The Intemperate
Rainforest

The Intemperate Rainforest

NATURE
CULTURE
AND POWER
ON CANADA'S
WEST COAST

BRUCE BRAUN

University of Minnesota Press
Minneapolis • London

Earlier versions of chapters 2 and 3 were previously published in "Buried Epistemologies: The Politics of Nature in (Post)colonial British Columbia," *Annals of the Association of American Geographers* 87 (1997): 3–31. Sections of chapter 2 also appeared in "Producing Vertical Territory: Geology and Governmentality in Late Victorian Canada," *Ecumene* 7 (2000): 7–46.

Published by the University of Minnesota Press
111 Third Avenue South, Suite 290
Minneapolis, MN 55401-2520
http://www.upress.umn.edu

Library of Congress Cataloging-in-Publication Data

Braun, Bruce, 1964–
    The intemperate rainforest : nature, culture, and power on Canada's west coast / Bruce Braun.
        p.   cm.
    Includes bibliographical references (p.) and index.
    ISBN 978-0-8166-3399-9 (hc) — ISBN 978-0-8166-3400-2 (pbk)
    1. Nature—Effect of human beings on—British Columbia.   2. Human ecology—British Columbia.   3. Rainforests—British Columbia.   4. Rainforest conservation—British Columbia.   I. Title.
    GF512.B7 B73 2002
    333.75'09711—dc21
                                                                              2001005830

Printed in the United States of America on acid-free paper

The University of Minnesota is an equal-opportunity educator and employer.

12 11 10   10 9 8 7 6 5 4

FOR MY PARENTS

Contents

Preface

British Columbia may be among the most difficult places to examine the matter of nature, if only because here the category seems so self-evident. With the possible exception of Alaska, no other region in North America is so predicated on the idea of nature's *externality.* From the province's forest industry to its burgeoning adventure travel market, nature is set up as a domain entirely separate from, and opposed to, society. It is a place to which one *goes*—the site of "resources," a stage for "recreation," a source for "spiritual renewal," and a scene for "aesthetic reflection."

The tenacious hold of the notion of nature's externality can be found elsewhere too, of course. This separation is a hallmark of capitalist modernity, a contradictory social formation that externalizes nature with a vengeance, even as it gives rise to worlds in which the cultural, the political, the technological, and the organic are tangled in ever more complex knots. A closer inspection of urban life in Vancouver makes these tangled webs evident: from log booms on the Fraser River to greengrocers on Commercial Drive, from traffic jams each Friday to raw materials and manufactured goods moving through its port, nature runs through every street and alley, every kitchen, restaurant, and office. Whether paper, tomatoes, oil, or electricity, life turns on the circulation of nature as both object and idea. But if nature infuses the city, so also the city infuses nature. Despite appearances, one does not leave culture behind when one crosses the city limits and travels into the surrounding hills and forests. From highly commercial resorts to so-called wilderness hideaways, from the managed forests of tree farms to the pressing matter of climate change, nature in BC is infused with social intent. It is made a site of fantasy and desire, manipulated as a source of energy, transformed through science and technology, managed as scenery. Far from two separate domains, there is but one, a hybrid realm crisscrossed by flows of energy and matter and the movements of animals, plants, people, machines, and ideas. Somewhat after the fact, we try to squeeze these tangled relations into separate boxes—nature on the one side, culture on the other.

This book presents a simple, yet counterintuitive, argument that nature's *externality* is merely an effect produced through the discursive and material practices of everyday life. This does not mean that mountains, trees, rivers, salmon, and grizzly bears do not exist. Rather, it calls attention to the ways in which BC's landscapes are shot through with language, meaning, and history, even as they are assigned to the category nature. As I argue in some detail about one of BC's most iconic "natural" entities—the temperate rainforest—what appears as primal nature is in fact far more social than we might first think. Indeed, I suggest that in BC the notion of "nature" hides a great many things from view. By "opening up" the temperate rainforest, as I do in the chapters that follow, a complex terrain of culture, politics, and power comes into view, within which the rainforest is continuously stabilized and destabilized as an object of economic, political, and aesthetic calculation and in which the future of many actors—both human and nonhuman—hangs in the balance.

As I explain in this work, it is perhaps First Nations who most directly bear the costs of the externalization of nature in BC, even as many others in the province derive great economic and personal benefit. This is in part because if nature is to be successfully constructed as primal, First Nations must be either erased entirely or collapsed into it. These sorts of cognitive failures have allowed many other people—the state, industry, environmentalists—to speak for the rainforest and its futures, while Natives have struggled to have their voices heard. One of my objectives, then, is to strive toward a new set of concepts that might inform a radical environmentalism that is attuned not only to the impact of people *on* the environment, but also to the relations of power and domination that infuse our environmental ideas and imaginations. Thus, I ask: What might a radical, *postcolonial* environmentalism look like? How do we need to renovate our ideas of nature and culture in order to achieve this? Ultimately, what might we gain by letting go of one of our most cherished notions—that nature is a site that somehow, magically, lies outside the messy world of history and politics. I introduce these questions in chapter 1 (at the end of that chapter can be found a brief synopsis of the rest of the book).

To write is to live in a state of indebtedness: intellectual, material, and emotional. Such debts cannot be quantified or adequately repaid. I owe much thanks to Derek Gregory, whose intellectual support and friendship sustained me throughout my years at the University of British Columbia. It was Derek who first suggested the book's title. Cole Harris, Dan Hiebert, Geraldine Pratt, Trevor Barnes, and David Ley were also valued interlocutors, as were many graduate students and friends in Vancouver who perhaps more than any others shaped my thoughts as I researched and wrote. Special thanks goes to Sarah Bonnemaison, Lynn Blake, Noel Castree, Dan Clayton, David Demeritt, and Matthew Sparke, all of whom contributed much through conversation and critical readings. Emmanuelle Arnaud, Trina Bester, Bill Burgess, Brett Christophers, Nicky Hicks, Jennifer Hyndman, Deirdre McKay, Yasmeen Qureshi, Ben Redekop, Ken Reid, Steve Rice, Magdalena Rucker, and Kamala Todd also contributed in countless ways and made it difficult to leave the city. A special thanks too to Ruby Willems, who provided inspiration and support.

In Berkeley I had the good fortune of meeting Donald Moore, whose generosity was exceeded only by his acute insight. Michael Watts kindly provided institutional support, and in turn the opportunity to work with a tremendous group of scholars. Sincere thanks to Julie Greenberg, Sarah Jain, Caren Kaplan, Jake Kosek, James McCarthy, Allan Pred, Jennifer Sokolov, Rebecca Stein, and Dick Walker. At the University of Minnesota, Ananya Chatterjee, Lisa Disch, Vinay Gidwani, Leila Harris, Helga Leitner, Tsegaye Nega, Eric Sheppard, Mary Thomas, Karen Till, and Joel Wainwright have been both wonderful friends and thoughtful colleagues.

Carrie Mullen has provided stellar editorial guidance throughout the project. Her patience as deadlines came and went was much appreciated. Sarah Whatmore, David Schlosberg, and Neil Smith made many constructive comments on an earlier draft. Mark Lindberg provided cartographic support, often on short notice. More people than I can name here agreed to interviews, gave access to records, and searched for files and photographs. Without them this book could not have been written. Research was supported in part through a Social Sciences and Humanities Research Council of Canada (SSHRC) doctoral fellowship. A SSHRC postdoctoral fellowship helped see it through revision.

Finally, much thanks to John, Mary, Brad, and Geoff. Not only have they come to terms with my itinerant ways, but all have at points contributed to the ideas expressed in the pages of this book.

Abbreviations

AAC	annual allowable cut
AIM	American Indian Movement
BC	British Columbia
CMTs	culturally modified trees
COFI	Council of Forest Industries
FoCS	Friends of Clayoquot Sound
GCMs	global climate models
GDP	gross domestic product
GIS	geographic information systems
GM	genetically modified
GSC	Geological Survey of Canada
IRC	Indian Reserve Commission
IWA	International Woodworkers of America
LUPAT	Land Use Planning Advisory Team
MB	MacMillan Bloedel
MEC	Mountain Equipment Co-op
NDP	New Democratic Party
NRDC	Natural Resources Defense Council
PPWC	Pulp and Paper Workers of Canada
REI	Recreation Equipment Incorporated
TEK	traditional ecological knowledges
TFL	Tree Farm Licenses
VAG	Vancouver Art Gallery
WCWC	Western Canada Wilderness Committee

1. The Intemperate Rainforest

The world we know is not this ultimately simple configuration where events are
reduced to accentuate their essential traits, their final meaning, or their initial
and final value. On the contrary, it is a profusion of entangled events.
—Michel Foucault, "Nietzsche, Genealogy, History"

Clayoquot Sound, British Columbia

During the summer of 1993, few Canadians remained unaware of the
drama unfolding on a remote logging road in Clayoquot Sound, British
Columbia. Here, for three months, protesters gathered daily in the pre-
dawn darkness to await vehicles carrying loggers to work sites deep in
the forest. At the first sight of approaching headlights, they would take
their positions on the road, a court injunction would be read, and members of the local Royal Canadian Mounted Police detachment would
begin the task of untangling limbs, lifting bodies, and carrying or dragging blockaders to buses contracted to transport them to the nearby
town of Ucluelet, where they would be charged and released.[1]

As brief as these encounters were—often less than thirty minutes—
they had the desired result. Images of heroic environmentalists defending pristine nature from a rapacious forest industry flashed across TV
screens in Canada and abroad. This was high drama, and in a matter of
days, Clayoquot Sound was placed firmly on the political landscape, not
only within Canada, but internationally as well. As the summer wore
on, protesters came from farther afield—France, England, Germany,
Australia, the United States—and began to include prominent figures
such as Robert Kennedy Jr., members of the Australian rock band Mid-
night Oil, and Green Party members from the European Parliament. By
early fall, when the onset of winter rains brought the logging season and
protests to a close, more than eight hundred people had been arrested,

1

making the blockade one of the largest acts of civil disobedience in Canada's history.[2]

It was hard not to be drawn in by the courage and passion of the protesters, and by the success of their tactics. As national and international coverage expanded and threats of consumer boycotts grew, logging in the region ground to a halt. As someone who frequently travels to the Sound to enjoy its sheltered inlets, mist-enshrouded forests, and remote beaches, I cannot say that I was saddened. Yet, I watched the protests with an uneasy mix of admiration and alarm. The passion and the principled commitment of the hundreds of people who risked arrest and jail deserved respect. Here, on a remote dirt road, a story constitutive of capitalist modernity—that nature exists merely as a stock of resources for the accumulation of wealth and for nation-building projects—was refused for its violence to plants, animals, and people alike. Many protesters made great personal sacrifices in order to highlight and disrupt the sacrificial logic of global capitalism, a logic that deems certain landscapes and communities expendable in the name of nation, profit, and progress.[3] Yet, I also had reservations. In the popular press, and in the rhetoric used by key actors, debate over the future of these forests was often cast in terms of a binary logic (pristine nature/destructive humanity). This, I feared, presented the complex politics of the rainforest in far too simple terms, and in a manner that stood in the way of a progressive ecopolitics attuned to the profusion of entangled events, actors, and practices that constituted BC's "war in the woods." Not only did this binary logic authorize certain actors to speak for nature's defense or its management (environmentalists, transnational capital, and the state), it risked marginalizing others (local communities, forest workers, First Nations) who understood, and related to, the forest in very different ways. The slogan on posters distributed by the Western Canada Wilderness Committee—"Wild beaches. Wild rainforests. Wild forever"—may have been wildly successful among urban North Americans schooled in what Neil Smith (1996) describes as "an almost instinctive romanticism," but at what price? What cultural and political subjects could not appear within its comforting frame? In what ways might the *defense* of nature, and not only its *exploitation,* be complicit in forms of erasure and abjection, mirroring similar displacements achieved by industry and state? In short, how might the "temperate rainforest" be understood as a site of irreducible difference rather than as transparent and singular?

The Intemperate Rainforest examines the forest as an epistemic, cultural, and political space. This departs from most treatments of forest politics in BC, which assume the forest to be an unproblematic identity, and therefore see conflicts over the forest only in terms of competing interests and diverging opinions about how it should be used, preserved, modified, or restored. In these accounts, the forest is self-evident and exists in a space outside politics; it is simply the object over which politics happens. This book, in contrast, asks how something called the "forest" is made visible, how it *enters history* as an object of economic and political calculation, and a site of emotional and libidinal investment. The purpose of such an investigation is not to contest or dilute the profound ecological critique of industrial forestry that emerged in the last decades of the twentieth century. Rather, it is to situate this critique in a wider field of cultural and historical practices—and relations of power—through which these forests have been invested with layers of cultural and political meaning. It is to suggest that the natures we may seek to save, exploit, witness, or experience do not lie external to culture and history, but are themselves *artifactual*: objects made, materially and semiotically, by multiple actors (not all of them human), and through many different historical and spatial practices (ranging from landscape painting to the science of ecology).

Necessarily, this work is *genealogical* in character, a concept I draw from the writings of Michel Foucault. For Foucault (1977), genealogy disrupted identity thinking, it approached things—such as the body, sexuality, government—as effects of shifting configurations of discourse and practice, rather than innate properties found in the world. This does not deny the materiality of such objects, as if they merely exist in our heads. Rather, it points to the historical, cultural, and political conditions through which objects attain legibility. Thus, instead of accepting an identity such as the "temperate rainforest" as self-evident, and thereby an unproblematic ground for politics and policy, genealogy "breaks it apart" and inquires after its emergence as an object of economic, political, and aesthetic calculation. The advantages of such an approach are many. It allows us to recognize understandings *of* the forest—and our interests *in* the forest—as historical rather than timeless and partial rather than objective. It also alerts us to the play of power. One of the aims of this book is to recognize the rainforest as a contested domain, where epistemology and politics are not separate. To question the forest of forest politics is therefore also to attend to the subjugated histories and buried epistemologies—often *colonial* epistemologies—that are hidden by, or within, the terms and identities through which forest politics in the province is organized and understood. Finally, and perhaps most important, to break apart the forest is also to potentially open space for a more informed—and inclusive—public debate over what sorts of futures we want, a debate not constrained by the binary terms of contemporary forest politics.

If this seems somewhat abstract, let me make it more concrete by turning to two events in BC forest politics. The first occurred at the beginning of the 1990s, when I traveled with a friend to the forest industry town of Port Alberni, in order to participate in a weekend meeting designed to draw together different "stakeholders" in what was then already referred to as BC's "war in the woods." The meeting was a fascinating event that provided a window into an unfolding political drama. At few other occasions could one imagine such a diverse assortment of individuals: members of the province's forest workers' unions (PPWC and IWA),[4] representatives from various Nuu-chah-nulth communities,[5] non-Native residents of Tofino, Ucluelet, and Port Alberni, environmentalists,[6] tourism operators, fishers, members of the New Democratic Party (NDP) that formed the official, social-democratic opposition in the province's legislature,[7] and even a professor of social work from a nearby university. Nor would one expect to find a group with a more ambitious objective: to locate common ground over the fate of the province's temperate rainforests.

At the time of the meetings, I was studying local attempts to challenge the hegemony of state foresters and transnational forest companies (which had cutting rights on public lands), and related attempts to develop alternative community-based planning mechanisms. This was an important topic, and still is. But, as is so often the case, the event provided an occasion to reflect on very different, but equally pressing, questions. During the meetings, we occasionally sat in circles in the hope that this would facilitate open and frank discussion. It did, but with vastly different results than the organizers had imagined. As the weekend progressed, tempers rose and fell, participants left and then were convinced to return, and speakers berated each other for their intolerance and inflexible positions. The "temperate" rainforest, I learned, was a terrible misnomer, and when the simmering dispute over Clayoquot Sound exploded three years later, few who attended the meetings were surprised.

It was commonly thought that the meetings failed because stakeholders had divergent interests. Reflecting on the event sometime later, however, I began to wonder whether the language of "stakeholders" and "interests" was, in important ways, inadequate. I was struck by the spatial organization of the sessions, with chairs set in circles. Circles assume a center, and the physical arrangement of chairs implied that, despite the varied economic and political "interests" of participants, they were all contemplating and discussing the same object. The meetings pre-supposed an epistemology. Thus, it followed that if individuals could only look beyond self-interest, they might find points of convergence, reconcile differences, and come to a unified position over the fate of this thing called the rainforest. By this view, the problem was one of bias, something that could be overcome through the application of reason.

This book finds one of its origins in the failure of consensus in this small community center on Vancouver Island. Where participants had sought unity, they found only disparity; in the place of accord, dissent. Instead of a single vision of the rainforest, conflicting images and in-commensurate differences clashed in a chorus of angry voices. I came away convinced that merely bringing people into the same room was not sufficient for a radical environmental democracy. What was needed was a sustained examination and appreciation of the multiple ways that the forest was invested with meaning and value. What was required was an examination of our passionate attachments to these forests, attach-ments whose histories were not merely personal, but political, not mere-ly local, but carried from afar.

The weekend was not completely wasted. Respect for difference, after all, comes at least in part from having one's preconceptions chal-lenged. Indeed, despite my concern to highlight questions of and about difference, I still believe in the value of efforts to achieve consensus. Not because I believe that these arenas somehow provide the possibility of undistorted communication, where differences are resolved in the clear light of reason—there can be no one Archimedean point from which to arbitrate contesting visions of the forest, no position free of the play of discourse and power, no forest imaginary without remainder. Nor be-cause I retain a melancholic longing for a unified Left. Nostalgia gets in the way of the hard political work of living with, and working through, difference. Indeed, as I demonstrate in this book, it is precisely when multiple voices are made to speak in the name of the One that we need to be most alert to what has been left out. Rather, I support such events in spite of their shortcomings because I remain committed to struggles for ecological justice rather than the sacrificial logic of the market, com-mitted to efforts to build social natures rich with possibilities for people, animals, and plants to flourish, rather than positing an external nature that can be exploited or saved by a select few. For these reasons, I feel that we cannot do without one of the central assumptions of a radical

ecological politics: that if we listen attentively, and critically, we can locate affinities, build coalitions, and imagine other, better, ways of being
together that do not reduce all of nature, and all of culture, to the logic
of the commodity. It is the terms through which I see these events that
have changed. Where I once imagined that consensus could be reached
by stripping away ideological preconceptions, I now realize that if any
consensus is to be reached, it will have to be something built rather than
found, provisional rather than final, always open to examining its own
constitutive exclusions.[8]

A second event can sharpen our focus as well as draw attention to
some of the methodological issues involved in the present project. Three
years after the fateful Port Alberni meeting, the BC government released
its now infamous "Clayoquot Sound Land Use Plan." This plan divided
the forests of Clayoquot Sound into various zones, each dedicated to
different uses. Accompanying the plan was a map (Figure 1.1) that was
promptly reprinted in countless newspaper and magazine articles. The
map turned on a color scheme. Areas colored red (black in reproduc-

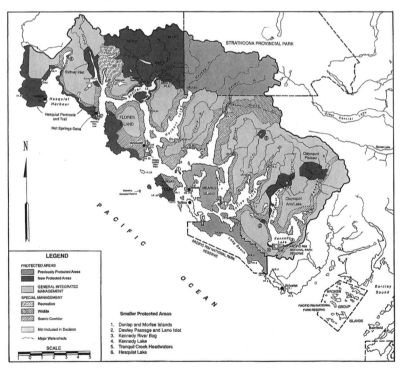

Figure 1.1. Clayoquot Sound land-use decision, Government of British
Columbia, May 1993. Scale in kilometers.

tion) were to be set aside for preservation; those (disingenuously) colored green (light gray in figure) were to remain open for logging. Other colors and codes identified areas that were to be managed for different activities and qualities: recreation, wildlife, visual scenery, and so on. This map is of interest for various reasons. First, for what it did *not* show. The map did not take into account the spatial, environmental, and economic practices of the various Nuu-chah-nulth groups living in the Sound.[9] This was alarming, given that apart from the predominantly white tourist and forestry towns of Tofino and Ucluelet, Nuu-chah-nulth communities composed almost the *entire* population of the Sound. It appeared even more alarming when one considered that the lands and waters of the Sound were central to issues of identity and economy in these communities, and that at no time—either through treaty or conquest—had the Nuu-chah-nulth ceded their traditional territories to the Canadian state (see Figure 1.2).[10]

For supporters of Native land rights, this land-use map raised troubling questions. How was it that in 1993, after years of conflict between

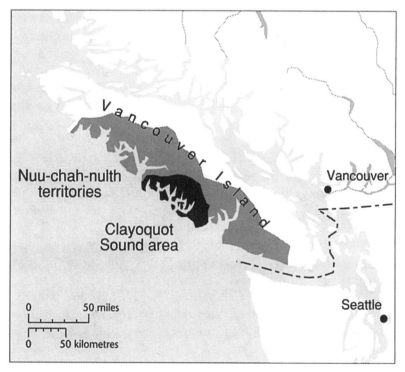

Figure 1.2. Location of Vancouver Island, British Columbia, showing Clayoquot Sound, Nuu-chah-nulth traditional territories, and U.S.–Canada border.

First Nations and the Canadian state (at both federal and provincial levels), it still seemed entirely natural that the Sound be conceived in terms of wilderness, working forests, recreation zones, wildlife reserves, scenic corridors, and so on, with no reference to the spatial and economic practices of the Nuu-chah-nulth? What authorized this erasure? And why was the press—and the public—so willing to understand the conflict in Clayoquot Sound in these terms? The lengthy front-page story that ran in the *Toronto Globe and Mail,* Canada's so-called national newspaper, provided readers with a lengthy guide to the various actors in the conflict, but none of these were Nuu-chah-nulth (Matas 1993). In turn, many environmental groups responded to the plan by vigorously debating the *quantity* and *quality* of lands preserved, tacitly accepting the constitutive erasure of the Nuu-chah-nulth despite their rhetoric of support for Native land rights. That few Natives joined protesters on the blockades is a topic that has still not received the attention it deserves.

What the land-use plan—and its aftermath—revealed were "environmental imaginaries" that somehow excluded, or at least narrowly circumscribed, relations between First Nations and their surrounding lands, and that limited the sorts of epistemological, political, and territorial claims that the Nuu-chah-nulth could make to the rainforest. How this erasure was achieved historically, and its reiteration and contestation today, is a topic that I explore most directly in chapters 2 and 3. The marginalization of the Nuu-chah-nulth in forest politics has not only been a result of how the forest was conceived, of course; it has also been achieved through legal and political means, backed by the coercive power of the state. But there can be little doubt that particular concepts of nature, culture, indigeneity, modernity, and progress have been deeply implicated in institutional and state practices in the region, and that a series of cognitive failures and discursive displacements have made it immensely difficult to recognize the political presence and environmental practices of the Nuu-chah-nulth, or to register hopes and plans for modern social and environmental futures that diverge from non-Native Canadian norms. Epistemic erasures are not innocent; they justify political and territorial erasures (Gregory 1994).

This brings me to one of the major arguments of this book: if the forest on the West Coast is a deeply cultural and political space, then it is so in ways that bear the continued imprint of colonialism. To speak of the rainforest is thus necessarily to speak of colonialism and its aftermath, to follow the trace of the past in a so-called postcolonial present. I will have more to say about the fraught career of the term *postcolonial* in a moment. Suffice it to say that in BC postcoloniality must be rendered in an ironic, rather than triumphant, mode. Before discussing this, let me return to the same planning map in order to address some methodological issues. In a corner of the map a legend relates colors and patterns to specific land uses. Because maps claim to represent a separate reality, legends are weighty things, for they tell us how to properly read the map's code. For a map to make sense, its legend must be clear and unequivocal. In other words, legends stabilize meaning. To question a legend is thus also to *destabilize* meaning, and to point to its ideological functioning (see also Belyea 1992; Harley 1992; Sparke 1995).

As I explain at more length later, one of the objectives of this book is to disturb the machinery of (neocolonial) meaning making. Somewhat playfully, then, I treat the map's legend not as the unequivocal guide to proper interpretation that its authors meant it to be, but as an invitation to experiment and thought. By translating the legend into a menu, such as found on a computer screen, and taking each box to be an icon, other possibilities come into view. Instead of the legend fixing meaning, it can now be used to open meaning to the question of its contingency. One can imagine clicking on any box, and having additional screens and menus appear, each leading to a further constellation of discourse and practice that are constitutive of the land uses that the map presents as self-evident. In this way not only does the map bring together and mediate various ways of constructing the temperate rainforest, but each construction now appears as the product of multiple practices and myriad histories that come together to give the present its singular shape. The result is a far more messy, provisional, and political mapping, where the forest and its meanings are not timeless and fixed, but instead continuously in flux.

This is *serious* play, for in these histories and practices we can begin to see that what counts as the forest is an effect of power. Ultimately, it may be possible to see that at the heart of these innocent colored codes that signify wilderness, working forests, or visual resources are constitutive absences and colonial histories whose disavowal is necessary for the fiction of natural/national forests to exist at all, but that also threaten to return as an excess that cannot be contained within the order of the code.

Nature/Culture/Power: Toward a Critical Environmentalism

Although this study is concerned with the poetics and politics of nature in a particular place—Canada's west coast—it stands at the intersection of two broader projects. The first is an effort to develop a robust, critical environmentalism that recognizes the intertwining of social, cultural, technological, and ecological relations in the worlds we inhabit. The second is an effort to dismantle relations of domination set in place during European colonialism, but that continue to infuse the so-called post-colonial present. In BC, I argue, these two projects must be thought together. For these efforts to succeed, however, requires that we learn to think differently about a number of keywords such as nature, power, resistance, politics, indigeneity, and history. In the remainder of this chapter, I examine a number of these at greater length, before concluding with an outline of the book as a whole.

Social Nature

The arguments in this book turn on the concept of *social nature*. I use the term to indicate the inevitable intertwining of society and nature in any and all social and ecological projects (Fitzsimmons 1989; Harvey 1996). In academic circles, this is a commonplace assumption, evident in the cascade of metaphors—hybrids, cyborgs, networks, knots, assemblages—that seek to capture this intricate mixing of the material, textual, cultural, political, and technological.[11] Not unexpectedly, the notion that nature is socially constructed has been met with skepticism by some Western environmentalists who fear that it undermines efforts to "save" nature, and gives free rein to the transformation of nature by humans (Soule and Lease 1995). Against this view, I argue that, properly understood, the concept of social nature is an important source of analytic and political hope in the face of the radical social and ecological displacements effected by postcolonial capitalisms. Although there can be no doubt that the concept challenges some of Western environmentalism's most cherished notions (pristine nature, wilderness), it does so only because it finds these terms far too deeply invested in a nature–culture dualism that hinders more than it helps our ability to think critically about our relations to, and responsibilities for, the environments we relate to in our everyday lives (cf. Cronon 1995; Smith 1996; Castree and Braun 1998). Conceiving of ecopolitics as the project of saving external nature from its destruction at the hands of humanity, I argue, risks leaving us wholly unprepared to imagine *how* we might responsibly inhabit our complex socioecological worlds. The argument of this book, then, is informed by the belief that understanding nature as social—seeing the world in terms of nature–culture "hybrids"—presents the best way to understand and interrogate present-day social and environmental conditions and to imagine steps toward ecologically sustainable and socially just futures.

For my purposes, the concept of social or hybrid natures is useful for what it brings into focus within the context of BC's war in the woods. On the one hand, it reminds us that almost everywhere—including Canada's west coast—nature is socially produced, in the sense that what we see as "natural" internalizes not only ecological relations but social relations too (see Smith 1990). This opens important analytic possibilities. If nature is produced, rather than simply given (even if humans are understood as merely one among many actors in this drama), it becomes possible, if not imperative, to identify *the specific historical forms* that nature's production takes, and to locate the specific *generative processes* that shape how this occurs (see Harvey 1996). Radical geographers, for instance, have drawn attention to the ways in which nature's transformation follows historical shifts in modes of production, each remaking nature according to very different social and ecological imperatives. By this view, contemporary environments are analyzable as in part the outcome of capitalist social formations. As Noel Castree explains (1995), capitalism's abstract framework of market exchange, with its imperatives of competition and accumulation, brings all manner of environments and human labor processes together so as to quite literally produce nature anew.[12] That nature is increasingly remade in the image of the commodity is clear for all to see at the beginning of the twenty-first century. From genetic engineering to biodiversity prospecting, the commodity form now reaches ever more extensively and intensively into the fabric of nature (cf. Haraway 1997; Katz 1998). This is as true for Canada's west coast as anywhere else; the temperate rainforest is in important ways an artifact of market relations, whether through forestry, tourism, or preservation (the latter still turns predominantly on questions of "value"). Crucially, the concept of social nature is also germane to *earlier* moments in the region's history, even *prior* to the extension of European power and market relations, when these forests were not primeval, but only governed by, and an effect of, a different configuration of cultural, political, economic, and ecological relations.

I will return to this point later. First, let me note some other ways that the concept of social nature helps us rethink environmental politics. Numerous writers have suggested that the concept of social nature usefully undermines the *ideological* role that nature plays in support of relations of domination. Neil Smith (1990; see also Schmidt 1971) argues that, in the capitalist West, nature is commonly understood in two contradictory ways (external and universal) and that these work together to naturalize existing social relations. The notion of nature as external, for instance, denies any social relation with the environment. This, in turn, authorizes neo-Malthusian arguments about intractable barriers in nature to which humans must submit, a rhetoric popular in studies of population and resources in the 1970s and widely repeated during the heyday of "deep ecology" in the 1980s and 1990s (Ehrlich 1968; Devall and Sessions 1985). Lost in this language of limits are the social and political causes of scarcity, as well as humanity's creative capacity to transform nature. To be sure, this appeal to the *malleability* of nature is deeply anthropocentric, and has been heavily criticized for being so, although, as David Harvey (1996) rightly notes, there can be no "valuing" of nature—even by ecocentrists—that is not in some respect grounded in human concerns.

The view of nature as universal, on the other hand, lends itself to the naturalization of social relations: if humans are like any other animal, then human relations can be seen to be as fixed and timeless as natural processes (a position that feminists have so clearly shown authorizes forms of domination based on biology). To say an aspect of human social relations is "natural" (i.e., patriarchy) is to say that it is pointless to try to change it. In reality, these dual representations often work in tandem: "external" nature, for instance, posits scarcity as arising *in* nature, while "universal" nature naturalizes the very social relations that *produce* scarcity in the first place. In this sense, Fitzsimmons (1989, 106) writes, nature functions as a "concrete abstraction" that has "a mystifying role . . . in social and intellectual life."

This takes us back to a point already touched on—that the concept of social nature challenges *wilderness* as the privileged term of Western environmentalism. As I argue throughout this book, notions of "pristine" or "primeval" nature—quite apart from their sexualized tropology—posit nature as something that lies *outside* history, and thereby denies *other* histories of nature's occupation and use, specifically those of indigenous peoples (Cronon 1995). Indeed, the rhetoric of wild, or external, nature can easily become complicit in the displacement of people for whom these places were once or still are home. As Donna Haraway (1992, 296) explains, efforts to preserve wild nature often "remain fatally troubled by the ineradicable mark of the founding expulsion of those . . . for whom the categories of nature and culture were not the salient ones." As I explain in chapter 3, on Canada's west coast such displacements are often an effect of environmental discourse and occur in two ways: through erasing signs of existing human modification, thereby producing nature as pristine, or by collapsing indigenous cultures *into* nature to form a premodern harmony that must be protected *in its totality* from a threatening modernity.

Perhaps the most important reason to shift our vocabulary away from external nature to social nature is that it provides resources for thinking about how to live responsibly *in* nature (Cronon 1995; Castree and Braun 1998). The notion that nature is socially constructed, rather than a pure identity external to society, forces us to take responsibility for *how* this remaking of nature occurs, in *whose* interests, and with *what* consequences (for people, plants, and animals alike). It brings together ecology and social justice. Further, it allows us to see modified landscapes—cities, farms, parks, second-growth forests—as worthy of as much, or even more, attention from environmentalists as "wild" areas. Indeed, if nature is always already hybrid—constituted by the actions of humans and nonhumans alike—then the lines drawn between humans and nature are no longer the ones that must be anxiously policed (see chapter 6). This will undoubtedly make some people uncomfortable. From the latter perspective—arguably the dominant view in environmental circles—the task of ecopolitics is to defend nature from humans, with the result that battles lines are drawn and immense resources spent to preserve the last vestiges of nature's pristine character. Yet, as Cronon (1995) has so eloquently argued, this may return us to the wrong nature. It reinforces the same nature—culture dualism that has authorized the view that nature exists merely as a resource for the self-fashioning of humanity, and diminishes our sense of responsibility for those areas that do not fit this cherished image of nature (urban ecologies, agricultural lands, managed forests).

Social nature demands a different politics, one that does not traffic in purity but instead understands responsibility or ethics in terms of *relationality* (Whatmore 1997). As I explain in this book's Conclusion, metaphors such as "networks" and "assemblages" are helpful precisely because they force us to think in terms of a web of relations that encompass humans, animals, and machines simultaneously, and in which specific identities and forms are always contingent rather than fixed. What "responsibility" means in such a world is certainly more complicated. At the very least, it demands more attention to temporal and spatial connections, including those between humans and nature, and less to policing these boundaries. And, most certainly, it refuses to divide political issues into "environmental" and "social." A radical green politics, I argue, must recognize them as intertwined. By conventional measures, this is a scandalous environmentalism, but it is not, as some argue, an apology for the rampant, and irresponsible, transformation of existing ecological systems (Soule and Lease 1995).[13] Accepting nature as inevitably marked by humans does not mean that any and all human environmental practices are the same or equally desirable. There is no reason why ecology movements founded on a notion of social nature should be any less concerned with the health of the planet and its many inhabitants, human and nonhuman alike, than any other ecopolitics.[14] Much like the ecological sciences, the concept of social nature places the focus firmly on connections and relations; it simply includes rather than excludes humans. If we take the notion of social nature seriously, we are faced with the task of understanding the complexity of ecological systems, learning to trace the impact of specific practices throughout the networks that bind animals, plants, and humans into complex totalities, and evaluating actions based on their consequences across and through these networks. Social nature leads to an ethics and politics of "dwelling" rather than separation.

Far from blunting its critical edge, then, social nature may potentially lend itself to a reinvigorated environmentalism.[15] It brings society and ecology together into a single analytic field, allows us to critically examine and evaluate the many ways that nature is socially produced, and draws attention to the ways in which nature's production—including its preservation—is always entangled with much more than nature, including questions of class, race, gender, and sexuality. It does not dictate to us what future natures *should* look like, nor does it provide a template for developing normative statements about nature and its transformation; these are open-ended questions that will be decided by the play of historical forces and political struggle.

The Cultural Politics of Nature

The social construction of nature refers to the imbrication of society and nature. But the phrase can also refer to the fact that our images and ideas of nature do not simply reflect a preexisting reality, but, in important ways, constitute this reality for us. In other words, nature's "social construction" refers to an *epistemological* condition as well as an *ontological* one. This second sense of construction is also crucial to what follows.

In some respects, nature's *cultural* construction is self-evident. We are confronted daily with countless images of nature in films, advertising, zoos, and parks. Often we approach these with playful irony, much as we do most visual images in a culture saturated by them. At these sites nature is clearly "culturally mediated" (Wilson 1991). But the argument of this book goes much further. When we leave behind the theater, the simulated habitat of the zoo, or the well-ordered urban park and venture out into the countryside (or into the laboratory), our experience and understanding of nature are no less constituted through an array of historical and discursive practices that organize how nature is encountered and known. To paraphrase Derrida (1976), there is no place *outside* such cultural practices from which nature can be objectively known. Even when our relation to nature seems most immediate, it is profoundly shaped by the narratives, knowledges, and technologies that enable experience.[16]

There is no need to rehearse at great length arguments about this understanding of the social construction of nature. Feminists grappling with questions of gender and sex have made these debates commonplace (see Butler 1993), as have scholars working in the field of science studies, who understand knowledge as the outcome of specific, historically situated practices involving both human and nonhuman actors (see Latour 1987, 1999; Haraway 1997; Hess 1997). Much of this work draws on poststructuralist theory and its postfoundationalist critique of epistemology. My own path into these questions—by no means the only one available—goes through philosophers and historians such as Martin Heidegger (1962, 1977b) and Michel Foucault (1970, 1979) and feminist theorists such as Donna Haraway (1991, 1997) and Judith Butler (1993), for whom, in different ways, the social identities through which we make sense of the world are seen as historically constituted and, crucially, the effect of power.

For many poststructuralists, Heidegger figures centrally as the thinker who was most concerned with the world of appearances, or, most specifically, how things came to be present to humans as objects of calculation. Behind his writings in the early and middle decades of the twentieth century lay the suspicion that our knowledges about the world *concealed* as much as they *revealed,* a notion that surfaces again in Foucault's concept of "subjugated knowledges." Heidegger used the spatial metaphor of a forest clearing *(Lichtung)* in order to capture the temporal character of how objects were encountered by, or showed up for, human subjects.[17] Not insignificantly, *Licht* means "light"; thus, within the clearing, things came to light—they became visible. In some respects, Heidegger's formulation was consistent with a long tradition in Western thought in which truth is equated with illumination (which is why we speak of the "light of reason"). But, in important ways, Heidegger turned this tradition on end. While retaining the trope of illumination, he argued that the clearing within which things came to light was *historical* rather than timeless, thereby linking intelligibility with cultural practice (Rouse 1987; Dreyfus 1991).

For my purposes, the significance of Heidegger's forest metaphor lies in its challenge to an epistemological model that in Europe had held sway since the Enlightenment. Rather than locating "illumination" in the familiar Enlightenment story of a disembodied Reason contemplating the world, he introduced history—and, by extension, notions of *practice* and *politics*—to how things "show up" in the world.[18] In other words, Heidegger contested the notion that understandings of the world could be explained by reference to something formal or to a world that was abstractable from our everyday involvements with it and with one another. Indeed, Heidegger (1962, 1977b) viewed the latter view as a source of great danger. The risk inherent in modernity's subject-centered representational epistemologies, he warned, was that they forgot the historical character of *Dasein* ("being-there") such that human knowing too easily became dehistoricized and transparent, and the order of the world too easily understood as something simply discovered in the passive act of observation.[19]

As Joseph Rouse (1987, 173, 182) summarizes, Heidegger's contribution to epistemology was his recognition that "what things are, and what characteristics they can have, depends in part upon the practical configuration within which they become manifest. There are no essences independent of this configuration of practices and the language invoked within it. . . . We encounter 'nature' through our practices, as it fits in and is revealed intelligibly in that context." This introduces to our discussion of social nature a series of very different analytic questions. If nature comes into visibility through specific historical and discursive practices, then how does this occur? What is left invisible or unthought in our accounts of the world? How does truth constrain? Foucault's insistence that knowledge be understood as, in part, an effect of power, further sharpens our sense of what is at stake in the project of historicizing nature's appearances. What must be brought into thought is not nature per se but the construction of *spaces of visibility* in which nature— and our economic and political investments *in* nature—is constituted (Rajchmann 1988; Gregory 1994). These might include sites and practices as diverse as technoscience, nature documentaries, outdoor sport magazines, or projects of mapping resources such as those of the Canadian Geological Survey in the late nineteenth century. In each, what counts as nature (and culture) cannot be disentangled from practices of representation.

It is this sense of the culture and politics of nature that Donna Haraway (1992) captures so brilliantly in her discussion of nature as *trópos* (figure, construction, artifact, movement, displacement). For Haraway— like Heidegger—what counts as nature cannot preexist its construction; when we take nature to be self-evident, we simply mistake our discursive practices for the things they seek to describe. I want to draw from this three implications that are critical to the arguments in this volume.

First, the social construction of nature thesis demands that we rethink how we understand and locate the political in ecopolitics. No longer can struggles over nature proceed as if nature were a self-evident entity. Indeed, if what counts as nature is an effect of multiple histories and practices, then taking nature as commonsense risks concealing and reinforcing the relations of power that are constitutive of, and internalized within, nature's historical forms of legibility. It is this link between power and knowledge that lies behind Donna Haraway's (1994, 60) declaration that "queering what counts as nature is my categorical imperative." For Haraway, as for many other feminist, queer, and antiracist activists, nature is an ideological battleground.

It should be emphasized that the deconstruction of identities (such as the body or rainforest) does not signal the end of politics, but rather establishes *as political* the terms in which such identities are articulated (cf. Butler 1990). Political struggle occurs in and through the concepts and images we use to describe the world. Thus, as Butler (1993, 50) explains, "To call a presupposition into question . . . is to free it from its metaphysical lodging in order to understand what political interests were secured in and by [its] metaphysical placing, and thereby to permit the term to occupy and serve very different political aims." As I explain in this book, what counts as nature in British Columbia is both highly political and fiercely contested. Thus, calling the temperate rainforest into question is not an arcane intellectual exercise, nor a nihilism that seeks to undermine morality; rather, it opens opportunities for reconfiguring and redeploying what counts as nature in new ways, and perhaps for aligning ecopolitics with other political struggles, including the anticolonial struggles of Canada's First Nations.

Second, the two meanings of the social construction of nature merge together. One of the problems that haunts many Marxist approaches to social nature is their narrowly defined sense of the generative processes that shape nature's transformation. Economic processes tend to be emphasized, while much less attention is paid to cultural practices. Neil Smith (1990, 31), for instance, argues that the events and forces by which nature is "produced" are "precisely those that determine the character and structure of the capitalist mode of production." Under dictate from the accumulation process, he explains, "capital stalks the earth in search of natural resources. . . . No part of the earth's surface, the atmosphere, the oceans, the geological substratum or the biological superstratum are immune from transformation by capital" (49, 56). Ten years later, David Harvey echoed similar sentiments: "capital circulation," he wrote, "has made the environment what it is" (1996, 131).

Such arguments are compelling at a time when nature is everywhere remade in the image of the commodity, from genetics to nature reserves. Yet, in the years since Smith's *Uneven Development* was first published, the *singularity* of such accounts has come under attack. To draw again on Haraway (1997, 141), the production of nature "is not reducible to capitalization or commodification"; it involves heterogeneous processes that are joined and separated in multiple "reaction sites."[20] One of the important insights that Haraway and others have brought to the study of nature is a refusal to see the economic and the cultural as distinct ontological domains. This point is captured especially well by Arturo Escobar (1996, 46), who argues that we need to attend to "the connections between the making and evolution of nature and the making and evolution of the discourses and practices through which nature is historically produced and known." These do not operate at different levels but form part of what Gilles Deleuze and Félix Guattari (1987), in a direct challenge to such dualisms as thought/matter, humans/nature, culture/ economy, refer to as "assemblages." Indeed, Deleuze and Guattari's metaphor of assemblages is particularly apt for a discussion of social nature (cf. Kuehls 1996). In a formulation that resonates with, but also extends and radicalizes, ecology, they argue that it is not possible to separate things from their relations, whether these things be human bodies, machines, words, or animals. Rather than a fixed set of relations, these assemblages are characterized by a shifting configuration of flows and intensities, in which things are constituted by the play of forces that take possession of them (Deleuze, 1983). Objects and identities within these assemblages are not pre-given; rather, they are nodal points on a "plane of immanence," in which the totality of relations at any one moment is constitutive of things themselves. Clearly, within such assemblages, it is not only the distinction between nature and culture that loses salience, but also the distinction between discourse and matter: representational practices become, everywhere, insinuated into the very *fabric* of nature— including our own bodies (see Martin 1998).

This may seem unnecessarily abstract, but is not hard to see how the imbrication of thought and matter occurs in practice. As I have shown elsewhere, the discourse of geology in the mid-nineteenth century—itself the outcome of a complex configuration of ideas, actors, institutions, and spatial practices—at once deterritorialized nature and brought about new territorializations (Braun 2000a). The implications both for the space-economies of industrial capitalism and for emerging forms of governmentality in nations such as Canada and Australia were substantial, for the simple reason that once nature was "geologized," it could be made available to, and transformed by, forms of economic and political calculation. In such contexts, the discursive and the material do not just coexist—a notion that retains their essential difference—but implode into knots of extraordinary density. Rethinking the social construction of nature along these lines helps us to recognize the way in which cultural, economic, political, technological, and ecological relations are tangled together. The chain saws and bulldozers of industrial forestry are not merely machines; they embody cultural and economic relations too. Likewise, the forests that are produced in industrial forestry's wake are not simply a reflection of capitalist relations, but internalize heterogeneous cultural, political, and ecological conditions that shape how its production proceeds.

Third, the social construction of nature introduces new questions about the role and practice of critique. As is well known, Deleuze was most interested in assemblages as the wellspring of potentialities, or a vaguely defined becoming. His was an exuberant thought that sought to free desire from what he, as well as Foucault, understood as the bureaucrats of truth. For Deleuze, assemblages are to be celebrated as "productive machines" from which subversive possibilities are produced in unpredictable and potent conjunctions (Jacobs 1996b, Doel 1999). From amid the multiple mixings of socio/spatial/ecological events may emerge new, unforseeable possibilities. The future, for Deleuze, is radically open, an active experimentation. Deleuze is somewhat less helpful, however, for attempts to understand fixity and identity, or for attempts to systematically interrogate and question the present. Our social worlds may be open-ended, but they are not a realm of boundless possibility (Harvey 1996). Deleuze's thought is not altogether incapable of addressing this. He argued that there are always "knots of arborescence in rhizomes." Likewise, his insistence that things exist on a "plane of immanence" enables us to see objects and identities in terms of their constitutive relations, and this in turn suggests the possibility of producing a cartography of the web of relations within which *both* forms of closure and new possibilities occur.

For an investigation of the cultural politics of nature in a place such as British Columbia, however, Foucault's genealogical method is perhaps the most helpful. As already noted, genealogy provides a means to explode the self-evident character of things, in order to show their constitutive qualities, histories, and relations. Genealogy *denaturalizes,* it shows the conflicts, accidents, and practices that are constitutive of the present. As Foucault (1977, 142, 147) explains:

> If the genealogist refuses to extend his faith in metaphysics, if he listens to history, he finds that there is "something altogether different" behind things: not a timeless and essential secret, but the secret that they have no essence or that their essence was fabricated in a piecemeal fashion from alien forms. . . . What is found at the historical beginning of things is not the inviolable identity of their origins; it is the dissension of other things. . . . [Genealogy] disturbs what was previously considered immobile; it fragments what was thought unified; it shows the heterogeneity of what was imagined consistent with itself.

Like the fecund potentiality that Deleuze locates in his "assemblages," Foucault's "histories of the present" are equally attuned to a politics of possibility, or what Foucault (1984) elsewhere called "the undefined work of freedom." After all, if the present is the effect of accident rather than will or design, as Wendy Brown (1998) explains in her gloss on Foucault, then it is more *historical* and more *malleable* than would otherwise seem.[21]

Postcolonialism and the Cultural Politics of Nature

As evident in my interest in grasping points where change is possible and desirable, my interest in the social construction of the temperate rainforest is much more than an academic one. It is grounded in ongoing political struggles over what counts as nature and over whose voices should be heard in conflicts over land, resources, and environment. In particular, it responds to efforts by First Nations to gain a measure of control over the future of social nature on their traditional territories. *The Intemperate Rainforest* is, therefore, a local "site-based" reading of the cultural politics of nature, one that takes BC's postcolonial present as its point of departure, rather than beginning with a universal, once-for-all-time environmental ethic.

Locating the Postcolonial

Because the matter of postcoloniality runs throughout this book, some comments on the term are in order. Defined broadly, postcoloniality refers to the cultural, economic, and political conditions that exist in the aftermath of colonialism and names a desire to engage them in a critical fashion.[22] However, the term comes with a number of potential pitfalls. Perhaps the most serious is that the *post* in *postcoloniality* can easily be taken to refer to a historical stage that comes after, and thus transcends colonialism, thereby effacing important continuities between colonial pasts and so-called postcolonial presents (McClintock 1992; Shohat 1992). The obvious danger is that emphasizing the *post* in *postcolonialism* risks diverting attention from colonial and neocolonial relations in the present (McClintock 1995; Jacobs 1996). A second difficulty arises from the term's universal application. References to "the" postcolonial condition are all too common, yet these overlook the fact that colonialisms and their aftermaths are *local* in their effects and practices, even if they draw on, and are enabled by, universalizing discourses (modernity, reason, progress). As Stuart Hall (1996) notes, colonialism and decolonization have been, and continue to be, widely differentiated affairs. Not only were British, French, and Spanish colonialisms markedly different (Mills 1991; Gregory 1995), but each was also geographically differentiated. British colonialism in India worked through very different means, and had very different effects, than it did in South Africa, Canada, or Australia. Colonialism may have everywhere involved the institution of racial, political, epistemic, and economic systems that benefited colonial regimes (both in Europe and among elites within European colonies), but this rarely occurred in the same way twice. Finally, with its emphasis on the onset and effects of colonialism, the field of postcolonial studies risks recentering *Europe* as the (single) agent of history, while the history of other cultures is either relegated to prehistory or understood primarily in terms of its response to European colonialism.

If we are to speak of British Columbia as postcolonial, then, we must do so cautiously, and by specifying its particularity. Moreover, we must also recognize that postcoloniality is experienced differentially *within* the province. For its increasingly prominent South Asian communities, for instance, postcoloniality is experienced as diaspora; it is about living a doubled existence between "here" and "there," along with the renegotiations of cultural and national identity that come with this. That this is an effect of past colonial displacement is certainly true, but in ways that stand in marked contrast to other communities in the province.[23] The experience of First Nations is much different. British Columbia may have long ceased being a formal British colony, but its colonial relation to its indigenous peoples has continued to the present. In the years since 1871 when the province entered the Canadian confederation, west coast Natives have been segregated onto tiny reserves, had many of their cultural practices criminalized, seen their children placed in residential schools, found their traditional territories enclosed as private property or remade as public (Crown) lands, and been subjected to racism, government paternalism, and devastating forms of physical, sexual, and emotional abuse. Despite recent court rulings in favor of Native land rights, there is still great reluctance on the part of non-Natives to acknowledge, let alone address, these histories of violent displacement. Indeed, political parties at both the provincial and federal levels continue to bolster their electoral support by attacking Native land rights, and treaty negotiations proceed at glacial speed, if at all.[24] Although differences exist between the region's various First Nations, Native peoples as a group remain by far the most marginalized population in the province, economically, politically, and geographically. For First Nations, then, the *post* in *postcolonial* is at best used ironically; at worst it works insidiously to deny continuing forms of domination.

That *postcoloniality* refers to such an uneven and differentiated terrain was brought home in the 1999 election of Ujjal Dosanjh as leader of BC's governing party, the NDP, and thus also as the province's premier. That the province was the first in Canada to have a South Asian immigrant as its premier was widely reported—and justifiably celebrated—in the Canadian press. Even the *New York Times,* a paper that usually ignores its neighbor to the north, carried a lengthy story on Dosanjh's election. What went unsaid in these reports was that although it was now possible to have an Indian (South Asian) premier, the election of an "Indian" (aboriginal) premier was unthinkable. Nor, amid this doubly displaced "Indian-ness," is there any guarantee that a disaporic Indian would sympathize with an indigenous "Indian"; issues of ethnicity and class make such alliances unlikely. The point is that the vast differences between the experience of indigenous and diasporic communities in British Columbia must give pause to anyone who wishes to use the term *postcolonial* to describe the province as an undifferentiated whole. My use of the term is therefore both cautious and specific: in this book *postcoloniality* refers to the aftermath and not the transcendence of colonialism. I use the term to refer not to a period after the *end* of formal colonialism, but rather to that which follows after the *institution* of colonial relations, thereby allowing a focus on both continuities and discontinuities. Ultimately, places are postcolonial because colonialism "worked over" existing social, cultural, and political institutions and identities. Thus, although BC is, at least by these terms, clearly postcolonial, this postcoloniality is always a multiplicity. This does not mean, of course, that something called "colonialism and its aftermath" is the only game in town, or that the study of colonialism and its multiple displacements fully captures the experience of postcolonial subjects. This risks raising colonialism to the level of metanarrative (Gandhi 1998), and, as I already noted, recenters Europe as the only historical agent. Colonialism may have irrevocably marked life in British Columbia, especially for First Nations, but it certainly does not exhaust or account for all aspects of existence.

Nature, Culture, and Power in the Shadow of Colonialism

That forms of domination set in place in BC's colonial past continue to shape the present is thus a central theme in this book. For the most part, such continuities are not the result of deliberate and sinister strategies to exclude and silence (although at times this is no doubt the case). Rather, the political and geographical marginalization of subjects such as First Nations is as much a result of an array of discursive practices that both effect and justify their displacement. Often these simply reiterate other, earlier displacements that have receded from memory to become taken-for-granted elements in how people envision and speak about nature and culture on Canada's west coast.

Following Edward Said (1978, 1994), I assume that colonial power operates in and through discursive practices as well as juridical-political ones, and that this remains as true for forms of neocolonialism in the present as it was for colonialism in the past. Such power is often hidden, in part because discourse produces its own truth effects while simultaneously limiting the field of possible statements about a given topic. In the years since the publication of *Orientalism,* discourse analysis has been a preferred method for bringing to light the operation of colonial power. By reading texts against the grain, postcolonial critics have sought to show how they shaped European understandings of non-European peoples and lands, in a sense producing an "other" for Europe, which it could then both know and govern.

Like most work on questions of colonialism, the present study is indebted to Said's pathbreaking work. My interrogation of west coast forest discourses takes as axiomatic that power and discourse are inseparable, and that any challenge to contemporary inequalities in BC's forests must interrogate the discursive practices that sustain them. In contrast to Said, however, I understand colonial discourse to be a much more fractured, unstable, and ambivalent field than his accounts allowed. This is by now a well-worn critique, but it bears repeating. By presenting colonial discourse as monolithic, Said was unable to recognize the internal contradictions and gaps and fissures within colonial discourse that opened possibilities of subversion, or the manner in which colonial discourse failed to suture the totality of social and ideological fields (cf. Bhabha 1994). In Said's work, colonial discourse appeared everywhere to be *effective* to the point where it became almost impossible to imagine its contestation. Although discourse analysis is clearly important for understanding BC forest politics and the epistemologies that secure the hold of neocolonial power, it is ultimately unable to capture the degree to which the temperate rainforest, and its social and political relations, is unstable, contested domains, constituted within multiple and often contradictory discourses that cannot be collapsed into one.

Equally problematic is the associated assumption that resistance is located in a realm external to these discursive fields. The paradoxical effect of presenting colonial discourse in such totalizing terms was that there seemed to be no space either inside or outside Europe where one could locate an alternative political consciousness. One option that presented itself was to posit domains of cultural and political life that somehow remained protected from the incursions of colonialism (or modernity) and to find in these spaces resources for anticolonial politics (Scott 1985; Chatterjee 1993). While work along these lines reminded scholars that colonialisms—and their discursive formations—were uneven in effect and extent, such studies often assumed the existence of separate spaces (public/private, colonial/precolonial, modern/premodern) between which subjects consciously moved, taking on different identities at different sites. Not only did this rely on questionable assumptions about the autonomy of these spaces, and the consciousness of knowing subjects who moved between them, it also revealed a remarkable, and ultimately disabling, nostalgia for a "pure" subject of resistance who, in the name of authenticity, speaks from a position outside or prior to the social and discursive fabric of colonial modernities. Such quests were ultimately doomed to fail, because at some point authenticity is always betrayed by its contamination by the "other." As we will see in chapter 3, tropes of authenticity/inauthenticity infuse forest politics on Canada's west coast, and often stand in the way of the articulation of alternative postcolonialities.

There are no "pure" spaces on Canada's west coast. Claims to cultural authenticity in struggles over the fate of BC's temperate rainforests are at best strategic, at worst a vehicle for perpetuating relations of domination based on colonial stereotypes. Much depends on the context, and who is making the claim. As I demonstrate in several sections of this book, efforts to contest colonial natures rarely if ever do so from positions outside colonialism's discursive and political fields, but proceed instead by turning the terms and tools of colonial power against itself. This suggests the need to figure resistance in ways that do not require the assumption of spaces "outside," or that do not divide the world into binary spaces of power and resistance. Precisely because colonial discourse is *not* monolithic, contradictions and tensions within its terms can be exploited politically. Likewise, because the legitimacy of colonial power is not given once for all time, it must be continuously reasserted. In its reiteration lies the risk of failure; the regimes of truth that provide the basis for colonial authority might be rearticulated otherwise. The production of socially just, postcolonial natures on Canada's west coast is not determined in advance, but is open to the play of history and politics; struggles over the forest's cultural meanings are central to this.

Episodes

The five chapters that form the core of this book are intended as individual "cuts" into BC forest politics. Each begins with an event, artifact, or image drawn from contemporary struggles over the production and consumption of "nature" in British Columbia's temperate rainforests, and these provide points of departure for an exploration of the intersecting discursive, social, technological, and institutional relations that shape the way that nature in the region is constituted—and ultimately transformed—as an object of scientific, economic, political, or aesthetic calculation. In each case, I draw out the complex ways in which these natures are also entangled with histories of colonialism and present-day struggles to imagine alternative postcolonial futures.

Although individual chapters are related to others, I make no attempt to resolve them into a totality. If the juxtaposition of studies achieves an effect as a whole, my hope is that it draws attention to the simultaneity of many forest politics rather than one, and the need to attend to the many different arenas that a progressive environmentalism must engage. Each chapter points to important dynamics at play in the temperate rainforest, and seeks to develop new ways for thinking about the intertwining of past and present, culture and nature, discourse and power, space and place, local and global in the complex environmental struggles that criss-cross the region. By moving across different sites, and drawing on different histories and practices, I show how nature and culture, place and identity are continuously (de)stabilized on Canada's west coast. Indeed, the matter of nature's *multiplicity*—and how we understand it—is a central concern of this book, and one of the reasons it is written in this form. My intention is not only to move away from talking about nature or culture in the singular, but also to move environmental thinking away from the notion that landscapes can be understood as discrete, bounded places (or analyzed through the archaeological trope of layers). In and between these chapters, place is rendered dialectical, constituted at the nexus of multiple material-semiotic practices. These do not coalesce to form a stable entity, nor do they respond to a single logic. There is no one spatiality or temporality in the temperate rainforest. Despite the romantic notions of place that prevail in some environmental circles, places are inherently rhizomatic, the effect of spatial networks that are global in scope (such groups as Greenpeace recognize this explicitly in their political strategies) and temporal rhythms that operate at different speeds. This does not empty place of its meaning; rather, it suggests the opposite. Drawing on a phrase used by Kathleen Stewart (1996), all places are "occupied places," at once subject to the occupation of external forces, yet occupied—experienced, lived, hoped for—in their own right. It is for this reason that David Harvey's (1996) counsel that we attend to a dialectics of space, place, and environment must be followed. How we understand this dialectic, and the conjunctural moments that it produces, is partly what is at stake here.

Crucially, by insisting on the simultaneity of *multiple* spatiotemporalities rather than viewing the present as a unity, it becomes possible to imagine a vastly expanded political terrain. It has always been tempting to write the history of BC's forest politics in terms of a single narrative—extractive capital, colonialism, speciesism—but it is far more tangled than these narratives allow. The chapters that follow resist such simplifications, but at the same time they do not deny the centrality of certain processes. That market forces have radically reconfigured these forests, and our relation to them, is undeniable. The increased ability of capital to flow across national borders, the restructuring of production and management, the mechanization of the workplace, and growing competition from other timber-producing regions all impact the social production of nature. The pressure to keep annual timber harvests at unsustainable levels—thus risking the integrity of both ecosystems and communities—is based in large measure on economic imperatives, as well as the complex ways that the interests of local communities and workforces have been tied to the profits of transnational forest companies (see Marchak 1983, 1997; Barnes and Hayter 1997). But economic processes always work in concert with other social, cultural, and political processes; they do not exhaust or determine the dynamics informing nature's social construction, nor do they exist as the only, or always the most important, site of politics.

I begin chapter 2 by examining histories of displacement. My guide in this story will be Simon Lucas (Klah-keest-ke-uss), a Hesquiaht elder who in 1975 testified to his people's isolation in the face of the spatial and colonial logic of industrial forestry. Tragically, Lucas's words were neither heard nor heeded, and it fell on other Nuu-chah-nulth leaders to repeat them in 1993 in the context of the Clayoquot Sound Land Use Decision, this time backed with the threat of court action. My concern in this chapter is to interrogate how such displacements were effected and naturalized, both in the past and in the present. By turning to a pivotal moment at the end of the nineteenth century, for instance, I explore how BC's temperate rainforests came to be constituted as *natural* and *national* spaces simultaneously, but only through the constitutive erasure of indigenous territorialities. Present-day displacements, I argue, occur in part through the *reiteration* of these earlier erasures, further reinforced through the supplanting of moral reason by instrumental reason whereby forest politics in the 1980s and 1990s turned largely on what qualified as "rational" forest management. By calling attention to these displacements as *constitutive* of present-day claims of forest "custodianship" by state and corporate capital alike, I suggest a colonial logic that continues to lie at the heart of BC's forest economy.

If chapter 2 explores the cognitive failures that are necessary for industrial forestry, chapter 3 raises similar questions in relation to campaigns waged by local and global environmental groups to "save" the temperate rainforest. Here my guide will be a popular book of nature photography, published by the Western Canada Wilderness Committee, Canada's largest preservationist group. The book contains some of the most remarkable photos taken of the landscapes on the west coast of Vancouver Island, and was instrumental in constituting a community of interest both locally and globally. Such practices, I argue, are necessary in an age of globalized capitalism and ecological crisis. Yet they must also be attentive to the discursive displacements that enable the ability of some to "speak for" nature in political struggles while disqualifying others. By examining depictions of nature and culture in this key volume, I reveal a problematic, and in many ways contradictory, postcolonial eco-politics on Canada's west coast. I focus specifically on the anxious policing of the "pristine" and the "primitive," and show how staging the temperate rainforest in these terms is crucial to the Wilderness Committee's contradictory support for wilderness preservation and Native land rights. Finally, I contrast this with a very different vision of the rainforest set out by the Nuu-chah-nulth in a court battle over Native sovereignty: a place heavily trammeled rather than pure, used rather than pristine, and home rather than wilderness.

Chapter 4 shifts gear and turns to a very different facet of nature's production on Canada's west coast: adventure travel and ecotourism. Here I examine the ways in which large areas of Canada's west coast have come to be sites for the re-creation of self, body, and identity for (mostly white, middle-class) metropolitan subjects. I show how this enframes the temperate rainforest—and its residents—in a certain mold, but also how it opens new spaces for politics and identity for First Nations peoples, both in terms of relations *between* First Nations and non-Native society and *within* Nuu-chah-nulth communities. If, as I argue in chapter 4, adventure travel and ecotourism are particularly *modern* forms of travel, deeply invested in, yet also seeking to transcend, a great divide between civilization/nature and modernity/premodernity, then chapter 5 might be considered in part an analysis of how the west coast came to be positioned on one side of these divides as an anachronistic space *within* modernity. Turning to the landscape art of Emily Carr (1871–1945), arguably BC's most widely known artist, I explore the degree to which her work fixed the west coast as a primitive space outside modernity, a space governed by natural history, not social history. Carr was central to constructing the coast as a place of fantasy and desire for Canadians, and is partially responsible for the mythic character of the rainforest. By exploring the circulation of her work in the present, I also raise questions about how a colonial past continues to irrupt in a postcolonial present, as well as how her landscape imaginaries have come to be challenged, and perhaps even displaced, by a new generation of First Nations artists who seek to present a very different image and aesthetic of the forest.

In chapter 6, I return to the present, and turn from the apparently subjective domain of art and aesthetics to the supposedly solid ground of science, in order to explore how competing claims about the forest's "disappearance" are constructed. A Sierra Club map of the "disappearing rainforest" provides an entry point into a discussion of ecology as a contested field, one that, despite its claims to mirror nature, fails to provide the unambiguous answers that we look to it for. By blurring the boundaries between science, culture, and politics, it becomes possible to see science as itself one of the constitutive elements of nature's social production, rather than an objective arbitrator of environmental politics. The Conclusion returns to the science of ecology in order to expand one of its central tenets—relationality—to include humans, and to delink it from romantic notions of unity and balance. Instead, the chapter asks a much different question: In a world in which both nature and our knowledge of it are everywhere entangled with human affairs, what does a radical postcolonial environmentalism look like? What does it mean to "save" the rainforest? By redefining the terms in which we approach questions of nature, culture, and politics, it may be possible to lay the basis for a reinvigorated ecopolitics that is at once more expansive in its scope and more reflexive about its claims.

2. Producing Marginality

Abstraction and Displacement in the Temperate Rainforest

Focusing attention on the presence of the colonial imagination in today's post-colonial society is not a gesture of ahistoricism—on the contrary. Problematiz-ing historical distance and analyzing the way streams of the past still infuse the present make historical inquiry meaningful.

—Mieke Bal, "The Politics of Citation"

The Passion of Simon Lucas

Speaking before a Royal Commission on Forestry in 1975, Simon Lucas, then chair of the West Coast District Council of Indian Chiefs, outlined his people's frustrations: "We feel more isolated from the resources to which we have claim than at any time in the past," to which he added simply, "this is becoming more so."[1]

Lucas's poignant statement articulated the experience of many First Nation peoples during and after the rapid expansion and consolidation of British Columbia's coastal forest industry in the 1950s and 1960s.[2] In the twenty years preceding the commission, traditional "Nootka" (Nuu-chah-nulth) territories on the west coast of Vancouver Island had been rationalized within the regional and global space-economies of industri-al capitalism, reterritorialized according to the spatial and temporal log-ics of "sustained-yield" forestry, and incorporated into forms of eco-nomic and political calculation in the interests of corporations, state institutions, and non-Native communities far removed from their vil-lages. Incarcerated on tiny reserves imposed late in the nineteenth cen-tury, the Nootka increasingly found that they had almost no access to the resource wealth of their lands, except on the shifting margins of the white wage labor force.

Two decades later, it was not clear that much had changed. In sub-missions to a 1991 BC Task Force on Native Forestry, Native groups ex-plained that they continued to find themselves marginalized in their own territories (Derickson 1991). As workers they remained under-represented in an industry where seniority and Native seasonal practices were at odds. Few First Nations held forest tenures. And, as would become evident again in events leading to the Clayoquot Sound land-use plan of 1993, decision making still occurred from a distance—both spatial and institutional.[3] The result was that in the 1990s, as in the 1970s, Native communities witnessed landscapes historically tied to their communities remade by industrial forestry, often at an accelerating pace.

This chapter (along with chapter 3) explores the marginalization of First Nations in conflicts over environment and resources on Canada's west coast. As I discuss at more length later, this marginalization has not gone unchallenged. From the Nuu-chah-nulth in Clayoquot Sound to the Haida in the Haida Gwaii and the Gitksan and Nisga'a along the Skeena and Nass rivers in northern BC, Native groups have fought determined battles to be heard amid the often rancorous debates in non-Native society over the fate of the forests and forestry in the province. Yet, as the task force discovered, Native voices—if heard at all—have often been incorporated as simply one among many special interests within a system of forest management founded on productions of *colonial space* in the nineteenth century that separated and segregated Native reserves from Crown lands. Given that the forest system had as one of its enabling conditions the political fiction of an empty and homogeneous national space, it is perhaps unsurprising that when BC forest politics exploded onto the national and international stage in the early 1990s, First Nations found themselves once again on the margins. Despite the massive national and international attention given to conflicts over BC's old-growth forests, forest disputes were routinely framed in binary terms: environmentalists pitted against industry in a struggle over the fate of the nation's primeval forests. As I noted in chapter 1, the lengthy article in the *Globe and Mail* that broke the Clayoquot Sound story to a national audience profiled key individuals on both sides, yet at no point mentioned that the Sound lay within the traditional territories of the Nuu-chah-nulth. This seemed incidental to the political drama unfolding; it was as if Simon Lucas had never spoken.

How do we account for the passion (emotion/suffering) of Simon Lucas? And why, more than two decades after Simon Lucas spoke so eloquently of his people's increasing isolation, should forestry issues still be discussed in the terms first laid down in the 1945 Sloan Commission, which divided the province into "sustained-yield units" with no regard to existing Native territorialities?[4] What itineraries of silencing are at work in BC's temperate rainforests, and how do these reenact earlier colonial displacements that occurred in the late nineteenth century? Ultimately, how does a nation founded on principles of liberalism manage to disavow these constitutive erasures, even as it proclaims the centrality of indigenous peoples within its myths of origin?[5]

This chapter and the next focus on what I call, for want of a better term, technologies of displacement in the temperate rainforest. Specifically, they focus on how the temperate rainforest has been constructed as a domain separate from the cultural geographies of individual Natives and their communities, and subsequently resituated within very different geographies: the "nation," the "market," and more recently, the "global biosphere." The significance of these displacements, both in the past and in the present, is that they authorize very different people to speak for the forest than before. These are no longer traditional "owners" and "stewards" (such as those whose authority resides in the Nuu-chah-nulth system of *haḥuulhi*),[6] but, rather, forestry corporations, professional foresters, economic planners, and environmentalists.[7]

Before proceeding, two issues demand attention. First, my intention is not to speak for First Nations, nor to recover a unified Native voice of resistance to colonial forest practices. There is a long history of paternalism in Canada where state officials, religious leaders, and academics have claimed for themselves the position of representative of Native peoples, perhaps the most insidious and effective form of colonial power, because it appears to be no form of power at all.[8] Nor, in any case, is there a single, unified "Native" voice waiting to be uncovered by the researcher or ethnographer, a form of consciousness shared by all BC First Nations with origins prior to the displacements and dislocations of colonial modernities on Canada's west coast. There is no untouched kernel of culture lying dormant and awaiting renewal, no domain of authenticity hidden behind modern Native life; rather, there are only multiple Native cultural and political identities that assume their present form *within* and as *effects of* contemporary struggles over land, resources, and environment. Even less is it my intention, as some interlocutors have erroneously thought, to privilege Native land and resource uses as somehow more ecological than those modern forms of resource use that have emerged at the intersection of national development, global capitalism, and technoscience. This sort of romanticization of the ecological Indian is fraught with problems, not least its demand that the Native fill a slot in a Eurocentric primitivist imagination. Although Native communities occasionally fill this slot for strategic reasons, they always do so at the risk of being declared not authentically indigenous at a later time if they happen to step outside its bounds, as they inevitably must.[9] Rather, my intention is to respond—critically and ethically—to calls by Native peoples that, as non-Native Canadians, we learn to re-cognize the postcolonial present as in important ways *continuous with* colonial pasts, and that we learn to identify the ideologies of nature, culture, and nation that secure and sustain relations of domination. These chapters begin the task of reworking our critical capacities in order to bring into view how colonialist practices not only are part of an "ugly chapter" in Canadian history, but are still endemic today, inscribed into the very ways that we visualize and apprehend the world around us.

Second, my argument is informed by theoretical and epistemological positions outlined in chapter 1. To investigate the persistence of colonial relations in BC's forests, it is necessary to examine a series of forest discourses that at once delineate a *field of politics* and *authorize certain actors* within it. These range from the intertwined discourses of "conservation" and "nationalism" to environmentalist constructions of "wilderness" and the "indigenous." In political and analytic terms, what such an approach offers is an analysis of the cognitive failures that have contributed to the marginalization of Native voices in present-day resource development and land-use conflicts. At a more theoretical level, it contributes to a growing and productive encounter between poststructuralism and political ecology. One of the hallmarks of poststructural theory is that social and political identities are constituted historically—that they do not preexist their articulation in specific events, but are constituted and reconstituted in and through them. Accordingly, I argue that concepts of the forest, indigeneity, the nation, and so on are not given once and for all, but are themselves critical sites of political struggle. In this sense, identities are *performed* rather than static, not in some voluntarist fashion in which individuals freely choose between alternatives, but in the sense that these identities must be continuously reenacted and stabilized within the discursive practices that give them their legibility if they are to retain their political and ideological force. As noted earlier, this raises important questions concerning the nature of power and the site of politics. Here I echo Edward Said (1994, 7), who noted in his *Culture and Imperialism* that the "struggle over geography" in nineteenth-century imperialism was not only about "soldiers and cannons," but also about "ideas, about forms, about images and imaginings"—in other words, about representational practices that organized what was visible (and what remained invisible) and that, in turn, authorized the administration and regulation of foreign lands and people (cf. Mitchell 1988).

Said's comments apply today as much as they did to the nineteenth-century world he was describing. To claim that BC's "war in the woods" turns in large measure on questions of *representation* does not reduce politics to philosophical or literary concerns. On the contrary, it insists on the political significance of representational practices, even more so today in a political climate where Native groups must not only seek redress for past colonialist practices (dispossession of lands, the physical separation and segregation of Native peoples on reserves, the paternalistic administration of "Native affairs" through the Indian Act), but also confront a growing backlash among non-Native residents that has girded itself in the seductive rhetoric of liberalism in order to question why Natives should be granted "special privileges" to which everyone else is not entitled, or rights that might limit the individual freedoms of others.[10] In the face of this double erasure, the struggle over geography necessarily becomes a struggle over which relations are visible in the temperate rainforest and which are not.

Abstracting Timber, Displacing Culture: The Making of Natural/National Resources

Let me return to the matter of nature, capitalism, and postcolonial modernities on Canada's west coast. At least since Marx modernity has been characterized as a maelstrom of change, whereby social life is continuously made subject to ever-new forces of "creative destruction" and is experienced as "perpetual disintegration and renewal" (Berman 1982, 15; see also Harvey 1989). If this is true of *social* life, then its corollary must certainly be that *nature* itself has been subject to violent change, made and remade in the image of commodity production and technological change. This occurs today on a global scale. From industrial agriculture and suburban sprawl to global warming and acid rain, the material landscapes of advanced capitalism have been reordered to such an extent that at the end of the 1980s alarms were sounded about the end of nature (McKibbon 1989), and, in one of the great ironies of the period, the "preservation" of nature increasingly came to follow the logic of the commodity form, such that "ecological reserves" were produced either as the negative image of capitalist production or as themselves highly commodified spaces (cf. Katz 1998).[11]

Canada's west coast differs from this general claim only in the predominance of a *single* commodity: timber. Perhaps captured best by an abstraction called the "normal forest" (Figure 2.1), sustained-yield forestry's focus is first and foremost on the rationalization of *timber* production amid the heterogeneous, unruly, and culturally infused temperate rainforest. Indeed, the production of nature on the west coast of Canada centers so completely on the practices of scientific, industrial forestry (or, conversely, on the preservation of its opposite, "pristine nature") that forest conflicts have tended to focus only on the proportions of the landscape dedicated to each.

In the face of this continual remaking of nature, radical ecologists have persuasively argued that in advanced capitalism specific entities—produced as commodities for exchange—are violently abstracted and displaced from their ecological surrounds, threatening the continued viability of ecosystems and the wildlife populations they sustain. Indeed, this has become perhaps the most compelling green critique of capitalist modernity—that nature becomes displaced into systems of meaning, production, and exchange that have no intrinsic relation to an underlying ecological order.[12] Without reducing the force of this critique (although its terms must also be examined), I want to extend its central insight in a different direction. If industrial forestry on Canada's west coast has been guilty of abstracting the commodity from its *ecological* surrounds (with devastating consequences for local ecosystems, including hydrological cycles, fish stocks, and wildlife populations), then it has equally been involved in the abstraction not merely of "timber," but also of the "forest" as a unit of production, from its *cultural* surrounds

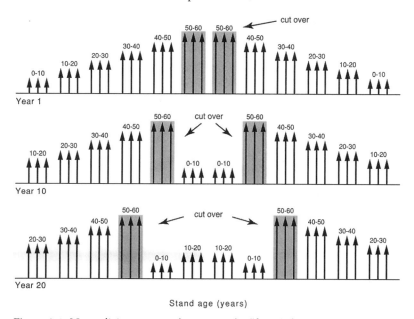

Figure 2.1. Normalizing nature: abstracting the "forest" from its surroundings. Reproduced by permission of David Demeritt.

(with equally devastating effects on local Native communities). With this in mind, let me return to the "war in the woods" in BC in order to trace some of the itineraries of silencing that have both enabled and hidden these displacements.

Custodians of the Forest

How are we to understand the isolation of the Nuu-chah-nulth? We can begin by turning to one of the most prominent forest corporations in BC during the 1980s and 1990s, MacMillan Bloedel (MB).[13] In many ways, MB epitomized the corporate face of industrial forestry in BC during the period, and partly for this reason it was the subject of intense scrutiny by local and international environmental groups.[14] In response, MB staged an aggressive publicity campaign in order to legitimate its authority as the forest's custodian, with television and newspaper ads, visitor centers, public-relations pamphlets, forest tours, and news releases constituting its most common methods.[15] In using these tactics, MacMillan Bloedel was far from alone; other companies engaged in similar strategies, and presented their cases to the public in comparable terms, often coordinated by the BC Forest Alliance, a prominent industry lobby group.[16]

For my purposes, MB's efforts to establish its authority in BC's temperate rainforests offer an important window into contemporary processes of displacement in the temperate rainforest.[17] What interests me here are the terms through which MB's authority is staged, and the erasures and disavowals that enable its claims to be self-evident. To say this differently, by exploring MB's rhetoric of "custodianship" we can begin to see how the cultural and epistemic space of the "forest" is also a space of abjection, where certain elements are cast off.

One of the company's early publications from this period—a booklet titled *Beyond the Cut*—can be used to illustrate the point.[18] Graced on its cover by a photograph of a new seedling growing out of the stump of a logged tree, the booklet is meant to convey a comforting story of scientific management and sustainable forestry, and to support MB's continued access to the "public" forest. Carefully crafted, it combines glossy photographs, graphs and diagrams, and a simple, accessible text, to form an attractive, easy-to-read document. For my purposes, its public-relations format—now ubiquitous—is less important than the terms, and strategic silences, through which the company establishes its authority. Essentially, this takes the form of an invitation to evaluate the company's forest practices in terms of three criteria: expertise, efficiency, and responsibility.

Expertise, arguably the most important term in the triad, is demonstrated through conventional, but remarkably effective, tropes of scientific and technological progress, a strategy that MB used repeatedly in its campaigns during the 1990s (Figure 2.2). MB's various booklets and brochures, for instance, routinely depict an army of scientists, resource managers, and "environmental specialists" engaged in fieldwork, laboratory research, and computer modeling. In one brochure, an entity called LUPAT—MB's Land Use Planning Advisory Team—is profiled, and described as a crack team of "environmental specialists" with expertise in "soils, wildlife, fish, water resources, and growth and yield projections." Elsewhere, tables and graphs present readers with extensive technical information about the forests and forest management, accompanied by photographs depicting "high-tech greenhouses" growing "genetically superior offspring."[19] Readers are also told that with the development of

Figure 2.2. Exhibiting expertise: forestry issues are often framed in technical rather than social or ethical terms, thereby shifting attention away from contentious issues such as land tenure. Source: MacMillan Bloedel.

new information technologies, MB is now able to "simulate the growth of present and future forests for 200 or more years and examine the results of different constraints" (MacMillan Bloedel, n.d., 6), and is thus able to plan far beyond the time frame of a single human generation.

Similar strategies could be found in visitor centers that the company operated in Port Alberni and Tofino.[20] Here, exhibits and interactive displays described the company's forest management cycle (Figure 2.3) and invited visitors to test their knowledge of temperate rainforest ecosystems. One display—an especially instructive computer game—combined the two, challenging visitors to take on the role of a forest manager whose objective was to maximize the forest's productivity. As players quickly discovered, this was possible only through *rational* management practices based on an *expert knowledge* of nature, something that all players eventually discovered they lacked.[21] In the case of the company's prominent forest management cycle display, the focus rested firmly on the scientific expertise needed, documenting the careful assess-

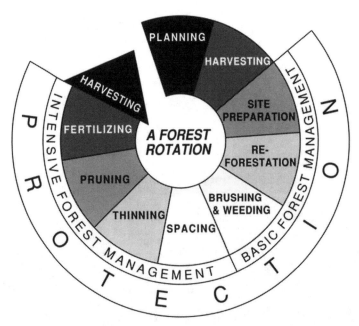

Figure 2.3. MacMillan Bloedel's forest management cycle as "cognitive failure." Discourses of scientific management efface the many ways that British Columbia's forests are contested spaces, as well as the historical practices by which the forest was produced as a domain separate from people. Source: MacMillan Bloedel.

ment and monitoring necessary to develop ecologically sensitive, site-specific plans. As explained in one display panel, after harvest "a site treatment plan is prescribed based on consideration of soil depth and texture, the amount of woody debris, and the species and health of small trees already growing on the site." Like a kind father guiding a child through the bewildering spaces of the city, MB guided its urban visitors through the alien spaces of the "working forest." Indeed, to complete the reality effect, visitors were invited to sign up for tours of the company's cutblocks in order to witness outside the centers what was so carefully exhibited inside!

These are familiar tropes that turn on discourses of modernization and development, and that rely heavily on the figure of the disinterested scientist. Neither requires much additional elaboration. The former carries the promise of ever-increasing rationality and progress, whereby practices based on such apparently "irrational" foundations as myth or tradition are progressively replaced by practices based on reason. Not only does this assume a single teleology to development with Europe as its model, but, as Jürgen Habermas (1972) persuasively argued in his early work, the discourse of scientific rationality effectively displaces questions of political legitimation from the realm of values (moral reason) to the realm of technique (instrumental reason), such that technical interests come to be established as values in their own right (see also Habermas 1987).[22] Donna Haraway (1991, 1997) has similarly argued that within this story of progressive enlightenment, it is the (male) scientist who seems least tarnished by error, because he appears to stand in a space of reason located outside the polluting influence of culture and politics.[23]

It is partly through enrolling a discourse of "technology" and "progress," rather than relating the details of specific practices, that MB's publications achieve their political effects. What matters is less whether MB's scientific credentials are actually solid (most readers have no way of evaluating this) than whether the company appears properly *scientific*. In short, science operates as the sign of the modern, a story that been repeated globally (see Prakash 1999). In turn, instrumental reason becomes a surrogate for moral or political reason. Placed beside highly aestheticized displays of forest renewal (provocatively inverting the before and after photos that the environmental movement has used so effectively in its campaigns against logging—see chapter 6), images of modern rationality shape a comforting story of scientific management that contains an unmistakable message: left to the company, the forest will be renewed, if not improved, for the benefit of future generations.

Despite the increasingly "reflexive" character of European modernities (Beck 1992), science and technology still remain privileged signifiers. In MB's literature, however, they work alongside another objective:
convincing readers and viewers that the company is managing the province's forest resources in the best interest of the "public." This follows
two paths: (1) showing that the company obtains the greatest value from
the resource (efficiency) and (2) demonstrating that the company is responsive to nontimber forest values (responsibility). The first is accomplished by drawing direct links between MB's activities and the consumer demands and the economic health of the province. Thus, MB
explains that its operations are necessary because it is meeting society's
basic material needs by "grow[ing] and harvest[ing] trees and turn[ing]
them into quality forest products that help satisfy society's need for
communication, shelter and commerce" (MacMillan Bloedel, n.d., 2).
At other points, MB emphasizes its contributions to employment and to
government revenues, showing itself to be indispensable to the economy
(itself an effect of the statistical tables, charts, and graphs that produce
it). Likewise, in TV ads that ran in the mid-1990s, the company responded to criticisms that it was more interested in achieving windfall
profits (by liquidating forest resources and selling them as raw material
or as primary manufactured goods) than it was in contributing to the
further development of the province. By focusing on new value-added
products such as Microllam and Parallam (brand names for laminated
veneer lumber, and parallel strand lumber, respectively), which use fiber
from "less-than-perfect trees," the company demonstrated that it was
committed to obtaining ever new and increased value from the forest resource.[24] The company, viewers were assured in a statement whose
double meaning was certainly not intended, was "making the most of
a renewable resource."

A final element in the company's strategy was its demonstration of
corporate citizenship. Its literature noted that MB facilitated and took
into account public input, had opened "its" forests to multiple uses, and
far exceeded its legislated responsibilities for the preservation of wildlife
habitat: "The forests of British Columbia are a great source of pride and
concern for the people of the province. No one wants to see them deci-
mated or devoted exclusively to timber production" (ibid., 12). To this
end, the company's literature explained how it cooperated with government agencies to preserve examples of old-growth forests in areas of special beauty or critical wildlife habitats. Thousands of hectares of forest-
land on Vancouver Island, British Columbians were told, had been
transferred from MB ownership or licensed tenure to parks and ecologi-

cal reserves, while logging in other "sensitive" or "aesthetic" regions had been deferred indefinitely. Under the guise of corporate citizen, MB appeared as the disinterested manager of a public resource, possessing the most objective knowledge and advanced technology, and therefore in the best position to mediate between the claims of interest groups clamoring to have their *particular* visions of the forest recognized: "As custodians of the forest, MB protects, cares for, and renews this great resource for the benefit of present and future generations. . . . The company's forestry policies are based on achieving an *optimum balance* for all users taking into account economic, recreational and environmental factors" (p. 2; emphasis added). In a world of competing demands and uncertain economic and ecological futures, only MacMillan Bloedel knew best.

Public Fictions and Fictional Publics

My point is not to dismiss the value of science; in a province where the forestry industry is so central, the rational management of the region's forests has immense implications for individuals and communities alike.[25] Nor is my intention to pass judgment on MB's claim that it has been a good steward of a public resource. To do so would be tacitly to accept the very terms that have organized forestry politics in the province, when it is precisely these terms—or, more to the point, the political geographies that these terms presume—that need closer scrutiny.

Rather, I have focused on questions of "scientific management" in order to suggest that this rhetoric relies on a more fundamental disavowal of *colonial history* that is constitutive of the company's claim to custodianship (and behind it, the state's claim of territorial sovereignty). These texts, and forestry discourse more broadly, must be read for what they do *not* say—the silences that enable the transparency of their message. Thus, to accept MB's invitation to evaluate its forest practices as the basis for establishing authority—even if, as in the eyes of local and international environmental groups, the evaluation proves negative—is to be complicit in the same historical amnesia that sustains the company's claims.

This becomes clear when we note that MB's rhetoric of enlightened custodianship pivots on the mobilization of a potent political fiction—the public. MB's authority appears legitimate because the company is seen to be meeting the standards of rational management (expertise), economic development (efficiency), and ecological sustainability (responsibility) that the "public" demands for "its" forests.[26] But this fictional public serves a crucial ideological function in postcolonial BC. By positing "a" public, both MB and the state assume an undifferentiated national space, situate readers within it as individual citizens, and thus construct a single, shared interest in the forest. This permits debates over forestry to proceed on the basis of a founding assumption—that the health of the resource is by extension the health of the provincial "body" and its constitutive elements (citizens), and thus that legitimacy is solely an issue of who is the best *manager* of the resource.

Not only does this discourse of a public interest flatten out differences within BC society along lines of class, race, and gender (which can enter the arena of forest politics only as special interests), it also erases histories of colonial displacement. By displacing the forest into the conceptual space of the "public," MB abstracts the forest from its specific cultural and political contexts and relocates it within the abstract, rhetorical spaces of the province and the nation. Such an abstraction is possible, of course, *only if the forest is first made to appear itself as an unmarked, abstract entity emptied of any social and cultural content.* Indeed, in its promotional literature, and in the Tree Farm License maps and the satellite photographs found on the company's visitor-center walls, this displacement is effectively achieved: the forest appears as a purely natural object devoid of any cultural and political geography (see Ingram 1994).

Herein lies the seductive magic of MB's representational strategies, or indeed *any* nationalist discourse of resource conservation that relies on the fiction of the "people." Ultimately, MB's forests are both any forest and no forest at all! Emptied of cultural history—displaced into the empty homogeneous space-time of the nation—the forest becomes a unit governed by natural history alone, incorporated within a national (or provincial) discourse of sustainable resource management, and tied to the administrative spaces of the state rather than to the local life-worlds of its Native inhabitants.

It is this absence of cultural history, rather than the positive discourse of scientific management, that in the final analysis makes MB's claim to custodianship transparent. Framed solely as resource landscapes without any competing territorial and political claims, such places as Clayoquot Sound are at once *de*territorialized and *re*territorialized as the province's forest, divided into units (Tree Farm Licenses), allotted to leaseholders (such as MB), and subjected to rational management (computerized models, scientific and economic rationalities) so as to produce sustained yield through rationalized forest rotations, all part of the administration of a national population and economy. In one of the many ironies found in BC's forests, foresters and economists today refer to this forest as the "normal" forest (see Figure 2.1).[27] While ecologists have complained that this has the effect of rendering the "natural" forest abnormal, my point is that it also *normalizes* the territorial claim of the nation-state and inscribes this claim into the very fabric of west coast natures.

In sum, MB's rhetoric assumes a priori the juridical, political, and geographical space of the nation-state and ignores its historico-geographical constitution. By staging the space of the nation-state as given (rather than something that must be continuously reiterated in order to retain its force) the Tree Farm Licenses that MB holds, and the "normal" forest it manages, are rendered common sense and thereby removed from the domain of politics and resituated in the domain of ontology. But this is most assuredly a political ontology; in the context of postcolonial BC, assuming the fixity of national/natural resources is a bad epistemic habit, one that incorporates colonial pasts by rendering invisible the historical and discursive practices through which these spaces have been constituted and naturalized, and that in turn authorizes certain voices—resource managers, bureaucrats and nature's "defenders"—to speak *for* nature. In a neat symmetry, then, what MB authors, authorizes MB.

Genealogies of "Wilderness": Unthinking Neocolonial Cultures of Nature

To argue that assuming the fixity of natural/national spaces is a bad epistemic habit is to suggest the possibility of other forest epistemologies not so tightly bound to a colonial or, later, national imagination. It is now commonplace to see the suturing of a (singular) national space or identity as provisional and unstable. Indeed, as we have just seen, the construction of the temperate rainforest as a natural/national space follows a performative logic: it is something achieved through specific material practices (mapping, discourses of scientific management, etc.) rather than given in advance. These practices, in turn, are iterative in character; they internalize and repeat earlier displacements, which over time take on the appearance of common sense. It is this performative, temporal aspect of the "nation" that, Homi Bhabha (1994) argues, introduces a space for political agency, an assertion I explore at the end of chapter 3, where I examine efforts by the Tla-o-qui-aht (a Nuu-chah-nulth First Nation) to articulate a different environmental/political geography.

Here I want to press another claim—that the self-evidence of any natural/national geographics can be disrupted by bringing into view its contingency. In the preceding pages, I argued that in the midst of BC's war in the woods, MacMillan Bloedel's authority was built, in part, by constituting the forest as a natural and public resource. Here I argue that this was not merely a cynical and calculated move on the part of a savvy corporate actor; it was also made possible by, and reiterated, histories of "seeing nature" on Canada's west coast that were colonial in character, and, perhaps more to the point, part of the historical operation of colonial power. In other words, the authority of the state and corporate capital in BC's forests today can be related to practices of imagining, representing, and purifying "natural" landscapes in the past. By taking our cue from Foucault (1977) and "listening to history"—that is, by refusing the notion of timeless forms and instead showing the *historical emergence* of these rhetorical and physical spaces named "nature" and "culture" on Canada's west coast—it may be possible to find something altogether different behind things, and thus challenge the political claims built on the temperate rainforest's constitutive absences.

It should be noted that the very notion of genealogical critique implies that it can never be complete, for it assumes that identities have no single origin but instead are "fabricated piecemeal from [many] alien forms" (Foucault 1977, 142). In what follows, I am necessarily selective; in order to begin the task of unthinking the (neo)colonial assumptions found in MB's texts, I explore at some length the writings, sketches and maps of George Mercer Dawson, a remarkable geologist and amateur ethnologist who traveled the coast with the Geological Survey of Canada in the 1870s and 1880s. By reading Dawson's journals and reports against the grain, I show that the fixity of national/natural spaces in British Columbia today—and the ongoing representation of British Columbian landscapes as a nonhumanized hinterland and thus an empty "national" space—is in part the effect of an array of historical practices that are reiterated in the present. Hence, what appears today to be nature's essential difference from culture—in other words, its identity as "wilderness"—can be rearticulated as an effect rather than an ontology.

Nature's Modest Witness: Science, Representation, and Politics

For students of west coast history, Dawson presents an intriguing figure. Born to a prominent Montreal family (his father, Sir John William Dawson, was principal of McGill University), Dawson spent his childhood in the circle of an emerging English cultural and intellectual elite based in Montreal. As a child he also contracted Pott's disease—a form of tuberculosis of the spine—which left him with a twisted body and the stature of a boy of fourteen. He would nevertheless be one of the most intrepid European travelers on the west coast at a time when its physical terrain and cultural geographies were only beginning to be known, and where forms of colonial power extended only as far as military force, European networks of communication, and modern regimes of power/knowledge reached (Harris 1997). His travels covered vast regions that until then had seen few Europeans, including the Yukon, much of northern British Columbia, and large areas of the coast and interior of the province. About each of these regions Dawson produced important works, including celebrated geological and ethnological studies.

For my purposes, Dawson merits attention for reasons that go beyond biography. His travels coincided roughly with the years that the federal Indian Reserve Commission (IRC) was allocating and demarcating a system of small "Indian reserves" across the province.[28] This makes Dawson's texts especially valuable. To date, it has been the reserve commissions themselves that have most attracted the attention of historians (Fisher 1977; Tenant 1990; Harris forthcoming). It was through them, after all, that colonial relations were most effectively etched into the territory of the new province, bounding, within a quasi-legal and cartographic discourse, a "primitive" space of Native villages, and beyond this positing an empty and unclaimed nature open to settlement or enterprise.[29] More than any other single action, the activities of the reserve commissioners set out a *political cartography* that future settlement and resource development in the region would follow; after these commissions the division of west coast territories into two distinct domains— one natural/national, the other cultural/indigenous—became increasingly naturalized as a way of seeing, and as a political fact. Yet, although the activities of commissioners, the political tactics of the myriad individuals involved in the land question, and resistance among First Nations in the years that followed have received attention, less notice has been paid to the representational practices, scientific discourses, and imaginative geographies that underwrote the commissions' cartographic imagination. Gayatri Chakravorty Spivak (1990) reminds us that administrative practice presupposes an irreducible theoretical moment. What I take this to mean, in part, is that cartographic practices like those of the IRC—reiterated a century later by British Columbia's forest ministry and corporations such as MacMillan Bloedel—did not simply occur as a result of administrative fiat but were made possible by a series of other discursive practices that made legible to power a space of administration, and that in turn invited and legitimated the actions of state administrators.

Dawson's work is thus doubly intriguing. Placed alongside the mapping strategies of the IRC, it brings into view the ways in which the cartographic inscription of colonial relations through the reserve system was prefigured and facilitated by a more general textualization of the west coast that rendered it a legible space, albeit in highly partial ways.[30] As is now commonly recognized, *legibility* and *instrumental reason* bleed into each other (cf. Braun 2000a). By keeping this firmly in view, we can move toward a more complete understanding of how the IRC was able to proceed with its national(ist) cartography, as well as how contemporary forest politics reiterates this pivotal colonial moment—a point lost on many commentators, who assume that the story *begins* with the administrative practices of colonial officials (or, after 1871, with the work of officials appointed by provincial and federal governments).[31] At the same time, Dawson's work must be read with care. Dawson purported to see these landscapes through the eye of science, rather than as a colonial apologist. In contrast to the maps of the IRC, which can be situated within the realm of politics with somewhat less effort, Dawson appears as a "modest witness" simply recording the truth of west coast natures and Native cultures, operating objectively in a sphere that appears insulated from the intense social and political struggles that swirled around these same places.[32] Yet, "legibility" is not something *in* nature awaiting discovery by the disinterested observer; it is achieved through historically situated representational practices. As I will show, despite their avowed objectivity, Dawson's writings, sketches, and maps never left the sphere of history and politics. Instead, they provide an exceptionally vivid picture of the colonial regimes of vision that underwrote the production of distinctly colonial spatialities in British Columbia.

Displacements: Bounding the Native, Producing Nature

Dawson was a prolific writer whose interests ranged widely. It was not uncommon for him to shift from speculation over the role of glaciation in the formation of British Columbia's coastal landforms to an analysis of folklore and language among Native communities. Although his writings appeared in such diverse places as the the *Canadian Naturalist* and *Harper's Weekly*, it is his work for the Geological Survey of Canada (GSC) for which he is most acclaimed. In his official capacities as a field geologist, and later director of the GSC, Dawson contributed numerous studies to the Survey's annual reports. Among these were detailed surveys of various regions of British Columbia, fulfilling one of the conditions that the colony of British Columbia attached to union with the Dominion of Canada in 1871. The importance of the GSC and its surveys has been noted by a number of scholars. Morris Zaslow (1975) has argued that the Survey was a prime instrument in "pushing back the frontiers" of the Canadian state. Suzanne Zeller (1987) tied the formation and activities of the GSC to imagined geographies of a "transcontinental" nation, and also to more utilitarian concerns for national economic development. Both writers make important links between the GSC and nation building. However, my reading of the Survey, and of the work of George Dawson in particular, departs from theirs in important ways. For the most part, Zaslow and Zeller remain committed to a realist epistemology. Although the GSC is shown to be an important arm of the state, both accept that the role of the Survey was to *enumerate* the wealth of the new nation, leaving aside the relations of power that both infused and enabled its representational orders. Both also retain a model of state rationality that assumes both a notion of sovereign power and an instrumental relation between the state and knowledge of its territory. By this view, the Survey was merely concerned with a long—albeit consequential—process of documentation in the interests of a state whose power existed prior to, and apart from, how its territories were rendered legible.

The GSC surveys were certainly occupied with the enumeration of the qualities of the state's territory, but to tell the story in this way—as merely the objective documentation of a pre-given landscape—underestimates the role that the surveys played in *constituting* these landscapes as intelligible domains. The previous story assumes too much. On the one hand, it invests the state with occult powers to know in advance the utility of scientific knowledges that were then only just emerging, or that would be reshaped by the very practices of institutions such as the GSC and its workers. On the other hand, it assumes that the sovereign power of the state preceded the knowledge-producing practices of its agencies, rather than in part emerging from the very ways that the state's territory came to be represented and known. To return to an earlier point, what is of interest is not how the GSC's surveys served the interests of a state concerned to possess and occupy its lands, but how they constructed spaces of visibility (and spaces of invisibility) that were at once both partial and of immense social, ecological, and political consequence (Rajchmann 1988; Gregory 1994). In other words, at issue are not questions of instrumentality and interests, but of enframing and power.[33]

We can trace this conjoining of enframing and power in Dawson's work at a number of levels. Most immediately, Dawson's reports provided west coast landscapes with the appearance of order and totality; his surveys—along with other knowledge-producing practices in the period—gave form to a geography that had previously been known only in a fragmentary manner. This can be seen clearly in Dawson's report on his 1878 explorations in the Queen Charlotte Islands (Dawson 1880b). Like most GSC reports, it begins with a general description of the islands—a bird's-eye view that situates them in relation to the rest of the nation, and provides a general outline of their physical geography: coastline, harbors, rivers, mountains, and so on. Subsequent chapters and appendices located and described the islands' geology, zoology, botany, and Indians. While appearing to merely document the landscape, the very organization of Dawson's survey was more than just incidental to the operation of colonial power. Its general overview, for instance, gave to readers in distant centers of administrative power a *cartographic* orientation, one that permitted the islands' constitutive parts to be placed within a larger whole, and the islands, in turn, to be situated within a wider national geography. At the same time, Dawson's report divided the islands into discrete domains; plants, animals, rocks, and Indians were documented separately, as if unrelated entities, a practice that had its antecedents in eighteenth-century natural history. In turn, each domain could be further subdivided, with objects described according to their specific characteristics. Geological specimens, for instance, were divided into a number of classifications: Triassic; Cretaceous coal bearing; Tertiary; glaciated and superficial deposits, and so on. Likewise, Dawson analyzed the Haida in terms of their physical appearance, social organization, religion and medicine, the potlatch and distribution of property, folklore, villages, and population.

As I will explain shortly, this division of west coast landscapes into discrete knowledge domains was far from innocent. As Mary Louise Pratt (1992, 31) explains in reference to natural history in the eighteenth century, such practices "extracted specimens not only from their organic or ecological relations with each other, but also from their place in other people's economies, histories, social and symbolic systems," displacing them into systems of knowledge and forms of economic and political calculation unrelated to the practices of local inhabitants. First, however, it is worth noting that Dawson's surveys, far from a neutral documentation of the qualities of the state's territory, contained within them a remarkable anticipatory vision that saw landscapes not in terms of their contemporary cultural geographies, but in terms of a future national development. It must be remembered that at the time he undertook his Queen Charlotte trip, the white settler population in BC was still outnumbered by Natives. Further, whatever European settlement had occurred was clustered almost entirely at the extreme southwest corner of the province (Galois and Harris 1994). Beyond its extent, the land was still primarily known and experienced in terms of indigenous knowledges and Native spatiotemporalities, a historical condition largely erased in Dawson's reports, but clearly evident in his journals, where Native guides, workers, and informants—usually unnamed—routinely enter and exit his daily log.[34] Despite this, and in the face of considerable evidence of the vitality of native territorial, cultural, and economic practices, Dawson and other members of the GSC viewed these territories through an implicit national teleology that assumed that the blank spaces on maps of the new nation would eventually be filled in two mutually reinforcing ways: with ever more complete knowledge, and with the enterprise of white settlers.

Dawson shared this telos of science, nation, and capital with many others. Speaking of his journeys across Vancouver Island, fifteen years before Dawson traveled farther north, the explorer Robert Brown set out the metaphorical terrain that would underwrite the nation's territorial claims: "It was the intention . . . that we should strike through the unexplored sections of the Island, carefully examine that tract as a *specimen,* and thus form a *skeleton* to be filled up afterwards" (quoted in Hayman 1989, 9; emphasis added). Later, Brown described the findings of his explorations as "tests of the whole," by which the regions between his traverses could be judged (Brown 1869). On numerous occasions, he fantasized its future transformation at the hands of settlers, presenting this as a fait accompli and linking knowledge of this "body" to its transformation: "The trail from Victoria to Comox crosses the Quall-e-hum River close to the coast, and an extension of this would form a transinsular road connecting coal miners of Nanaimo and the farmers of Comox with the wild savage of Nootka, Klay-o-quot [Clayoquot] and Barclay Sound" (Brown 1864, 25). Dawson shared many of the same assumptions.[35] Writing of a particularly promising region in the Queen Charlottes, he imagined a land of rich resources and routes of communication: "before many years extensive saw-mills will doubtless be established. . . . The quality of the spruce timber is excellent, and beside the immediate shores of the harbour, logs might probably be run down the Naden River from the lake above" (Dawson 1880b, 38). Like Brown, Dawson enacted the bounded space of the white settler nation, naturalized the sovereign claims of the Canadian state within these borders, and reproduced in a fantasy of European (dis)possession what had already been accomplished elsewhere in the Americas. Much more than enumeration, Dawson's writings were quite literally a means of national *incorporation,* constructing and filling in the internal structure (skeleton) of the nation (specimen) and inscribing this corporeal fiction onto west coast lands.

Significantly this in-corporation of the nation-as-body and the visualization of the body's internal structure were enabled by, and hinged on, the very divisions and displacements noted previously. Here we can begin to trace a relation between knowledge, power, and colonial spatialities, one that not only underwrote the anticipatory vision of Dawson and his contemporaries, but also continues to inform discourses of environment and resources in the present. As noted earlier, at the same time that the skeleton of the nation was metaphorically given flesh, it was also anatomized or divided into its component parts, whereby specific entities—minerals, trees, Indians—were apprehended entirely apart from their cultural and ecological surrounds and displaced and re-situated within other systems of signification. These processes of division and displacement are clearly on display in Dawson's journals. At one level, these journals are unremarkable, documenting the mundane activities of scientific fieldwork. In them Dawson recorded his observations and kept a daily account of his movements, including descriptions of the social and technical mediations that made his movement and his scientific observations possible: people he met, how he traveled, where he stayed, who acted as his guides, instruments used, measurements made, and so on.[36] Yet they also give us a window into the ways that late-nineteenth-century European Canadians saw and represented these complex cultural and physical landscapes. On the reverse side of his journal pages Dawson occasionally wrote additional notes. At times these were little more than details that he had neglected to include in his daily entries. But as often, he used the back of his journal pages to compose a remarkable second text that discussed the regions he visited at more length, elaborating on the physical landscape or on aspects of Native culture.

Much of the material in these sections was later incorporated into Dawson's official reports for the GSC. It is the *organization* of these parallel texts, more than the specific details of their content, that concerns me. Reading these passages, we find a world meticulously divided into two. Some passages dealt exclusively with geology or botany; others dealt only with Native culture. One example will suffice. From Dawson's journals we learn that between 8 August and 10 August 1878, the geologist, accompanied by his brother Rankine, an Indian guide named Mills, and the crew of the schooner *Wanderer,* traveled from Skidegate to Masset, along the east and north coast of Graham Island (see Figure 2.4).

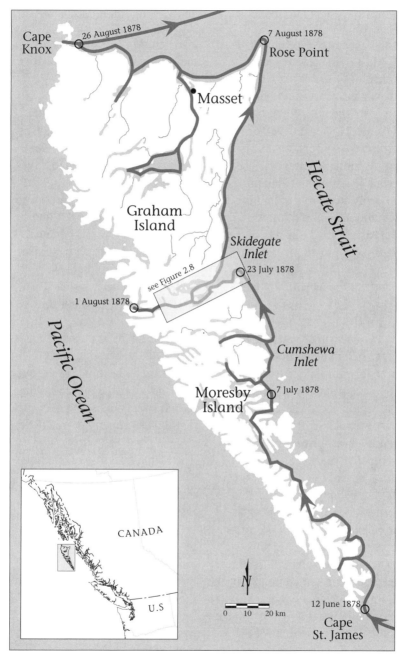

Figure 2.4. Route of George Dawson's journey to the Queen Charlotte Islands, 1878. Adapted from Cole and Lockner (1993).

Text 1: Physical landscape	Text 2: Cultural landscape

The Coast between Skidegate & Masset, in some respects resembles that between Cumshewa & Skidegate. A bare open stretch with no harbour & scarcely even a Creek or protected bay for Canoes or boats, for long distances. The beach is gravelly & sometimes coarsely stony to a point near windbound camp of track Survey. Beyond this it becomes sandy, & though not without gravel continues generally of Sand, all the way to Masset.

Lawn Hill is evidently Caused by the outcrop of volcanic rock described in field book, is probably Tertiary. Beyond this for some distance, & including the region about Cape Ball, cliffs, or low banks of drift-clay, & sands characterize. They are generally wearing away under the action of the waves, & trees & stumps may be seen in various stages of descent to the beach. In some places dense woods of fine upright clear trees, are thus exposed in section, & there must be much fine spruce lumber back from the sea everywhere. Very frequently the timber seen on the immediate verge of the cliffs, & shore is of an inferior quality, rather scrubby & full of knots.

Potlatch. Mr. Collinson gives me some additional light on this custom.

When a man is about to make a potlatch, for any reason, such as raising a house &c. &c. he first, some Months before hand, gives out property, money &c., So much to each man, in proportion to their various ranks & standing. Some time before the potlatch, this is all returned, with interest. Thus a man receiving four dollars, gives back six, & so on. All the property & funds thus collected are then given away at the potlatch. The more times a man potlatches, the more important he becomes in the eyes of his tribe, & the more is owing to him when next some one distributes property & potlatches.

The blankets, ictus &c. are not torn up & destroyed except on certain special occasions. If for instance a contest is to be carried on between two men or three as to who is to be chief, One may tear up ten blankets, scattering the fragments, the others must do the same, or retire, & so on till one has mastered the others. It really amounts to voting in most cases, for in such trial a mans personal property soon becomes

Text 1: Physical landscape *(Cont.)*	Text 2: Cultural landscape *(Cont.)*

The soil is generally very Sandy where shown in the cliffs, or peaty in bottom places where water has Collected. Sand hills or sandy elevations resembling Such, are seen in some places on the cliffs, in section, & there is nothing to show that the Soil away from the Coast is universally sandy, but the fact that the upper deposits of the drift spread very uniformly & are of this character. Further north the shore is almost everywhere bordered by higher or lower sand hills, Covered with rank Coarse grass; beach peas, &c. &c. Behind these are woods, generally living though burnt in some places. The trees are of various degrees of excellince, but most generally rather undersized & scrubby. This part of the coast is also characterized by lagoons, & is evidently making, under the frequent action of the heavy South East sea.

exhausted, but there an under-current of supply from his friends who would wish him to be chief, & he in most popular favour is likely to be the chosen one.

At Masset last winter, a young man made some improper advances to a young woman, whose father hearing of the matter, was very angry, & immediately tore up twenty blankets. This was not merely to give vent to his feelings, for the young man had to follow suite, & in this Case not having the requisite amount of property, the others of his tribe had to subscribe & furnish it, or leave a lasting disgrace in the tribe. Their feelings toward the young man were not naturally, of the Kindest, though they did not turn him out of the tribe as they might have done <u>after</u> having atoned for his fault.

<u>Totems</u> are found among the indians here as elsewhere. The chief ones about Masset are the Bear & the Eagle. Those of one totem must marry in the other.

Figure 2.5. The double vision of colonial science: parallel texts found on the reverse pages of George Dawson's journals. Source: Cole and Lockner (1993, 57–59; emphasis in original).

On 11 August, the day following his arrival in Masset, Dawson records that he attended church, dined with the missionary Mr. Collinson, read recent newspapers, and "wrote up notes." All these events are duly recorded, part of the daily log of his journeys, itself a fascinating document that reveals the complex sociopolitical networks that Dawson negotiated in order to successfully complete his scientific research.[37]

On the reverse, two remarkable texts are found (Figure 2.5).[38] In one we find an enumeration of the "wealth of nature." Here the sciences of botany and geology occupy center stage. Specimens are located and related in space, physical processes are described, and possibilities for establishing communications (or lack thereof) are noted. In the second, Dawson describes Native peoples, their customs and behavior (on still other occasions Dawson described Haida villages in great detail). Potlatches are explained, traditions related, and social organization explicated. There could hardly be any example that more clearly revealed the "double vision" of European travelers and colonial science. It is as if Dawson had simply turned his gaze from one domain of objects to another one entirely unrelated to the first: on the left primeval nature, on the right Native culture. The same separation is found in his photographs: Native villages and individuals on the one hand, displayed as a spectacle of primitive culture; geological sites and landscape vistas on the other, set out as the "natural" domain of the nation (see Figures 2.6 and 2.7).[39] Thus, while indigenous peoples were described in great detail—their physical features and cultural forms documented and enumerated—they were simultaneously detached from the surrounding landscape, which, accordingly, was encountered and described as devoid of human occupation. In other words, Dawson distilled the complex social-ecological worlds through which he traveled into neat, unambiguous categories: primitive culture and pristine nature. Despite the great attention Dawson paid to both, no relations were drawn between them. The former was contained within the village, fixing a sedentary Native presence "in" place, while the blank spaces beyond the bounds of the Native villages were filled with the colored spaces of geological and botanical maps (Figure 2.8). Indeed, his reports on the Haida would eventually circulate separately from his geological studies.[40]

Dawson's representations of the Queen Charlotte Islands (Haida Gwaii) disclose the peculiar colonial visuality that lay at the heart of the production of space on Canada's west coast. To be sure, this vision was rarely singular or without contradiction. Although colonialist double vision pervades his "scientific" writing, Dawson's "political" opinions

Figure 2.6. *Skedans Indian Village.* Photograph by George Dawson, 18 July 1878. In his photographs, as in his journals, Dawson divided the land into natural and cultural landscapes. He was fascinated by the totems at Skedans and took five photographs of the village; the following day, he examined the surrounding landscape for mineral deposits. Source: National Archives of Canada, PA38148.

Figure 2.7. *Basaltic Columns, Tow Hill.* Photograph by George Dawson, 9 August 1878. As a geologist, Dawson was trained to evaluate British Columbia's landscape in terms of geological formations, revealing an "environmental architecture" that could be mapped entirely separately from the cultural landscape. Source: National Archives of Canada, C51371.

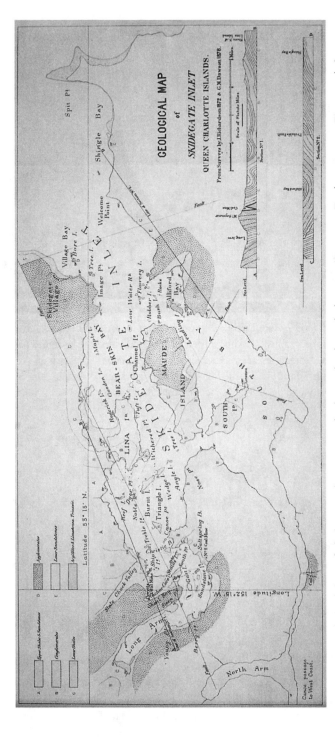

Figure 2.8. Geological map of Skidegate Inlet, Queen Charlotte Islands, including cross sections. For location, see Figure 2.4. Source: Dawson (1880).

were more ambivalent. At times he subscribed to the thesis that Natives were destined to decline and ultimately disappear before a superior European modernity, yet elsewhere he argued strongly that the Haida, if not others, would survive, and that steps would need to be taken soon to deal with questions of land title. Dawson also felt that the Haida could be trained as productive workers in the resource industries that were certain to spring up across the island. His descriptions of the material culture of the Haida, moreover, contained evidence of precisely that which his spatial and epistemological divisions disavowed: a Haida artifice that suggested territorial and ecological practices that did not fit Dawson's neat epistemological divisions. Many of the same elements of Haida culture that so captivated Dawson—their large houses, totem poles, artistic abilities, and elaborate rituals—spoke of much more than an unexpected cultural sophistication among primitive peoples; they also told of intimate contact with the surrounding territory. Yet although Dawson's enframing of culture and nature is troubled by spatial practices that clearly exceeded its terms, he was unable to register this at either a practical or an epistemological level. Amid the rugged, fog-enshrouded landscapes of the Haida Gwaii, a colonial visuality held sway. Dawson divided into two domains—nature and culture—what in the everyday lives of the Haida remained as one.[41]

Mere Surfaces

Dawson's colonialist visuality was likely reinforced by his schooling in nineteenth-century stratigraphical geology. With the shift from mineralogy to stratigraphical geology in the late 1700s and early 1800s, geologists had become resolutely global in their outlook, even as they focused on the most local of landscapes. This paradox can be explained by reference to changes in how geology understood the objects and purpose of its analysis of the earth. The emergence of "strata" as an object of study, and the possibility of correlating vertical sequences from disparate geographical sites, enabled geologists to develop a visual language (and visual technologies) that made local landscapes legible as part of a larger *global* order.[42] This had important implications for our story. With the visual language of stratigraphical geology, faraway places could be related and compared as never before. Correlation gave geology (and empire, with which geology was closely associated) a global reach; if the presence of coal or a certain mineral was associated with a sequence of strata at one site, for instance, it was reasonable to expect the same presence at another site where these strata were found.[43] Thus, as the nineteenth century progressed, landscapes were increasingly read not as complex, integrated cultural and material landscapes, but instead for signs of their inner architecture, a prior structure that appeared entirely unrelated to local sociopolitical contexts (cf. Stafford 1990). When Dawson looked at exposed strata on the Queen Charlottes (Figure 2.9), he did not see them as related to the islands' inhabitants, but as part of a separate natural order. Local inhabitants might appear, as in the case of the small boat pictured in Dawson's sketch, but only to mark such things as scale.

Equally important was geology's depreciation of mere human history in the face of an almost immeasurable geological time. Human history, it was becoming evident, was but the most recent—and remarkably brief—chapter in a far longer story of the earth. This dovetailed neatly with the double vision of Europeans on the coast, further prying an underlying geology from the inhabitants of a given territory. The transient details of human occupation, etched only temporarily on the earth's surface, paled in significance to the depths of geological time, which told stories of natural processes, and of time horizons, that could not be measured in terms of merely human time. Indeed, the sailboat in Dawson's sketch can be read in this manner. More than mere decorative embellishment, yet much less than the recognition of a dissenting Native claim to the land, the boat signifies the transient character of *all* human endeavors, Native and European alike. Yet, in British Columbia, as in other colonial contexts, this had uneven effects, merging with imperial ideologies of the inevitable decline and disappearance of primitive

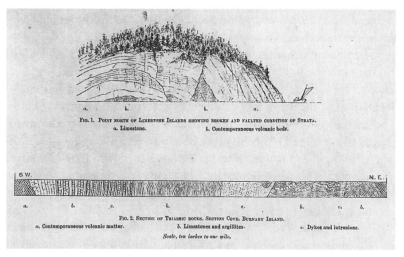

Figure 2.9. Nature's "architecture": diagram of strata north of Limestone Islands and section of Triassic rocks, Burnaby Island. Source: Dawson (1880).

peoples, regardless of Dawson's own beliefs or political and ethical positions. Destined to disappear—whether measured in minutes, days, or years—the boat marks the fleeting presence of the Haida at the moment of the arrival of a European modernity.

Finally, geology, as a way of knowing, served to reinforce for Europeans their sense of difference from, and superiority to, premodern cultures. Whereas Dawson wrapped his geological descriptions in a rhetoric of certainty, he viewed Native knowledge of the landscape as untrustworthy. Often in his texts Native knowledge is treated as rumor. Further, as a people without Geology, they had no legitimate claim to the resources of the land, because it was assumed that they were unaware of what was there. In contrast to Dawson's systematic and disciplined observation, Native communities were thought to possess a transient, undisciplined gaze. They knew only that which they stumbled upon by accident, and thus had no real knowledge of the tremendous riches that lay on or beneath the "mere surface" of the country. Indeed, the sailboat in Dawson's sketch might be read in yet a third way. Passing before the cliffs, its occupants appear completely unaware of the ways in which the rocks could be read so as to reveal the landscape's hidden secrets. Only Dawson, floating further offshore, could properly read the signs, supplanting the irrevocably partial vision of local inhabitants with a universal and objective gaze.

Through its displacement of local landscapes into global orders, its erasure of human occupants, and its privileging of European science over other ways of seeing, geology served to annex new territories not only into domains of imperial science, but also into imperial forms of political and economic rationality (Stafford 1990; Braun 2000a). Dawson's geological maps exemplified this doubled imperial imagination, revealing an inner architecture that could be appropriated as yet another new frontier for capital. Enterprising settlers, armed with a rudimentary knowledge of geology, could now "read the rocks" according to an assumed plan, and indeed were encouraged to do so.[44] Dawson himself would go on to write texts about Canada as a "field for investment" (1896), and create provincial maps of the region's important resources (1880a)—important not for Native inhabitants, but for nascent resource industries.

In short, what we find in Dawson's writings, sketches, and maps is the unveiling of nature's plan, a plan which both preceded and lay external to a Native presence and that would be fulfilled only through the judicious mixing of European (Canadian) capital and labor. It was a world divided in two: on the one hand, an anachronistic space dense with the last vestiges of a primitive culture; on the other, an empty, natural space awaiting its inevitable transformation as part of a modern nation-state.

The Appearance of Natural Order: How Colonial Power Works

Here we might pause in order to draw out some of the implications of returning to historical texts such as Dawson's in the midst of present-day struggles over land, resources, and environment in British Columbia. Dawson's texts suggest the possibility of writing genealogies of unmarked categories such as nature, the land, and the nation, genealogies that find in these inviolable identities numberless histories and hitherto marginal events. They also help clarify how colonizing power works, and how colonial pasts continue to infuse the present. The illusion of representations such as Dawson's surveys, journals, and maps was that they appeared to be entirely without illusion: they were faithful to the things represented, promising complete and certain knowledge (even if this was continually deferred, as Robert Brown [1869] noted, leaving details to "more minute afterinspection"). This promise allowed readers (and writers) to view west coast landscapes as if their order emanated from nature itself, rather than from the ordering of appearances in representational practices.[45] Reading the survey only as a more or less accurate record within a story of progressive European acquaintance with west coast lands obscures the manner that the survey enframed the land within regimes of visibility. It is important to be clear about what is meant by this. At issue is not the question of accuracy. Dawson did not *misrepresent* the lands and peoples of the Queen Charlotte Islands. He did not get it wrong. Nor is it an issue of whether his surveys were complete or incomplete, as would be the case if we approached his work as no more than the description and cataloging of the qualities of the landscape. Rather, his surveys produced an *effect* of truthfulness, an effect that obtained its force through a metaphysics that assumed that behind representation lay the true structure of the world. Even if Dawson himself did not get it all completely right, his readers would have assumed that behind his sketches, maps, and texts lay the truth of these landscapes, which his representations to a greater or lesser degree captured. Through the hold of this metaphysic, Dawson's surveys, and the writing of countless other European travelers, could be taken as approaching nature itself, effacing the particular representational practices through which they were produced and finding in the order of representation the order of reality itself. In this sense, Dawson's texts were *colonizing*, not just *colonial*.

This point bears underlining: Dawson's surveys and journals did not invent objects and landscapes in flights of fancy. These were material practices that engaged material worlds. In rendering the land visible, the surveys constructed from what was encountered an ordered scene that could be read. Such practices were highly material in two important respects: they were social practices embedded in particular historical and institutional settings, and they underwrote nature's social transformation. More than mere texts, they did not leave the land untouched. Instead, they actively displaced and resituated landscapes within new orders of vision and visuality, authorizing particular activities and facilitating new forms of governmental rationality. It was only after the land was staged as a theater of nature, after all, that it could be made available to the sorts of political and economic calculation that Dawson and others anticipated.

Significantly, the production of nature in colonial discourse did not occur through a straightforward erasure of Native presence. Dispossession did not hinge on ignoring Natives; it hinged on *how they were described and incorporated within orders of knowledge.* In Dawson's texts (as in others of his time) indigenous people were identified and described in great detail. This presence, however, was contained in a discourse of primitivism and what Anne McClintock (1995) describes as "anachronistic space": a realm that remains anterior, and thus has no place in the unfolding history of the modern nation. At the same time that Dawson placed Native peoples such as the Haida on view before interested readers and state officials, he displaced them both temporally and geographically from their surroundings. Concurrently, Dawson described a national (physical) landscape consisting of certain geological and botanical entities, containing certain landforms and waterways, and subject to particular climates and meteorological phenomena. The result was a textual and spatial separation of the tribal village from the modern nation, a double vision that marks BC politics to the present.

Conclusion: The Cultural Politics of Tree Farms

Only subsequent to these displacements—and their institutionalization in the practices of the Indian Reserve Commission—did it become a matter of course to represent British Columbia's natural landscape with no regard to its original inhabitants. More than sixty years after Dawson's surveys, Justice Gordon Sloan (1945) wrote his *Report of the Commissioner Relating to the Forest Resources of British Columbia,* a document widely seen as defining the present spatial and temporal organization of forestry in BC. In it the displacements of the late 1800s were disavowed and the erasure of Native peoples was incorporated as a constitutive element of forest policy and politics in the province. For Sloan—and for later commissioners such as Peter Pearse before whom Simon Lucas made his impassioned appeal—the forest stood as an entity entirely separate from its inhabitants, subject to the administrative and political objectives of distant actors.[46] Sloan saw the forest as a single, although geographically differentiated, *natural* system. In this system, Native reserves appeared, if at all, only as small, anachronistic—and ultimately inconsequential—islands. Accordingly, it seemed entirely reasonable to Sloan to generalize and extend a model of sustainable forestry—and its accompanying tenure system—across the entire extent of the province's known exploitable forest resources, without any appreciation for *competing* cultural and political claims. Perhaps more than any other single act since the separation and segregation of First Nations on a reserve system in the late 1800s, this would work to isolate people such as Simon Lucas and First Nations such as the Nuu-chah-nulth from their physical surrounds.[47]

In the decades after Sloan's report, forest management seemed far removed from the province's colonial past. It was about trees, about extracting the most value from them, and about doing this in the "best interests" of the public. In many ways, this was the mandate of the 1975 Pearse Commission before which Simon Lucas spoke his words of isolation, and in large measure it continues to define a cartography of politics in which certain issues (local versus state planning, private versus public tenure, regulation versus deregulation of forest practices, how to determine levels of harvest, appropriate royalty rates, and so on) dominate public debate, while others (sovereignty) remain on the margins.

What Lucas's poignant words brought into such sharp relief was the *colonial logic* underpinning British Columbia's forest regimes. The strong reaction of the Nuu-chah-nulth Tribal Council to the 1993 Clayoquot Sound Land Use Decision, on the other hand, reminded British Columbians of the remarkable *durability* of these colonialist foundations. Indeed, the reception of Lucas's testimony before the Pearse Commission is telling in this respect. Despite Lucas's appeal, Pearse's report remained silent on questions of Native rights. At the time, there was simply no place available for Lucas to fill, discursively, politically, or geographically. In its 381 pages, the only comments addressed to Native concerns appeared in a short section that discussed the possibility of new allocations of timber licenses. "Native Indian reserves," Pearse wrote, "present another potential source of forest land that might be combined with [nearby] provincial Crown land into sustained yield units, under band management" (118). Although this seemed promising, Pearse noted elsewhere that precious little Crown land *remained* from which to form the core of any new licenses. His acknowledgment of Native concerns in a sense came too late; moreover, it simply assumed the prior production of colonial space (no land remained) and therefore reinforced the terms of Native marginalization. In the eyes of the commissioner, the province remained divided between the "primitive" space of Indian reserves (amounting to approximately 0.4 percent of the land area of the province) and the "modern" space of the nation-state (the space of citizenship, where no prior claims other than those of the sovereign state could be broached). Outside their postage-stamp–sized reserves, Natives had no territorial claims—they would be treated like any other tenure holder. In other words, by stepping outside the narrowly circumscribed positions defined for them within the spatial and cultural imaginary of BC resource politics, Natives stepped into a space of citizenship that resolutely refused to acknowledge the validity of Lucas's challenge to nationalist cartographies.[48]

In this chapter I have sought to uncover some of the buried epistemologies that shape BC's postcolonial present. I wish to conclude with three points. The first concerns how we understand marginality. As I have shown, margins are not foremost a *spatial* category (as in writing or speaking *from* the margins). To conceive of marginality as a spatial category risks essentializing the margin as a site that exists outside or prior to its discursive production, and that is then subsequently excluded from, made subject to, or incorporated within forms of power. Rather, marginality is produced in, and integral to, forms of colonial power; it is that which *must be excluded* from conceptual frames in order for identities—such as "nature" or "nation"—to appear coherent and complete. This does not mean that spatial practices are unimportant. As evident in the writings of George Dawson, and in contemporary forest practices, marginality was produced precisely *through* the representation and production of space. It was, after all, through the spatial separation of Native peoples from their lands that wilderness was able to appear as an object in and of itself, an identity whose constitutive disavowals were made visible in the figure of Simon Lucas.

Second, marginality is not something that originates only in juridical-political practices; it is something produced discursively as well. As important as public policy—and legal institutions—may be for implementing colonialist practices (even today, after the end of "formal" colonialism), such policies cannot be fully understood, or contested, without attending to their conditions of possibility. The cartographic exercise of mapping Indian reserves in the late 1800s was founded on a colonialist discourse that made it appear common sense to divide BC landscapes into two, immensely unequal, parts. Likewise, forest policy today obtains its force, in part, through the reiteration of these discursive and cartographic displacements. This has important implications for where we locate the production of marginality, as well as how we imagine its interruption. To focus on administrative policy alone is to overlook the theoretical moment presupposed in it.[49] Or, stated in a different way, it is to overlook the ways in which colonialism and neo-colonialism are related to, and enabled by, practices of signification. As Nicholas Thomas (1994, 2) explains:

> Colonialism is not best understood primarily as a political or economic relationship that is legitimated or justified through ideologies of racism or progress. Rather, colonialism has always, equally importantly and deeply, been a cultural process; its discoveries and trespasses are imagined and energized through signs, metaphors, and narratives; even what would seem its purest moments of profit and violence have been mediated and enframed by structures of meaning. Colonial cultures are not simply ideologies that mask, mystify, or rationalize forms of oppression that are external to them; they are also expressive and constitutive of colonial relationships in themselves.

Rather than focus only on state policy or legal judgments to explain why, in the words of a witness to the Task Force on Native Forestry, First Nations today live as "prisoners in their own lands," it is equally as important to interrogate the representational practices that have made this possible.

Finally, this chapter has been concerned centrally with the question of temporality. The relation between the past and the present is a complicated one, made more difficult by historical-geographical narratives of the inevitable unfolding of a European modernity and its specific cultural, political, and spatial forms, such as the nation-state. Canadian history is often rendered as a single thread that runs unbroken from Jacques Cartier to the present. Not only does this problematically assign pre-contact periods to prehistory, it also internalizes two other assumptions that are disabling for thinking *territoriality* differently. The first is that the present *transcends* the past, such that "error" in the past (colonialism) is overcome by "enlightenment" in the present (postcolonialism); the second is that the end point of this linear history is the (singular) nation, in which all differences are incorporated as the same (multiculturalism being a somewhat ambivalent instrument of this incorporation of difference). In this chapter I have sought to trouble these narratives in a number of ways. First, I have questioned the assumption that colonial pasts are simply "left behind" as unfortunate stages in the nation's development. As I have sought to show, the present is iterative in character: it necessarily draws upon past practices, such that the past continues to shape the present. Second, I have questioned the notion that history is somehow singular and cumulative (or linear). If the past infuses the present, this occurs in many different ways simultaneously. The temporality of the present is decidedly multiple, as different histories punctuate the present in ways that that do not add up to a single, unified totality.

Here we can see the significance too of rereading historical texts such as Dawson's. Amid today's forest politics, First Nations have struggled to articulate their rights to land and resources. These have been immensely difficult struggles, in part because non-Native Canadians have such a difficult time recognizing histories and territorialities that are different from those of the nation-state. In legal struggles, and in negotiations with the federal and provincial governments, First Nations have been forced again and again to show "evidence" for their claims. The starting point in these disputes has been the *validity* of the sovereign territorial claims of the Canadian state; when these are taken for granted, First Nations become special interests like any other. In such a context, countering colonialism's cognitive failures—both historical and in the present—is crucial. Indeed, this was perhaps *the* central task faced by the Gitksan and Wet'suwet'en in their early 1990s land-claims case (see Monet and Skan'nu 1992; Solnick 1992; Sparke 1998).[50] Rereading texts such as Dawson's is therefore doubly important, for if the rhetorics and (de)territorializations that underwrote Canada's dispossession of First Nations in the late 1800s are left unexamined, then past colonial authority appears legitimate, and, by extension, so does the authority claimed today by the state and transnational forest companies. By noting that dispossession occurred not simply through legal pronouncements, but also through a visuality that at once geographically located and spatially contained Native presence (and therefore authorized European claims to an empty land), the reiteration of these displacements in the present can be challenged, opening a conceptual space within juridico-political discourses in which past and present forms of Native territoriality and possession might again be visible.

3. "Saving Clayoquot"

Wilderness and the Politics of Indigeneity

> Nature is not a physical place to which one can go, nor a treasure to fence in or bank, nor an essence to be saved or violated.
>
> —Donna Haraway, "The Promises of Monsters"

The Spatial/Temporal Logics of the "Normal Forest"

In his 1945 *Report of the Commissioner Relating to the Forest Resources of British Columbia,* Justice Gordon Sloan reinforced colonial displacements that occurred at the end of the nineteenth century. As I explained in chapter 2, his "forest" achieved its coherence as an object of state political and economic calculation only through a series of cognitive failures that erased existing forms of Native territoriality. In Sloan's report, nature and the nation coincided, the former produced rhetorically as the prehistory of the latter. Sloan is better known, however, for introducing sustained-yield forestry to the province, and it is the particular order of time and space that his plan inaugurated that will be our focus here. As we will see, the crisis of the 1990s—and the protests and blockades that shook the province—were in important ways a logical outcome, produced by the system itself.

Sustained-yield forestry works on a simple principle: each year the amount of tree fiber removed from the forest should equal the amount of fiber added through growth. Or, stated somewhat differently, the number of trees harvested in any given management unit should be such that by the time one rotation through the unit is completed, the areas harvested first will be ready to be cut again. The resulting forest—what foresters call the "normal forest" (Figure 2.1)—would therefore consist of an equal distribution of trees of different ages. In practice, of course, this proved difficult to achieve. Determining the optimum rotation period alone depends on many factors, from adequate knowledge of inventories and growth rates, to calculating a variety of economic, ecological, and technological variables that determine at what age and at what rate it is desirable to harvest.

The "normal forest" is a model of rational management. On paper, its clear logic and regular order has a certain aesthetic appeal, but it also carries a number of important consequences. Sustained-yield policies, for instance, assume that over time the *entire* "working forest" will be re-made in the image of a single commodity: timber. This makes it difficult to imagine the integration of *other* forest uses, a perennial problem that has resulted in endless conflicts over rights of access, and intense struggles over the legitimacy of competing claims made by various social and cultural groups for whom the forest is the site of very different emotional, political, or economic investments. It is immensely difficult, for instance, to manage a forest for timber values *and* aesthetic values, not to mention other economic uses of the forest, such as the harvest of mushrooms, or the production of reliable sources of clean drinking water. To take into account competing uses threatens the very spatial and temporal organization of the "normal forest," with significant consequences for the communities that after 1945 found their fates ever more tightly bound to this forest system. The "normal forest" also carries enormous ecological consequences, although, as we will see in chapter 6, these are fiercely debated.

It is not surprising, then, that sustained-yield forestry in BC has many critics.[1] But it is the spatial and temporal logics of sustained-yield forestry that arguably have had the most to do with *where* and *when* voices of protest have been heard. In most sustained-yield units on the coast, harvesting began with the most accessible stands. As these were depleted, networks of roads were extended into outlying areas, and more remote forests brought into production. This left distant areas untouched until the final stages of the first rotation. By the 1980s, fewer and fewer uncut areas remained on Vancouver Island, and most of those were slated for logging. It was no surprise, then, that as the decade proceeded, forestry conflicts intensified. With the extension of industrial forestry into these last regions, the "end of nature" appeared imminent. Borrowing a phrase used by environmentalists, industrial forestry appeared on the verge of "liquidating" the last vestiges of the "ancient rainforest." If this presented itself as a new problem, it was in many ways the predictable outcome of the solution devised for a different problem that appeared critical in the postwar era: how to rationalize forest production so as to produce a rational and stable society. The forest system set in place in 1947 and reaffirmed in 1955 was consistent with the high-modernist planning of the period, which had as its objective an ordered, rational society and assumed that it could be achieved through scientific management. Implicit in the forest system established in the period was the assumption that by organizing the province into spatially bounded "working circles" organized around manufacturing centers, the logic of the "normal" forest (with its regular and predictable levels of production) would contribute to the sedentary workforce, lifelong employment, community-centered life, and traditional family structures headed by a male wage earner that were assumed at the time to be society's "normal" form (Braun 2000b).

It is important to note, then, that what appeared in the 1990s as a new crisis (the end of nature) was in fact presupposed in Sloan's recommendations. These remote forests were not now suddenly coming under the sway of industrial capitalism; they had been an integral part of the economic and social calculations of sustained-yield forestry in British Columbia from the very beginning. Without including these distant forests, the spatial and temporal patterns of sustained-yield forestry proposed in the 1940s made no sense, nor did the social and economic objectives predicated on them. Eventually, the circle would have to be closed. Predictably, as the final stages of the forest's normalization approached, valley-by-valley conflicts between communities dependent on the full completion of the forest cycle and environmentalists concerned to preserve the last "intact" forests intensified, for with each tree cut, each *remaining* tree became that much more important for all involved.[2]

Speaking for Nature

In the 1980s, logging intensified on the "wild side" of Vancouver Island.[3] With each additional valley slated for logging, conflict increased and previously remote and inaccessible sites—Carmanah, Walbran, Tsitika—became household names, linked to media images of protesters chained to logging equipment or perched precariously in the limbs of trees. Clayoquot Sound was one of those sites. In the early 1980s, residents and activists learned of plans by two forest companies—MacMillan Bloedel and International Forest Products—to extend logging into the region, which contained some of the largest remaining tracts of "intact" temperate rainforest on the island.[4] Protesters responded by blocking MB workers from landing at Meares Island in 1984, and stood in front of road builders at Sulphur Pass in 1988. For the next few years, conflict over these forests simmered as the government explored ways of reaching an agreement between the various parties regarding the timing, scale, and methods of logging in the area.[5] Efforts at compromise proved futile, and the provincial government unilaterally imposed its land-use plan in April 1993. Far from bringing the issue to a close, it added fuel to the fire, as protests quickly spread beyond the province's borders and thousands of activists from across North America, Europe, Australia, and New Zealand flooded into the region in an effort to halt logging activity in the Sound.

As noted in chapter 1, the protesters' blockades delayed loggers by only a matter of minutes, about as long as it took for the Royal Canadian Mounted Police to drag the protesters off the road and into waiting buses. Yet this was in many ways immaterial, for the blockades' success was measured in terms of publicity. Schooled in globalization, the protest's leaders were savvy political actors who knew that images traveled widely and instantaneously in late-twentieth-century mediascapes, and that political identifications were not necessarily contained solely within the boundaries of place, or the nation-state (Appadurai 1996; see also Thompson 1995).[6] By staging dramatic confrontations and by bringing the protests to centers of consumption in North America and Europe, environmentalists rapidly built a constituency in and across disparate locations, strategically mimicking the circuits of capital and commodities that characterized the forest industry.[7] Their ability to capture media attention remains virtually unmatched in Canada's history, as evident in the front-page coverage given the protests by Canada's leading news magazine (Figure 3.1), and in the clear anxiety expressed in its worry that the *world was watching*.[8]

My interest in this chapter lies less in the media tactics of protesters than in the manner, and by whom, the temperate rainforests of Clayoquot Sound came to be represented. An image from the 1993 protests can help clarify my argument and the stakes involved. In this scene, flashed across the nation's TV screens, the camera followed a protester as he was pulled off the barricades by members of the Royal Canadian Mounted Police. In the brief moments that passed as his limp body was dragged to the waiting bus, he cried out in a voice clearly audible above the chaos: "Here I stand for the wild things who have no voice in our courts, our boardrooms, and our politics. I speak for the wolves, the trees, and eagles." This was clearly a passionate and principled ethical statement, which spoke of an obligation to non-human nature, and contested the hubris of industrial societies that ruthlessly objectified nature as mere resources for the accumulation of wealth or for nation-building projects. The man physically bore witness to the ecological violence of capitalist modernity and raised questions

Figure 3.1. "The World Is Watching." Cover of *Maclean's* magazine, 16 August 1993.

about its anthropocentric moral and political horizons. Around what do we organize our moral responsibility? How do we define these boundaries? To what does our "democratic imaginary" extend?[9]

Yet, as important as his statement was, it struck me as deeply problematic, and not only for its obvious masculinist heroics (a perennial problem among radical ecologists). Why was it *this* man, rather than someone else, who spoke for nature? What made it possible for protesters like him—predominantly white, middle-class professionals (or their children)—to stand before the machinery of the forest industry and speak as nature's defenders in a region claimed by the Nuu-chah-nulth as their traditional territories? And does not "speaking for" nature presuppose a "nature" to be spoken for? If nature is "without voice," as the man claimed, who decides the words in which nature speaks?

The attempt by environmentalists to "save Clayoquot" opened up as many issues as it resolved, including the ways in which boundaries were drawn between nature and culture, tradition and modernity, the wild and the tamed. But perhaps the most significant issue that emerged in these conflicts was the ambivalent place of indigenous peoples within environmental discourse. As will become clear as I proceed, at the heart of the environmental imaginaries that informed many Canadian and international environmentalists working in Clayoquot Sound lay much more than what William Cronon (1995) has described as a thoroughly cultural discourse of wilderness; it was mediated by and through a discourse of indigeneity. As we will see, this brought First Nations peoples onto the stage of ecopolitics, but at the same time narrowly circumscribed the terms in which their appearance could be understood. In this chapter, I argue that to the questions already raised must be added others: How is "indigeneity" defined and by whom? What has indigeneity come to *signify* within discourses of modernity? Ultimately, what accounts for this category having such currency in contemporary ecopolitics, and what are the political implications in postcolonial British Columbia?

Sightings/Sitings

In what follows, I address these questions by relating and comparing two specific sightings/sitings of Clayoquot Sound landscapes where what counts as nature and indigeneity is at stake.[10] The first is a popular and widely distributed book of photography published by a local environmental group in the context of its fight to save Clayoquot Sound from industrial forestry. Although dating from the early 1990s, I focus on it at some length because it established many of the rhetorical and visual norms that have since governed the representation of west coast landscapes and its Native inhabitants by prominent environmental groups in the region. The second is a map of "culturally modified trees" taken from an archaeological study produced as evidence in a land-claims trial centered on Meares Island, one of the most contested sites in Clayoquot Sound. This map was part of a pathbreaking study, and since its production many Native groups along the coast have engaged in similar projects, now aided by powerful geographic information systems (GIS). Neither gives us the full truth about Clayoquot Sound—both are partial visions enabled by particular technologies of vision, ideologies of nature, and definitions of indigeneity. Critical inquiry attends to how the differences matter.

As we will see, the presence of First Nations in British Columbia's temperate rainforests poses ethical, political, and ideological problems for local and international environmental groups seeking to preserve west coast forests, and has resulted in representational practices that are both complex and at times immensely contradictory. These call out for analysis if a radical postcolonial environmentalism is to be successfully articulated. Clearly, there are risks involved in such a critique, especially at a time when a backlash against radical environmentalism has resulted in state environmental deregulation across North America and led to reduced public interest in environmental issues (Rowell 1996). Theoretical purity can be a dangerous game if it undermines political action. Yet, given that ecopolitics can never be just about nature on Canada's west coast, interrogating the terms on which ecopolitics proceeds is a pressing analytic and political task. Here it is important to restate that deconstructing environmental discourse—showing its aporias, cognitive failures, and enabling fictions—is not a detached or disinterested practice, nor is it a nihilism that willfully sets out to destroy the basis for any and all ethical statements. Rather, it engages with that which is close at hand (and heart) from a position that is made scrupulously visible: in this case, support for Native land rights.[11] Far from seeking to discredit or undermine ecopolitics, this chapter examines how its truths and coherency are attained, and thus attends to the critical absences that haunt its achievements. The objective is not to tear things apart; it is to engage in the work that must be done in order to locate other, more just, worlds where current practices of marginalization have less purchase. In this light, the angry responses by some environmentalists to recent "blasphemous" critiques of the wilderness idea say much more about their unwillingness to take questions of justice and equality seriously than about their credentials as radical ecologists (see Foreman 1995; Noss 1995; Sessions 1997; Willers 1997). Blasphemy, Donna Haraway (1991, 149) writes, "protects one from the moral majority within, while still insisting on the need for community. Blasphemy is not apostasy."

Defending Nature: *Clayoquot: On the Wild Side*

In 1990, Canada's largest wilderness preservation group, the Western Canada Wilderness Committee (WCWC), published a book titled *Clayoquot: On the Wild Side* (Dorst and Young 1990), which quickly became one of the most popular items available for sale at the Committee's store in Vancouver's Gastown district, a trendy heritage area heavily visited by tourists. Filled with more than 160 photographs, the book presented an awe-inspiring image of the Sound, bringing its snow-capped mountains, wave-battered headlands, and majestic forests to the attention of urban, middle-class audiences in British Columbia and beyond, many of whom had little experience or knowledge of the region. Although it does not exhaust the complex, shifting political relationships and alliances between First Nations and BC environmental groups, the book provides a starting point from which to think about questions of power and politics in postcolonial environmental discourse. Indeed, the very *success* of these kinds of publications—by now a time-tested formula for mobilizing support for nature preservation—demands that we learn how to read them critically.[12]

The first thing worthy of note about *On the Wild Side* is its format. For many North Americans weaned on *National Geographic* or the Discovery Channel, the genre is familiar (cf. Lutz and Collins 1993). Although the volume has considerable text (written by Cameron Young) it is the book's photography (by Adrian Dorst) that takes priority.[13] This primacy of vision over language, of course, is a hallmark of modern empiricism, and is part of what makes the book effective as a political tactic. Critiques of ocularcentrism have been well rehearsed and need not be repeated at any great length; suffice it to say that what gives vision its privileged position is its apparent transparency—the eye is thought to merely passively record what is there to be seen, rather than to actively constitute it (cf. Crary 1990; Jay 1993). Yet, we must proceed cautiously. It would be a mistake to assume that this realist fiction of unmediated vision has been held uniformly, or that it has remained unchanged through time, even within the West. As a number of scholars have shown, notions of unmediated vision ("seeing is believing") have often worked in tandem with other discourses that emphasize the materiality or embodiment of vision. Jonathan Crary (1990), for instance, has shown how the *eye* itself became an object of scientific inquiry in the nineteenth century, rather than that which guaranteed the *truthfulness* of empirical science. In other words, already early in the nineteenth century individual variance in the physical act of seeing was an issue, although, as McQuire (1998) explains, the concern at the time may have been less to recognize subjectivity in seeing than to contain its perils (thus eyes could be trained, corrected, and submitted to normalizing principles).

However, if in the nineteenth century the eye increasingly came to be viewed as embodied—merely human, and thus potentially flawed—the camera arguably came to embody the unmediated vision once claimed for the eye. This was owing in part to the camera's technical status as a mechanical seeing that could transcend the suspicions cast on individual perception (McQuire 1998). The camera was, in André Breton's words, a "blind instrument" of vision, an optical device that witnessed and recorded with the disinterestedness that only a machine could achieve. Mixing the primacy of metaphorical vision with the promise of mechanization (the removal of a more subjective perception), photographic representation in the twentieth century was often thought to not only assume parity with, but actually transcend, direct perception. Yet, the same drift of perspective that troubled the eye in the nineteenth century also came to trouble photography, for if the eye lost its privileged position with the introduction of questions concerning its *embodiment,* the camera came to be haunted by questions concerning its *operation.* As an instrument, it is clearly something used. This raises the issue of intentionality, or, more to the point, the positionality of the camera in time and space. Likewise, as a machine that captures things in an instant, it is profoundly disinterested in what comes before and what comes after. And, as a device that limits the visual field, it raises the spatial question of framing. Further, as evident in our desire to add captions, photographs have promiscuous meanings that are not easily contained—the truth of the image does not precede its interpretation. Finally, with the introduction of digital technologies that allow the manipulation of the image, photography is today in ever-greater danger of losing its privileged position as the exemplary model of objective vision. No longer something to be blindly trusted (if it ever was), today we often *expect* artifice and playfully look for the ruse (Figure 3.2).

It would likewise be a mistake to assume that all photographic genres are approached equally. Documentary photography and photographic journalism arguably retain a privileged claim to objectivity, as the medium that most accurately "records" history. Likewise, nature photography, despite its highly aestheticized mode, continues to aspire to "let nature speak for itself." Yet neither entirely escapes issues of *positionality.* As John Tagg (1988, 187) explained:

Figure 3.2. "There you are." Billboard, Vancouver. Photograph by author.

> The photographer turns his or her camera on a world of objects already constructed as a world of uses, values and meanings, though in the perceptual process these may not appear as such but only as qualities discerned in the "natural" recognition of "what is there." . . . The image is therefore to be seen as a composite of signs. . . . Its meanings are multiple, concrete and most important, constructed.[14]

This recognition that, in the famous words of Ansel Adams, "[a] photograph is not an accident—it is a concept," bears directly on how we read *On the Wild Side*. As documentary, it makes a claim to a reality that lies outside the photographic device and its human operator; yet it reveals as much about the environmental discourses that provide the frame as it does about the nature represented. No less than MacMillan Bloedel's publications, this book and its photographs must be approached as profoundly ideological entities, although in very different ways.[15]

Before investigating this further, let me suggest two additional reasons why cultural productions such as *On the Wild Side* must be taken seriously. First, the book provides insight into the intertwining of culture, technology, and politics. As partially a product of one of modernity's most prevalent and celebrated optical machines—the camera—the book cannot easily be assigned to the realm of either culture or technology, as if these were mutually exclusive domains. Nor can these be opposed to a third domain called politics. Indeed, that Clayoquot Sound became such an intense site of emotional and political investment on the part of distant actors during the 1990s was in part related to modern technologies of vision that made it possible to bring it into local, national, and global public spheres. Second, photographs are *mobile*; they provide a means for the 'forest' to be translated into a two-dimensional form (an inscription) that can be moved to new sites without modification. The significance of books such as *On the Wild Side,* then, is that they make it possible to bring local natures to distant places, and by so doing to draw together a community of concerned readers. As is now well known, political actors are increasingly constituted or enrolled through images. To say this differently, *On the Wild Side* constructs its reader as nature's defender, and this is enabled in part through the technologies of vision that it deploys. It was no accident that the mass protests that occurred in the summer of 1993 followed the wide circulation of images, both in Canada and abroad, of British Columbia's temperate rainforests, and included for the first time large numbers of people from the United States, Europe, and Australia. Although not uniquely responsible for the entry of Clayoquot Sound into these global spaces of publicity, *On the Wild Side* must be understood as part of the wider circulation of texts and images by which geographical and ecological imaginations are constructed.

Wilding Clayoquot

The first three sections of Dorst and Young's book offer the familiar wilderness fare that readers are accustomed to finding in American nature photography.[16] The book's introductory photo-essay, for instance, is titled simply "On the Wild Side." Here Dorst's photographs frame a sublime, enchanting landscape, capturing nature's powerful forces and its intricate, even delicate, ecological relations (Figure 3.3). Straining to give to his language the same visual sense as Dorst's photographs, Young describes the Sound as a "showcase of environmental elegance and diversity" (20). Elsewhere, the Sound's natural environment is described as a "spectacle," a "massive tableau," and a "sculpture," and is said to give "virtuoso performances" that are "tightly choreographed." And, indeed, in this book the Sound *is* visually stunning: from fog-enshrouded, crescent-shaped beaches and wave-pounded coastal headlands to shoreline trees sculpted by fierce winter storms and the almost unimaginable silence and luxuriant growth of its inner forests, the book's photographs and text reveal to the reviewer a truly spectacular(ized) landscape.

This spectacularization of nature should not be passed over lightly.

Figure 3.3. "On the wild side": framing nature as harmony. Photograph by
Adrian Dorst. Source: *Clayoquot: On the Wild Side*. Reprinted with permission
of the photographer.

There are important strategic reasons for such a tactic. In the face of the
pervasive instrumentalism of the state and extractive capital, the appeal
to nature-as-spectacle has proved effective for an oppositional ecopolitics. Who would not wish to see these spectacular sites preserved? Quite
apart from the ecological significance of the forest, why risk despoiling
sites of such obvious beauty? What is viewed as aesthetically pleasing
changes over time, of course, and will again in the future. When he
explored the coast in 1792, for instance, Captain George Vancouver
(1798) found the landscape dull and monotonous. My concern does
not lie with the aestheticization of the landscape (nor do I subscribe to
the view that aesthetics degrades or displaces politics). Instead, it lies
with how Dorst and Young *enframe* the Sound. As we will see, the book
achieves its ideological and political effects in part through a number of
crucial cognitive failures; or, said differently, it gains its authority
through a series of discursive displacements that render the forest intelligible in ways that produce it as an object in need of their (political)
representation.

We can begin tracking these displacements by attending first to what is obvious—that for Clayoquot Sound to appear resolutely wild and un-humanized certain aspects of the Sound must be ignored or erased. For example, despite the daily activities of workers, residents, and recreation-ists that are so evident as one moves through the channels and inlets of the Sound, humans and human activity are almost nowhere to be seen. Indeed, this extends beyond the first photo-essay to the book as a whole. The few places where people do appear, they are presented as insignifi-cant to the larger drama of wild nature and natural forces, often present-ed in the guise of passive observers, or, if female, conflated with the na-ture observed.[17] Crucially, only two photographs in the book depict scenes of labor (nobody, it seems, makes a living off these lands and waters, including its Native inhabitants). In its first sections, then, the book follows well-established rhetorical and pictorial norms for con-structing wilderness: the removal of nature from history, its construction as a domain entirely external to, and set against, culture, and the erasure of all signs of human labor in the landscape. It is not unimportant that the absence of humans extends to the photographer himself. Not only in this introductory photo-essay, but in the book as a whole, the *act* of photography is erased. As in much photo-realism, the camera and the photographer are kept well outside the visual frame, allowing the fiction of unmediated vision to proceed unquestioned. More to the point, this absence is precisely what allows all *other* absences in the photos and text to appear not to be absences. The camera merely records nature; what it captures is what is there to be seen.

Young's text produces a similar effect. The title of the first chapter—"A Gift of Nature"—reinforces the apparently unmediated nature of the author's encounter with the landscape. It also affirms his position (and by extension, ours too) as nature's representative (to whom nature "re-veals" herself, echoing Francis Bacon's sixteenth-century articulation of the relation between the inquiring empiricist and a sexualized natural world).[18] This is evident in Young's text, which positions the viewer/reader as an explorer taking part in a journey of discovery: "This is still a virgin landscape lost in time and governed by the unequivocal laws of nature—a gift of nature to humanity . . . you can feel that you are wit-nessing the perpetual big bang of creation. . . . This is a truly anti-matter world that spontaneously erupts with playful displays of its evolutionary masterpieces" (20). Here is a virgin landscape awaiting the (visual) pos-session of its representative. It is not passive, mind you: it gives "playful displays." Such possession, of course, appears uncontested, for this is a "virgin landscape," a place devoid of human history.

Without a voice of its own, and emptied of human inhabitants who might lay claim to the land, nature appears to be in need of a representative against the encroachments of industrial capitalism. Accordingly, the second chapter is titled "Bearing Witness to the Wilderness." It is here—and only here—that two photographs of the photographer, Adrian Dorst, appear, and it is important to consider how his presence is depicted. First, Dorst is the only individual in the book identified by name (all other humans that appear are *indexical* rather than *historical*). Second, he appears not as a photographer actively framing the landscape; rather, much like George Dawson, he appears as nature's "modest witness" (Haraway 1997). In one example, Dorst plays the part of the heroic explorer, piloting his zodiac through one of Clayoquot Sound's many secluded channels (the citationality is clear, even if unintentional). In the other he is shown crouched beside his dog, looking out over a primordial Clayoquot scene. Described by Young alternately as a biologist, a wildlife manager, and a photographer, he appears a dispassionate observer, a witness to the spectacle of nature's powerful forces, one whose vision is made even more sure through his eschewal of the distorting influences of modern culture: "He is about as close as you can get to a modern pilgrim. Equipped with a camperized van, his trusty zodiac and a weathered cedar-strip canoe, Adrian has dedicated himself to seeking out the mysteries of the west coast wilderness. . . . [He is] Robinson Crusoe with a zodiac" (28, 31). Whether Dorst would recognize himself in Young's portrait is incidental to the place that he occupies in the book's drama. The book presents him as its masculinized hero, speaking for a feminized, virgin nature. His relation with nature is presented as unmediated and authentic (for white, middle-class environmentalists, the camperized van and cedar-strip canoe are easily recognized signifiers of antimodernism). The home of Dorst's dreams, Young writes, would be a driftwood cabin nestled beside a rotund Sitka spruce on an isolated rocky headland.[19] As the book proceeds, the distance between nature and its representative continually shrinks, until the latter is merely the voice of the former.

This is not a univocal text, of course. The presence of Dorst's zodiac, for instance, rests uneasily with Young's effort to present Dorst as an antimodern pilgrim. Yet, taken together, the first three sections—Dorst's opening photo-essay, and the chapters "A Gift of Nature" and "Bearing

Witness to the Wilderness"—establish the region as irrevocably wild, located external to, but threatened by, an industrial society. "Scanning the complex shoreline," Young writes, "Adrian sees an island paradise; but he also sees its impending destruction" (32). Within this frame, a politically engaged, scientifically sound, witnessing is authorized, and it is the concerned "scientific" pilgrim who will ultimately demonstrate the correct mode of dwelling in the land (even if this dwelling is precisely a nondwelling that "lets it be").

Preservation through Indigeneity

Clayoquot Sound is far from unoccupied. Today, more than 2,200 people live in the region, of which members of three Nuu-chah-nulth tribal groups account for about half.[20] The non-Native population resides almost exclusively at the south end of the Sound in the village of Tofino, while Nuu-chah-nulth communities are centralized at four village sites: Ahousat, Opitsat, Hesquiat, and Esowista.[21] All but the last are accessible only by boat or float plane.[22]

This Native presence complicates any "wilding" of Clayoquot Sound. Indeed, the authors are well aware of this Nuu-chah-nulth presence, and explicitly state their support for Native land rights. And, in one chapter of the book, they bring the Nuu-chah-nulth back into their picture of spectacular nature. The support given First Nations by BC environmental groups (particularly by the WCWC and the Rainforest Action Network) must be commended, but, as I show in what follows, the relation between environmentalists and First Nations is deeply ambivalent, and this stems in part from contradictions within environmental discourse itself. What interests me here is how the authors incorporate—and in so doing also contain—the presence of the Nuu-chah-nulth within in the text and photographs of their book. By focusing on this, we can perhaps begin to trace the outlines of a postcolonial environmentalism in BC, one in which First Nations are ambivalently situated.

I will return to the details of *On the Wild Side* in a moment. First, let me say more about some wider discursive and political shifts of which the book is symptomatic. During the 1980s and 1990s, environmental rhetoric became increasingly wedded to a discourse of indigeneity, not only in BC but globally. This has proved to be a complicated conjunction. On the one hand, it has expressed a growing recognition that the extension of capitalist modernities and their ecological displacements is a *human rights* issue too, because aboriginal communities are witnessing the rapid commodification and transformation of the environments that sustain them. At issue here is a socially produced scarcity visited upon aboriginal groups through the disruption of existing territorialities. On the other hand, this concern has at times been refracted through overtly romantic and even primitivist lenses that conflate the preservation of *cultural* diversity with the preservation of *bio*diversity. Although far from a consistent rule, this is often informed by one of two assumptions: that safeguarding indigenous cultures would help protect nature, because indigenous peoples are thought to have an interest and/or expertise in sustaining existing ecological relations; or, alternately, that the preservation of nature is necessary to preserve indigenous cultures, because they are seen to have a necessary relation to nature. By this view, transforming nature is seen as equivalent to destroying Native cultures. This perhaps explains how Colleen McCrory of BC's Valhalla Society could draw parallels between the loss of pristine forests and the loss of the "last shreds of Native cultural heritage" (McCrory 1993). This conjunction of ecology and indigeneity is perhaps most prevalent in settler societies such as Canada, Brazil, and Australia, where aboriginal groups often reside in, or have been pushed to, marginal sites that are now being opened to capitalist development. It is not uncommon in this context to see this conjunction articulated *by* aboriginal groups themselves, which consciously or unconsciously occupy this slot with considerable political effect. Articulations of indigeneity have also appeared in such places as Indonesia and India, although with somewhat different meanings (see Gupta 1998; Li 2000).

Clearly, the conjunction of ecology and indigeneity has been ambivalent. On the one hand, it has radically transformed the terrain of both environmental politics and anticolonial politics, providing a new cartography for politics and identity (Li 2000). But, like all openings, it also has been accompanied by great risks. Much turns on the terms in which indigeneity is framed, by whom, and in whose interests. As I will show, if indigeneity is abstracted from any specific historical context, it can very quickly become a cipher whose content is filled by the environmentalist or the state official, and thus a vehicle for very different political and ecological agendas from those of the people purportedly represented. This is a crucial issue, for in the conjunction of environmentalism and indigeneity on Canada's west coast—what I have named "postcolonial environmentalism"—ecopolitics can very easily become merely the latest in a long history of neocolonial incorporations, where indigenous identities are defined and contained *within* the environmental imaginaries of European environmentalists and the postcolonial nation-state.

These are strong claims, yet, as we will see in the remainder of this chapter, how the Nuu-chah-nulth are constructed as indigenous is of vital importance for the sorts of political-ecological futures that can and are being imagined for Clayoquot Sound. What must be foregrounded, then, is not indigeneity but the politics of its articulation.

I will have more to say about discourses of indigeneity later. Let me return to *On the Wild Side*. As mentioned previously, although the book begins with a rhetoric of wilderness, the authors are well aware that the Nuu-chah-nulth have long dwelled in the region and have staked political claims to its land and resources. And, like many environmental groups active in Clayoquot Sound, they recognize that linking environmental struggles with land-rights struggles makes for both good ethics and good politics.[23] In the middle of their book, the authors include a chapter that focuses explicitly on the Nuu-chah-nulth. This introduces a marked ambivalence that the authors must carefully negotiate: at the heart of what had been framed as a natural paradise lies an ineradicable cultural presence. How this is resolved is critical.

Before I say more, it is important to be clear about where my concerns lie. Dorst and Young support Nuu-chah-nulth land rights. Their book is dedicated to the Nuu-chah-nulth, and the text—at least in this chapter—is clearly sympathetic to Nuu-chah-nulth struggles against a colonial state that has consistently refused to recognize Native sovereignty while exploiting the resources on traditional Native lands. Yet, as is often the case, these anticolonial sentiments are contradictory, and nowhere is this more evident than in the very chapter in which Dorst and Young explicitly "bring in" the Nuu-chah-nulth. This ambivalence is evident in several ways. Placed where it is in the book, the chapter gives the impression of simply inserting Native people into, and as part of, a preexisting *natural* landscape. Whereas previous chapters revealed a spectacle of natural forces and later chapters focus specifically on the rainforest, wildlife, coastal ecology, and so on, this chapter turns its attention to Natives, effectively naturalizing them within the landscape. Like other organisms, Natives appear as merely one of nature's many elements, part of a natural history that can be objectively ordered and described.

Paradoxically, there is some space in this for the articulation of a radical anticolonial politics. Read through a somewhat different lens than the one the authors intend, this naturalizing of Nuu-chah-nulth culture gives weight to the argument that the historical (and present-day) practices of the Nuu-chah-nulth are partly *constitutive* of the temperate rainforest (and thus that contemporary Nuu-chah-nulth communities should be its custodians). This potentially explodes the book's rhetoric of wilderness, replacing it with a historical geography of *social nature* that places the Nuu-chah-nulth, and not just the unequivocal laws of nature, as central actors in the ongoing drama of nature's production. This reading, however, is quickly foreclosed. Although the authors celebrate Nuu-chah-nulth culture, they banish it to the past. It is *pre-contact* Nuu-chah-nulth life that receives most attention. Before the arrival of Europeans, Young writes, the Nuu-chah-nulth "lived on a grand scale," developing "a cultural philosophy and a life-support system [that was] subtle and complex" (41). To be sure, placed against the comments of British Columbia Chief Justice Allan McEachern, who claimed in *Delgamuukw v. Her Majesty the Queen* (1991) that pre-contact Native life was "to quote Hobbs *[sic]* . . . at best 'nasty, brutish, and short,'" Young's text presents Native life and culture in positive terms, deflecting to some extent the debilitating force of a primitivism shared by McEachern and the state.[24] Young also relates work by anthropologists and archaeologists who have estimated populations on the so-called wild side of Vancouver Island prior to contact to have been as high as seventy thousand, adding weight to arguments that the Nuu-chah-nulth, at least in the past, had clear proprietary claims to the land.[25]

Dorst's photographs mirror Young's ambivalent prose. Eight photos in the chapter display a Native presence, but again, this presence is narrowly circumscribed. Only one shows Nuu-chah-nulth people: unnamed members of the Tla-o-qui-aht band paddling a cedar canoe (Figure 3.4). All others focus solely on *artifacts*. A lone totem is found still standing in the forest (Figure 3.5); a dugout canoe is shown deposited on a local beach; yet another canoe is pictured floating offshore, silent and empty in the fading light of dusk. Other photos carry a clear message of decline. A fallen totem pole is depicted in the process of being reclaimed by nature. The decaying corner post of a traditional longhouse is shown with a young tree growing from its decaying wood. In another, more troubling image, a moss-covered skull is identified as the remains of a Nuu-chah-nulth.

Dorst is not alone in his nostalgia. For Young, the story of the Nuu-chah-nulth appears as tragedy: the *fatal contact* of modernity on a primitive people. This is conveyed strongly in the chapter's subheadings: "coastal ghosts" . . . "a nation in distress" . . . "where worlds collide" . . . "last remains." Much of the chapter uses the past tense: "once was the site"; "islands were home"; "abandoned" villages, and so on. Although it is true that in the last two hundred years Native communities in the

Figure 3.4. Framing native culture as traditional: Tla-o-qui-aht paddlers. Photograph by Adrian Dorst. Source: *Clayoquot: On the Wild Side.* Reprinted with permission of the photographer.

Sound centralized in response to population loss, economic and technological changes, political conflicts, and the intervention of a colonial state, and that this left numerous signs of former settlements scattered along the coast, the absence of any reference to *modern* Nuu-chah-nulth life results in an overarching narrative of decline. In a memorable passage, Young mourns the loss of an authentic Nuu-chah-nulth indigeneity:

> On one of the beaches, the last remnants of a rare west coast ruin lie concealed from view by a tangle of salal bushes. Hidden away here are the structural remains of at least eleven traditional longhouses, including several corner posts and a handful of beams. Low midden ridges encircle each dwelling, outlining all the house sites—the last dead imprint of a once vital village. Jim Haggarty believes the village may have been built during the 1850s. It was perhaps the last attempt by the land's original people to carry on their age old way of life.
>
> As Adrian Dorst bends low inside the salal thicket, he steps gingerly over a slumping cedar timber to study one particular house post with grooves chiseled all around its perimeter. Smooth and stately,

Figure 3.5. Lone totem in forest. Photograph by Adrian Dorst. Source:
Clayoquot: On the Wild Side. Reprinted with permission of the photographer.

this cedar post looks for all the world like the fluted marble columns that once supported the classical structures of Greece and Rome.

The light is fading on this long summer's day, and during that slow ebb into darkness, Adrian can faintly imagine the sounds of cedar canoes being hauled up on the beach, the chatter of fishermen unloading their halibut, and the strong smell of smoking salmon in the air.

For a brief moment Adrian is able to conjure up those ghostly images, and the beach seems to come alive. But out at Pacheena Point, evening sports fishermen have tired of riding the ocean swells and are racing back to Bamfield. The roar of their outboards drives the ghosts back into hiding. (44)

Native ghosts disrupted by a modern reality. Even in their most powerful assertions of Native presence, the overarching sense is not one of modern trajectories, but instead of loss: "the wilderness rainforest appears substantially undisturbed by human activity, yet for the trained eye, examples of *past* human use are everywhere." Admittedly, this is not "nature itself" as portrayed in the opening chapters, or in the publications of forest companies—evidence of human use is everywhere. But the key word is *past*.

Where the initial promise of this chapter gets diverted, then, is at the point that Native culture is assigned to the premodern, and Clayoquot Sound is relegated to an anachronistic space anterior to the present. Nowhere in the book's photographs, and only rarely in the text, are Native people depicted engaging in any activity, or using any technology that might be considered modern. The question that begs asking is, why must the Nuu-chah-nulth appear in this register? What accounts for their relegation to prehistory in a book that is explicitly dedicated to their contemporary political struggles? Why should *pre-contact* Native life be what is celebrated and its passing mourned, rather than Native culture as modern and technological? What governs this representational logic?

Romancing Nature

Although not all environmental groups in BC hold the same positions, *On the Wild Side* raises some important questions about the how First Nations are understood and incorporated within the region's postcolonial environmentalisms. Why should *this* be how the Nuu-chah-nulth are brought into portrayals of Clayoquot Sound? What authorizes these photos, and not others, to pass as a celebration of Nuu-chah-nulth life and culture? In turn, what enables them to appear as *support* for Native land claims? I will suggest that the ambivalent place of First Nations in environmental discourse in BC is the result of two reinforcing discourses that find expression in Dorst and Young's representation of Clayoquot Sound. The first, which needs little explanation, is romanticism (and the transcendental naturalism that lies at its core). The second is a discourse of indigeneity, which is articulated with romanticism in ways that give environmentalism on Canada's west coast its particular *postcolonial* form.

Neil Smith (1996) argues that in North America, romanticism is almost an instinctual reflex, one that thoroughly infuses ideas of nature and shapes the value placed on particular landscapes (see also Nash 1967). For William Cronon (1995), the pervasive appeal of romanticism—with its dualisms of modern/premodern, culture/nature, tame/wild, profane/sacred—goes a long way to explaining why wilderness is accorded such importance (see also Jasen 1995). Wilderness, Cronon argues, is taken to be that place where one can escape an inauthentic modernity (the relationship between urban life and moral decay has been an enduring theme in North American culture) and where one can glimpse the face of God. It is a place that lies anterior to modernity, a place before the "fall" into modern culture, a place of renewal and regeneration. In short, Cronon argues, wilderness is shot through with culture: from ideologies of rugged individualism and the assertion of various masculinities, to notions of moral order and spaces of spiritual rebirth. To speak of wilderness is to remain firmly within culture's orbit.

For my purposes, romanticism—and its particular North American expression in wilderness—is important for what it can tell us about postcolonial environmentalisms on Canada's west coast. Before I address this, however, let me sound a note of caution. Just as environmental discourse is not unified or singular, neither is "wilderness" univocal. Many environmentalists understand wilderness as an *ontological* entity (either something "out there" in the world that has certain characteristics, such as the absence of roads, or as something lost that we should strive to restore); for others it is more properly a *philosophical* category that refers to that which lies beyond, exceeds, or threatens to disrupt, the rationalization of social and ecological life in modernity (Light 1995; Rothenberg 1995). In this latter sense, the focus rests more on wildness than wilderness, where the "wild" is understood less as a place that one can travel to than a rhetorical place that illuminates the limits of Western rationality.

The "wild" in this second sense refers to modernity's unassimilable other—its remainder—and signifies an ethical position to the extent that it names a responsibility to humanity's "others."

Most commonly, however, wilderness is understood as an ontological condition. It refers to a place, and, in particular, to its physical character. What qualifies a place as wilderness is its ability to appear to lie *outside* human history. By this definition, wilderness is that place where nature is most pure, where nature is still untouched by humans. It is this last meaning that dominates Dorst and Young's text, as it does most mainstream North American environmentalism. In the postcolonial geographies of British Columbia, this sort of environmentalism is immediately fraught with problems. To begin with, it risks acknowledging only certain issues as properly "environmental." As environmental justice advocates have argued, so long as wilderness occupies a privileged position in environmental discourse, everyday environments at home and work are given less attention.[26] Likewise, in cases where environmental issues are intertwined with questions of race, class, gender, and sexuality, these are too easily dismissed as social problems, or as civil rights issues, and thus of no interest to environmentalists. More significant, for my purposes, is how wilderness discourse intersects with *other* land-based politics in BC, as well as the way that its organizing tropes narrowly circumscribe how First Nations peoples can appear as political actors in BC forest politics.

In chapter 2, I noted that a state discourse of natural resources turned on the *spatial* displacement of First Nations. In this chapter, I have shown that an environmental discourse of wilderness turns on the *temporal* displacement of First Nations. This merits further discussion. If we return to *On the Wild Side,* we see that the authors' preoccupation with wilderness (as the absence of culture) means that the presence of humans can be admitted to the frame only to the extent that it does not disrupt the landscape's ability to signify nature's premodern purity. In other words, humans can enter the picture only insofar as their presence does not cause these landscapes to slide across the great divide from nature to culture. Within this system of signification, the Nuu-chah-nulth can appear in only one mode: *as a natural culture* ("At home in the wild"). If present in any *other* mode—that is, as modern, technological peoples—the region loses its identity as wilderness. It becomes, instead, a "modified" landscape, and thus less worthy of our attention. Along the nature–culture continuum that governs mainstream environmentalism, this brings the Sound somewhat closer to New York (a place that interests few environmentalists) and somewhat farther from Eden (one that interests a great many more).

Thus, for the book to acknowledge Native presence, yet simultaneously retain the Sound as "wild," it has to establish and maintain an *equivalence* between Nuu-chah-nulth culture and nature. This is accomplished by introducing a second binary opposition that is then collapsed into the first (Figure 3.6). As is evident in Dorst's photograph of the Tla-o-qui-aht paddlers, this second opposition relies on the trope of "tradition," and turns on the presence or absence of technology. The outboard motors that Young decries are seen as an intrusion of a debased modernity into an Edenic paradise, disrupting the timeless harmony that binds indigenous culture to wild nature. In the absence of such modern technology, the Nuu-chah-nulth can be safely located *outside* history, and thus subsumed *within* nature. Although aboriginal cultures may "use" nature—cedar trees for dugout canoes, for instance—in the absence of modern technologies their use is not viewed as being any different in kind from the kinds of marks that animals might make. Ultimately, in Dorst and Young's book, nature incorporates the Nuu-chah-nulth, while modernity destroys both.

Indigeneity as a Residual Category

If *On the Wild Side* anxiously watches over boundaries between the traditional and the modern, and between what is properly and improperly indigenous, we must ask why these distinctions have become the salient

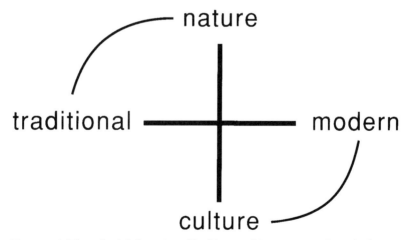

Figure 3.6. The colonial rhetorics of "wilderness." By mapping these dualisms onto each other (culture/nature, modern/traditional), native peoples are conflated with nature, and areas are seen to remain "natural" only if the cultures that live there remain "traditional."

ones. What accounts for indigeneity becoming such a privileged signifier in environmental discourse? Why has preservationist politics come to
the point where it is most effective when it works *through* a discourse of
indigeneity, rather than through ecology alone?

Before proceeding, it is important to recognize that, like wilderness,
indigeneity carries multiple meanings and is deployed in different ways.
At one level, it refers simply to a political identity arrayed against the
social, economic, and territorial ambitions of colonial—and especially
settler—states. In this sense, indigeneity emerges historically as a vehicle
for the political and territorial claims of communities displaced in the
formation of postcolonial nation-states. It is an identity defined in response to, and largely within, legal and political arenas in which the
truth of indigeneity—usually framed in terms of prior occupation and
use of land—must be demonstrated in order for political and territorial
claims to be accepted as legitimate. In other words, indigeneity emerges
in these arenas to fill a place defined though juridical-political discourses
that simultaneously limit the meanings it can hold (Clifford 1988;
Povinelli 1999; Li 2000).

In a somewhat different sense, the indigene is deployed in literature
as a rhetorical figure that disturbs nationalist narratives of the postcolonial settler state. Here, the indigene names that which does not properly
fit, that which has had to be expelled or erased in order for the settler nation to obtain its present form. This is not the place to examine this
meaning in great detail, except to note that in this case, as with the prior
definition, *indigeneity* does not name a *pre-given* political or cultural
identity. Rather, it is something that obtains a certain political and ideological standing within specific historical junctures (Li 2000).[27]

It is also possible to identify an additional cluster of meanings more
directly relevant to ecopolitics on Canada's west coast. Here I follow
Akhil Gupta (1998), who understands indigeneity as a *relational* term
internal to discourses of modernity, in a fashion analogous to the notion
of tradition. Both tradition and indigeneity, he argues, function within
this discourse as residual categories, in that they both name something
that modernity transcends or displaces. But, crucially, they have different valences. As Gupta explains, tradition stands to modernity in a relation of lack in the sense that what is deemed traditional is that which has
not yet been *sufficiently* modernized. This is a distinction familiar from
modernization theory, which sees modernity as the supplanting of belief, ritual, and traditional social structure by reason and rationality.
Tradition, then, is the name given to that which stands outside the mod-

ern, but that is best left behind (it is worth noting that the category has the perverse effect of consolidating the modern as a distinct historical stage). Significantly, the traditional can never be totally annihilated, for to do so would be to render modernity itself an incomprehensible notion; tradition, after all, provides the foil through which modernity defines itself (Gupta 1998, 180).

Indigeneity, by contrast, is that which *modernity* lacks. In other words, indigeneity refers to modernity's loss (understood as its failure rather than that which it has left behind in the march of progress). Essentially, then, tradition and indigeneity are both the Other of modernity, but whereas one is devalued by supporters of modernity, the other is privileged by modernity's critics. Gupta puts this well: "As the project of modernity has come under increasing attack, the evaluative scale has shifted: 'tradition,' long-conceived as the chief stumbling block to the arduous pilgrimage into consumer heaven, has been increasingly replaced by 'the indigenous,' the alternative, eco-friendly, sustainable space outside, or resistant to, modernity" (179). Understood along these ideological lines, Gupta argues, it becomes possible to see why indigeneity is able to refer to such a varied set of phenomena: it becomes a grab bag—a slot—in which to throw all that is not modern and that has the potential to resist modernity's incursions.

Gupta's account is helpful in a number of ways. It sees indigeneity less as something ontological (an identity that exists prior to its articulation) than as a potent *signifier* that obtains its meaning within a discourse of modernity. This in turn helps us think about indigeneity in terms of its point of emergence, in the sense that it achieves its prominence at a particular historical juncture. Gupta, for instance, traces the emergence of indigeneity to transitions in the global economy. Specifically, he points to the geographic expansion and restructuring of capitalist production in the last decades of the twentieth century, in which cultures in remote regions previously on the margins of the global economy were increasingly drawn into its circuits of production and consumption. By this view, indigeneity emerges at precisely the moment that these sites, and cultures, are seen to be under threat of disruption or displacement by global economic forces. It is worth noting that in this context Gupta sees indigeneity as a highly ambivalent identity: on the one hand, it can be articulated as a political identity in response to the displacements of global capitalism; on the other hand, it merely becomes another means of commodifying and thus incorporating these "remote" peoples into global capitalism (as evident in ethnic tourism and ecotourism).

Although this is helpful, Gupta's account is not entirely convincing. The most obvious problem is that he does not give adequate weight to earlier critics of modernity who also understood the supplanting of tradition through an optics of loss, but did not draw on notions of indigeneity. This conservative antimodernism has a long history, although arguably the "indigenous" contains a sense of active *resistance* that a term such as *tradition* does not. Struggles pitched in terms of the latter were ultimately doomed, because tradition is always eventually overcome. Equally problematic is the fact that Gupta fails to explain why it is *this* round of globalization—rather than any other one—that produces the indigenous as its privileged ideological term. Earlier rounds of globalization also involved the spatial extension of market relations. Yet, to a large extent, they were understood in the West in terms of a world divided in two: a Western modernity and a non-Western world in which tradition stood in the way. Why should present-day displacements be understood differently? Ultimately, we need to go beyond political-economic explanations. Here I wish to argue that indigeneity emerged not solely as a result of capitalist modernity's spatial extension into remote places (for this is a *constant* element in stories of modernization), but in a somewhat more conjunctural mode where these displacements occurred simultaneously with *(a)* a negative revaluation of modernity (that gathered force in the 1960s) and *(b)* a new metanarrative of "sustainability" (that took form in the 1980s). This does not throw Gupta's political-economic account into question so much as flesh it out. For instance, whereas past displacements were contained within a *positive* discourse of modernization-as-progress, present-day displacements are often viewed through a *negative* discourse of modernization-as-disruption. The indigeneous appears as that which resists incorporation into this global order. The shift has many causes, some internal to the West (a loss of faith in modernity's metanarratives, for instance), and some that come from beyond (such as critiques of development that emerged in the so-called Third World in the 1970s). Equally important to this story is the strengthening of an antimodernist environmentalism. This is evident in the terms in which modernity's Other comes to be understood in the 1980s. Whereas in the past modernity was set against tradition (figured as a belief system or a social structure), now modernity is set against (and threatens to tear asunder) something quite different: finely balanced cultural-ecological formations (in which the key terms are now a triad of place-ecology-knowledge). Significantly, modernity's Other now comes to be seen not as that which stands in the way of modernity, but as those people and cultures who hold the key to its sustainable future. To the extent that modernity threatens to displace indigenous communities, *it risks destroying itself.*

What I am suggesting is that cultural and ideological transformations as much as political-economic ones explain why indigeneity has such ideological and political purchase today. Before returning to Clayoquot Sound, let me make three additional observations. The first is that it would be wrong to see this as a development that is entirely *internal* to the West. Although it is certainly true that indigeneity fills a slot within discourses of modernity, this is no longer a modernity contained as European, but a heterogeneous and global modernity that allows for multiple indigeneities to emerge with rather different political and ideological effects (indigeneity means something different in India, Australia, or Indonesia). Moreover, so-called indigenous peoples are not without agency; discourses of indigeneity are not simply imposed from outside. Li (2000) notes that in Indonesia discourses of indigeneity opened new spaces for politics and identity that were quickly occupied by actors who are able to recognize themselves in the positions held out for them (she refers to this as the "tribal slot"). It is also the case, I think, that the shift in evaluative scale whereby modernity is increasingly defined in the negative owes a great deal to the interventions of those who do not properly fit the slots that modernity held out for them. In this sense, indigeneity emerges as part of the recoding of modernity that occurs at multiple sites, and not just within the West.

Second, although indigeneity displaces tradition as the key term that stands in opposition to modernity, this does not mean that tradition falls from sight, nor that it merely stands in the negative mode that Gupta assumes. Rather, as we saw in the case of Dorst and Young's book, the traditional reenters the discourse of modernity *as that which guarantees indigeneity's nonmodernity.* No longer that which must be overcome by modernity in its teleological story of progress, tradition becomes the term that mediates the pure from the impure (without a notion of tradition, indigeneity loses its meaning).

Finally, it should be noted that although indigeneity is arrayed against modernity, it does little to place in question either the linear, progressive history of modernity or its singularity. Although indigeneity emerges at precisely that moment when modernity is cast in the negative, it simply *inverts* modernity's value, rather than displacing it or rendering it multiple.

Authenticities and Incarcerations

We are now in a position to see how a discourse of indigeneity dovetails with environmental politics on Canada's west coast: through its temporal displacements it allows First Nations to be visible, but incorporates them within the terms of an antimodern preservationist politics. To the extent that the Nuu-chah-nulth appear properly indigenous, Clayoquot Sound can be situated in a mythical place *outside* modernity and thus a place that both deserves preservation and requires a modern representative to speak in its name.

In *On the Wild Side,* the presence of indigenous peoples does not threaten wilderness, it guarantees its ontology. This leads to an additional point: to the extent that ecopolitics hinges on a discourse of indigeneity, indigeneity itself becomes an ideological battleground (and potentially a prison house in which First Nations are discursively incarcerated). If to be properly indigenous is to be nonmodern, then indigeneity is something that potentially can be *lost*; at the point that indigenous peoples become too closely associated with the social, economic, and technological relations that signify the modern, they lose their claim to indigeneity. This is clear in much environmental literature in BC in which cell phones, chain saws, power boats, and Chicago Bulls T-shirts are kept outside the visual frame because they potentially disqualify the bearer's indigenous identity. Further, because indigeneity is defined in ways that bind together a triad of culture-place-ecology, these links must be carefully preserved. Far from being able to display the complex local-global relations that organize Nuu-chah-nulth life today, including the continuous movement of people between Clayoquot Sound and sites of employment elsewhere, the authors of *On the Wild Side* relegate the Nuu-chah-nulth to a timeless series of pure forms that lie outside history. The Nuu-chah-nulth are merely asked to fill a slot as modernity's Other. Only by doing this can the authors hold their contradictory political position of support for Native land claims and a desire for preservation.

There is a certain irony in this: as we saw in chapter 2, nineteenth-century colonial discourse simultaneously marked, yet spatially contained, a Native presence, allowing for a more complete spatial dislocation in the twentieth century. At the close of the twentieth century, a different representational logic emerged, one that gestures toward "giving back" to Natives their visibility, but only to ask them to speak the language of authenticity. A spatial incarceration is replaced with a temporal one. Iain Chambers (1994, 74, 81) argues that such temporal displacements have long underwritten the hegemony of the West and underwritten its more subtle forms of imperialism:

> The temporal (and teleological) distance between "primitivism" and "progress" . . . has consistently justified so much of the intellectual, political and cultural capital invested in the center-periphery model of culture and history. . . . Whether following religious or secular compasses, [this] has permitted evil to be externalized by relegating it to the "savage" and "heathen" peripheries of the world. Then, in a reverse image, that imperious gesture has more recently been extended to the mourning for the pristine culture of the primitive, that is, the past, the elsewhere, from within the perceived decay of the metropolitan present. . . . The notion of the pure, uncontaminated "other", as individual and as culture, has been crucial to the anti-capitalist critique and condemnation of the cultural economy of the West in the modern world. Such a perspective evoke[s] its surreptitious form of racism in the identification by the privileged occidental observer of what should (a further ethnocentric desire and imperative) constitute the Native's genuine culture and authenticity.

Who defines whose authenticity? As Chambers clearly shows, it is the observed who are once again spoken for and positioned, a problem that has surfaced often in environmentalist texts, where aboriginal peoples are asked to occupy a redemptive place for members of Western industrialized societies, rather than allowed to articulate a political future unconstrained by a discourse of authenticity (Emberley 1996; Sturgeon 1997). To talk of authenticity, Chambers (1994, 82) continues, invariably involves referring to tradition as something fixed, "as though peoples and cultures existed outside the language of time. It is to capture these in the anthropological gaze, where they are kept in isolation and at a 'critical distance', as though they do not experience movement, transformation and the disruption that the anthropologist represents: the West." It is likewise to hold modernity's Other in a geographical gaze, where a particular "dwelling-in-place" precludes the kinds of mobility of bodies, technologies, and ideas that characterizes the time-spaces of modernity. Perhaps tellingly for an environmentalism that proceeds *through* a discourse of indigeneity, the book gives the last word to Simon Lucas, the Hesquiaht elder who had appeared before the Pearse Commission in 1975. Here, however, he is identified by his Hesquiaht name, Khah-keest-ke-us: "The very survival of the Nuu-chah-nulth depends on the

survival of old-growth forests. Old-growth forests are our most important places of worship. Within forests we are completely surrounded by
life; within forests we can renew our spiritual bonds with all living
things." Located where it is, the quote appears to give tacit consent to all
that precedes it. Further, it allows only a *spiritual* relation between the
Nuu-chah-nulth and the forest, thereby legitimating the Wilderness
Committee's call for the region's preservation, for in preserving the forest, indigeneity is preserved too. This spiritual relation is clearly important to the Nuu-chah-nulth, but as we saw in chapter 2, this land is
clearly a *resource* landscape too, a place of work and livelihood.[28]

Despite the important support for First Nations land-rights struggles
by many BC environmental groups, the province's postcolonial environmentalisms are deeply ambivalent. Within its terms, First Nations often
find themselves with few options. For First Nations in BC, to forge a
modern future as participants in the region's resource economies is to risk
losing what many non-Natives consider authentic indigenous culture
and thereby also their right to speak *as* indigenous peoples for *their* lands.
Likewise, to leave their communities, or to develop a bifocal sense of
home, as so many Nuu-chah-nulth have in their search for work in the
province's cities, is to step across a temporal divide that separates indigeneity from modernity. On the other hand, to refuse modernization—
in essence to constitute identity around culture, place, and ecology, as
the environmental movement implicitly asks—is to risk remaining forever outside the economic circuits of the global economy, placed where
European cultures have so consistently slotted indigenous peoples: in nature, always outside modern forms of rationality, as undeveloped, primitive precursors to modern cultures, and, ultimately, in need of representation. Both rhetorics easily lead to an assumption that—as was noted
by a Hesquiaht woman—aboriginal people are "incapable of being main-
tainers of our own territory" (Charleson 1992).

Here lies the larger problem. By removing signs of *modern* Native
culture from the landscape, the same rhetorical maneuvers that enabled
the region to be seen as a resource landscape in the first place are again
deployed, only its terms are inverted: the hyperproductivism of industry
replaced by a transcendental naturalism in which First Nations are subsumed within a "natural" drama. The significance of this becomes readily apparent in a statement taken from the preface to *On the Wild Side*,
written by the renowned Canadian wildlife artist Robert Bateman: "The
world recognizes Canada as containing one of the last great remnants of
wilderness and we Canadians have always prided ourselves in *our natu-*

ral history. . . . This decade will see them saved or lost. Our generation must draw the line for all of them" (emphasis added). We are back where we began, with nature displaced within narratives of the nation. Indigeneity may be one of postcolonial environmentalism's most privileged terms, but the risk is that indigenous peoples merely become subsumed within the "natural history" being saved.

Nuu-chah-nulth Modernities: Writing Open Futures

Not all articulations of indigeneity are identical, nor do they all foreclose on politics in the same way. Indeed, what *counts* as indigenous, and how the relation between indigenous peoples and nature should be understood in west coast temperate rainforests, is something very much up for grabs. By turning to a second example from the political and legal struggles over the fate of the forests in Clayoquot Sound, it may be possible to sketch a competing vision of indigeneity and environment that refuses closure around either tradition or wilderness, but instead opens conceptual and political space for a uniquely *Nuu-chah-nulth* modernity.

In the 1990s, it became popular for First Nations to develop increasingly sophisticated maps of their historical and present-day territorial and environmental practices (see Beltgens 1995; Scott 1995). These efforts can be traced in part to a key project in the late 1980s in which a team of archaeological consultants produced a map of "culturally modified trees" (CMTs) as part of a multivolume work that examined Nuu-chah-nulth forest uses in Clayoquot Sound (Arcas Associates 1986; see Figure 3.7). The study was commissioned by lawyers representing two Nuu-chah-nulth bands (Ahousaht and Tla-o-qui-aht) in Clayoquot Sound, as part of their efforts to assert sovereignty over Meares Island. Here I explore the very different sense of indigeneity, modernity, and nature that it articulated.

Enrolling Allies

Although produced near the end of the twentieth century, the significance of this map lies in its response to dynamics set in place more than a hundred years earlier. As discussed in chapter 2, in the late 1800s Native groups in Clayoquot Sound were segregated onto small reserves, and all surrounding lands were declared the property of the Crown. At first this had little effect on Native territorialities because few areas of Clayoquot Sound were alienated as private landholdings. Thus, although boundaries between reserves and forests appeared on maps, these lines remained for the most part cartographic fictions. It was not until the

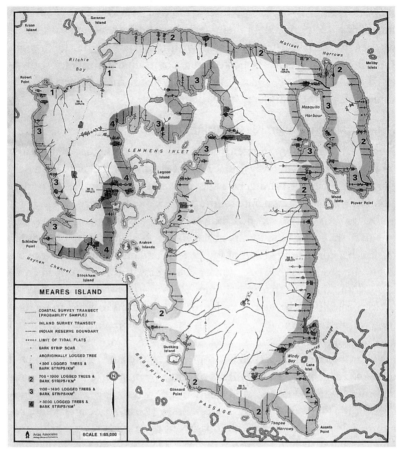

Figure 3.7. Map showing native tree use on Meares Island. Courtesy of Arcas Associates and *BC Studies: The British Columbian Quarterly.*

1950s that the Crown lands were divided into Tree Farm Licenses (TFLs) and the rights to the timber on them allocated to large forestry companies. By the 1980s, as more accessible timber elsewhere became depleted, MacMillan Bloedel submitted plans to commence harvesting on Meares Island, a prominent landmark in the Sound, and the site of the Tla-o-qui-aht village of Opitsat. In response, the Tla-o-qui-aht and Ahousaht (who also claimed traditional territories on the island), joined by non-Native allies in Tofino (which uses the island as its water supply), blockaded the island in order to keep the company's workers and equipment from landing. The result was a series of court struggles over who held sovereign rights to the island.

This context is important for how we read the map. Because the validity of Canadian sovereignty was assumed from the start—and thus what had to be disproved—it fell upon aboriginal groups to demonstrate the existence today of territorial practices that were continuous with those that existed prior to the arrival of Europeans. In other words, in Canadian courts aboriginal groups have had to demonstrate their indigeneity in certain terms: as bound to particular *places,* and, crucially, as *uninterrupted* (the assumption being that discontinuity was evidence of extinguishment). This explains why lawyers for the Tla-o-qui-aht and Ahousaht commissioned this study rather than some other. The legal context also helps explain the form that the study took. In any court trial, evidence must be produced that can pass the test of facticity. At every turn, then, the study emphasizes its scientific credentials. Accepted rules of archaeological research and ethnographic fieldwork are carefully explained, and then shown to have been followed meticulously by researchers in the field. Methods of locating and interpreting biological evidence of tree modification are presented in great detail. The authors explain, for instance, how their estimates of tree modification passed tests of "statistical confidence," emphasizing their systematic surveys (165 transects, each 30 meters wide, running inland through the coastal stratum, and covering approximately 10 percent of the territory) and drawing attention to their probabilistic sampling methods. The report also included discussion of what at first glance might seem curious topics for a court battle over Native sovereignty: the marks left on trees by certain tools, how to determine the age of trees, or how the growth of a tree may hide signs of human use. To address these questions, the authors included an appendix that provided criteria for identifying signs of cultural modification and provided extensive discussion of how the analysis of tree-ring characteristics can reveal evidence of modification even when the external signs do not point unambiguously to either cultural or natural causes (Figure 3.8). In short, the study had to educate the court on how to properly *read* the forest, and to do this the authors had to enroll as allies knowledges produced in the sciences of archaeology, anthropology, ethnobotany, and tree biology in what amounted to a "trial of strength" (Latour 1986) to determine which representations of Clayoquot Sound would prevail.

This brief sketch of the map's production allows a number of important observations. Like the wilderness photographs just discussed, for example, the map is assumed to bridge the gap between representation and reality, and to speak the "truth" of the forest. Both the photographs and the map draw on an empiricist tradition that assumes that if one looks carefully and without bias one sees what is really there. But they

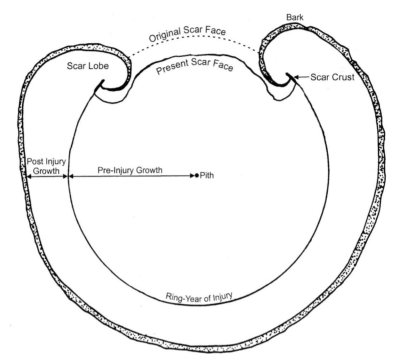

Figure 3.8. Locating culture in nature: diagram of tree scar from Native tree use. Courtesy of Arcas Associates.

reach this conclusion in different ways. In *On the Wild Side,* it is simply assumed that the camera lens mirrors reality. The archaeological study, by contrast, cannot tap into this faith in photo-mimesis, because the representations it brings to the court are clearly linked to the labor of fieldwork. Although both are the product of work, its "stain" is more evident in the latter. Thus, the authors find it necessary to discuss in great detail *how* they derived their truths, consistent with the protocols of "good" science, where proper methodology is seen as the guarantor of objectivity.[29] One of the side effects is that the study provides critics with a means of questioning the adequacy of its observations and conclusions, even as it tries to stabilize what the forest signifies in order to foreclose other interpretations of the landscape.

Contesting Tropes of Dispossession

My concern is not to identify this map as more accurate than other representations. A closer examination would find that its coherence *also* rests on a set of (successful) cognitive failures, not the least of these its masking of the ways in which these landscapes are contested terrains, not only between Natives and whites, but within Native communities too. Rather, my intention is to attend to the terms in which this study poses the relation between nature, indigeneity, and modernity, and, in turn, how this matters in struggles over the future of Clayoquot Sound, its forests, and its people.

We can immediately see important similarities and contrasts between this map and the photos in *On the Wild Side*. In both, signs of Nuu-chah-nulth forest use are present. For the nature photographer, however, these are unambiguously assigned to the past and contained within a discourse of primitivism. The effect is to secure the region's identity as wilderness. The map of CMTs, in contrast, presents the region not as dehumanized wilderness occupied by a primitive people living in harmony with nature, but as a hybrid natural/cultural landscape that is a product of ongoing and intensive use and transformation and thus as *emergent* rather than *timeless.* Indeed, the scale of Nuu-chah-nulth forest use is breathtaking. Through archaeological and ethnographic research (which included the participation of many Tla-o-qui-aht and Ahousaht actors) the authors located and mapped 1,779 CMTs on the island (ranging from felled trees to trees stripped for bark) and found that in places the density of modified trees exceeded three thousand CMTs per kilometer of coastline. Using their transect data as a base, the researchers estimated that the coastal stratum of the island likely contained twenty thousand CMTs, an estimate they termed conservative. This excluded CMTs that remained "archaeologically invisible" or that lay undiscovered further inland on the island, a region outside the specific parameters of the study, but not beyond the forest-use activities of the Nuu-chah-nulth. From this data the authors concluded:

> Archaeological evidence from the west coast of Vancouver Island and northwest Washington State indicate that tree use has been an integral part of Nootkan culture for at least 2500 years. The antiquity of human settlement and tree use of Meares Island is not known, but the enormous midden at Opitsat has probably been occupied for several thousand years. Our tree-ring dating showed that tree use on Meares Island goes back at least to the 1640s [Captain Cook arrived in 1778], and there is no reason to believe that this use does not go back further in time. Both the tree-ring dates and our ethnographic information indicate that this use has *carried on* to the present, *although there have been changes in specific tree uses in response to changing needs.* In a few cases, such as red cedar stripping, canoe carving,

and the collecting of traditional medicines, the present-day activity represents a direct *continuation* of a long-standing tradition. This *continuity* between the past and the present was illustrated during our research by a number of CMTs recorded during the archaeological survey which were later identified in the ethnographic research, and confirmed by the tree-ring dating, as instances of *very recent* tree use. (Stryd 1986, 35; emphasis added)

This was a significant intervention in the representation of the temperate rainforest. Disrupting the claims of transnational capital and the state, for which Clayoquot Sound was simply a resource landscape without any preexisting claims, the map asserted a substantial and transformative Native presence in the forest. First Nations in Clayoquot Sound did not simply wander aimlessly across the surface of the land living off its bounty; they actively used and transformed their environments in a purposeful manner.

This map takes on added significance when placed against the dispossession of Native lands by the Indian Reserve Commission in the late 1800s. As Paul Tennant explains (1990), dispossession at the time had been justified in part through recourse to Lockean notions of property that tied possession to signs of use.[30] If land showed no signs of labor—usually understood as permanent improvements—it could not reasonably be claimed as property.[31] The scandal of this map, then, is that it shows Nuu-chah-nulth land use to have been extensive and to have altered the landscape, not only appropriating a Lockean discourse of property and turning it against the state, but revealing the cartographic incarceration of First Nations achieved by the Indian Reserve Commission to have been based on its inability to correctly read the landscape.

The study also demonstrated continuity in land use between past and present. This turned in part on dismantling the "great divide" usually posited between pre-contact and postcontact periods in BC history. On the *far* side of this divide, such narratives assume, lay the presence of a timeless primitive culture governed by nature and mythology, whereas on the *near* side lie an acculturated people whose "primitive" past has been inevitably eroded by modernity (the introduction of European technologies, market relations, and the intrusion of distinctively non-Native economic and cultural practices). As we saw in *On the Wild Side*, in such accounts there is only one epochal event—contact—after which First Nations are ever in danger of losing their claim to indigeneity.

A further discussion of this assumption of historical rupture can help us draw out the full significance of the consultant's study. Yvonne

Marshall (1994) has argued that notions of rupture have been a consistent part of BC history writing, in which scholars have consistently divided the past into two parts: a static Native/natural prehistory and a dynamic European settler history. Anthropologists, she suggests, often divided history in the same way, evident in the work of such salvage enthographers as Boaz, Jacobsen, and Sapir, but also in much later work by ethnographers such as Philip Drucker, who published important studies of the Nuu-chah-nulth in the 1950s.[32] For my purposes, Drucker's writings are instructive. Despite writing more than 160 years after contact, and sixty years after so-called modern transportation links had connected the Nuu-chah-nulth with Euro-Canadians (steamships, telegraph, etc.), he remained focused on capturing the last vestiges of "authentic" Nootkan (Nuu-chah-nulth) culture. Drawing on a discourse of fatal contact, Drucker believed that in his time Nuu-chah-nulth culture had become cut off from this authentic past:

> The evil star of civilization dawned for the Nootkans and their neighbors on the Northwest Coast . . . when Cook stood in to King George's or Nootka Sound in the *Resolution* during his third voyage of exploration. . . . [Later,] when it was learned that in Canton the lush brown pelts were worth more than their weight in gold, the fate of the native culture was sealed. (Drucker 1951, 11)

Rather than a people actively initiating change, Drucker saw Native culture after contact as merely reacting to a modernizing world over which the Natives had no control. Thus, changing settlement patterns in the nineteenth century were viewed solely as responses to growing trade, rather than intricately tied to Native polity and society. Ultimately, all traces of "authentic" culture were destined to disappear:

> The final step in the transition to modern acculturation consisted of several parts: the establishment of Christie School at Clayoquot Sound in 1899; the increased white contact resulting from the establishment of canneries and other enterprises at Clayoquot, Nootka, and other localities; the white community at Tofino; and regular steamship service up and down the coast. (Ibid., 14)

Drucker (1955, v) would later write that "only fragments are to be found today of the aboriginal civilization described in these pages. Many of the Indians of the coast are nowadays commercial fisherman and loggers. Most of them are more at home with gasoline and Diesel engines than with the canoes of their forefathers."

These are remarkable passages not only because Drucker so consistently tropes the arrival of modernity as the death knell of indigeneity, but because some forty years later *On the Wild Side* repeats many of the same assumptions. Indeed, while both books assume that modern technologies—gasoline and Diesel engines—signal the end of indigeneity, the earlier volume at least registered the *presence* of modern, technological Indians, even if it then went on to deny them their authenticity. Such nostalgic tropes would matter little if they had no effect politically, but it is important to attend to what they license. Drucker's distinction between a "pure" past and a "corrupted" present allowed him to drive a wedge between Native cultural practices in the present and those in the past. "Aside from the fragments and people's pride in their identity as Indians," Drucker wrote, "Northwest Coast culture must be regarded as having disappeared, engulfed by that of the modern United States and Canada" (ibid.). By focusing on change as *rupture,* such statements legitimate the continued dispossession of Native lands. Indeed, as became evident in Drucker's later work, this notion of discontinuity was of a piece with his outright dismissal of Native land rights. Regarding the assertion of these rights in political and legal arenas, Drucker (1965, 229) wrote:

> The inspiration throughout was non-Indian, or by *sophisticated Indians long-removed from the native way of life and thought.* The techniques used were non-Indian—a petition drafted by attorneys, attempts to utilize British legal procedure, fund-raising campaigns to implement the legal contest, and the like. *The obvious conclusion is that Indian interest in land, outside the few heavily settled areas, was largely artificial.* (Emphasis added)

Sophisticated—that is, modern—Indians were no longer Indian. They used "non-Indian" techniques. Contact becomes rupture, and hybridity equated to inauthenticity.

Today, First Nations are placing increased emphasis on deconstructing this "great divide" between pre- and postcontact life, not in order to deny the significance of European–Native contact, but in order to question whether its effects were really so unidirectional and catastrophic, and whether the worlds before and after contact can be so neatly assigned to different eras. Along these lines, Marshall (1994) argues that if historical accounts are broadened out to include insights obtained from archaeology and oral histories, it becomes possible to see First Nations as having always already been adaptive cultures, reconfiguring existing social, political, and economic structures into ever new forms, while also transforming the land in which they lived. As she shows in her own archaeological research and in studies by other archaeologists and ethnographers among northern member bands of the Nuu-chah-nulth Tribal Council, change in Native culture did not *begin* when Europeans arrived on the shores of a land apparently lost in time.[33] Rather, the region already had a complex history of changing economic relations, cultural practices, and political structures that extended back several thousand years—long before the so-called historical period recorded in archival records—and reached forward into the present. These relations and structures were specific to particular regions, but set in a wider network of social and economic exchange that extended up and down the coast. Marshall suggests that one of the most important changes that occurred within "Nootka" culture and polity was the development of a whale-centered social-economic system around 1500 BP (and the concurrent reconfiguration of political relations). Few would argue that this shift brought about the end of a culture, and implicit in Marshall's argument is that European contact must be understood in the same way. Drawing on the texts of salvage ethnographers, for instance, she traces a story of continuity in which the shifting web of cultural, economic, and political practices that composed "Nootka" cultures in the years prior to contact is shown to have adapted to what was simply a *different* set of external relations (see also Clayton 1999). Indeed, Marshall views the current Nuu-chah-nulth political body—the Nuu-chah-nulth Tribal Council—as a political form that has antecedents in earlier coastal confederations that reach back before contact.

This is not mere quibbling over the past. The consequences of writing these alternate histories are highly political, as Marshall herself underlines:

> A "new past" from which an "open future" can be realized . . . requires a temporal frame broad enough to bracket and ultimately foreclose on colonization. Rather than start with the fur trade and end with the welfare cheque, [I have] chosen to begin with ancestors of the Nuu-chah-nulth whale hunters and end with an increasingly confident and increasingly self-determining Nuu-chah-nulth Tribal Council. (Marshall 1994, 340)

From this point of view, the only culture that has died was the one frozen in the accounts of early explorers, missionaries, and ethnologists.

If we return to the map of CMTs (Figure 3.7), we can see how it also contested colonialist narratives of disruption and inauthenticity. On the map, more than three hundred years of continual tree use on Meares Island were identified, dating from the period prior to contact and reaching into the present. This is not by itself a radical recoding of Clayoquot Sound landscapes, for, as we saw earlier, present-day practices can still be displaced temporally by casting them as traditional, or by understanding present practices as merely a response to *external* (i.e., European) influences. However, the authors of the study were careful to avoid this trap. Citing their own research and that of other ethnographers and archaeologists, they argued that "at the time of contact with the Europeans, Nootkan culture was characterized by massive woodworking (i.e. the use of large trees), and an extensive material culture in wood and bark" (Arcas Associates 1986, 3). Wood products were important items for internal cultural economies, but also as items of trade. In short, they argued that contact could not be used to fully explain either the existence or the scale of forest modification by Natives.

Equally important, although the authors admitted that with contact profound changes occurred, this did not bring to an end a preexisting tree culture, but rather resulted in transformations that were continuous with the past. The language here is primarily one of adoption and adaptation, not rupture.

> Nootka culture has changed profoundly since the arrival of the Europeans two hundred years ago. Nootkan adoption of Eurocanadian housing, clothing, containers, rope, boats, and other items resulted in the decline or end of many traditional tree uses. In other cases, the Nootka adapted their tree use practices to modern economic pursuits. European tools and felling techniques have been widely adopted by the Nootka in this century, and milling has altered the way in which the Nootka obtain their planks. (5)

Later, the study turns to how present-day Nuu-chah-nulth forest use has been adapted to more commercial pursuits, including the use of trees in fishing and sawmilling industries. Rather than a static triad of place-culture-ecology disrupted by modernity, what we find is a story of adaptation, adoption, and alteration in which modernization is merely part of a larger story. Contact brought changes, certainly, but the differences that matter in this account are no longer those of pre- or postcontact, nor is the language of rupture or discontinuity the most important one. The focus is not on a timeless culture, nor on policing authenticity. Changes in how or to what purposes trees were used are simply placed in relation to the evolving character of Nuu-chah-nulth society. Hybridity is neither inauthenticity nor loss.

With this modest map of culturally modified trees, the cultural politics of nature and indigeneity comes into clear focus. In contrast to the efforts of environmentalists, its authors refused to "wild" nature by "wilding" culture. The map depicted neither a pristine landscape nor a traditional indigenous culture living in harmony with nature. Its forest is a dynamic, trammeled ecology occupied by a people whose indigeneity is not defined as a residual category, but as constitutive of the region's modernity. Likewise, Native people do not lose their claim to speak for nature as soon as they begin to modify their environments, or as soon as they take up modern technologies. What this map allows is the recoding of the present in terms of a uniquely *Nuu-chah-nulth* modernity in which the future of the region's social nature is not determined in advance.[34]

Conclusion

The point of this chapter has not been to dismiss the political and media campaigns of BC environmentalists. In the face of the massive ecological displacements achieved in this era of globalized capitalism, we need more efforts like Dorst and Young's, not fewer. The brilliance of Dorst's photographs—and others like his—are that they remind viewers of responsibilities that extend beyond the bounds of the human. Viewed in this light, *On the Wild Side* served an important role as a catalyst for forms of global environmental citizenship that took seriously the need to challenge nature's commodification. This invariably involves speaking for *other* beings who are not accounted for in calculations of profit. In a world tied together by circuits of capital, commodities, and information, as well as carbon cycles and weather patterns, the thousands of people that journeyed to the Sound to put their bodies on the line must be commended. Ecological issues, as Thom Kuelhs (1996) notes, are not contained within the space of sovereign territoriality.

This imperative to speak out, however, does not place environmentalists beyond criticism. Indeed, it demands that we think carefully about how this speaking occurs. Despite support for Native land rights, BC's ecology movement has at various points found itself estranged from First Nations; and, when coalitions are formed between these groups, they are invariably tentative and fragile. It was not insignificant that in 1994 George Watts, a former chairperson of the Nuu-chah-nulth Tribal Council, accompanied then Premier Mike Harcourt on a European tour designed to head off the growing challenge to BC forestry by European

Greens. Before an assembly of German Greens and the German media, Watts accused the environmental movement of "neocolonialism."[35] Although dismissed by environmentalists as a lone Nuu-chah-nulth voice, and not representative of the current Tribal Council, Watts's decision to align himself with the state *against* environmentalists needs to be taken seriously.[36] Similar issues surfaced again in the summer of 1996 when, for a time, Nuu-chah-nulth groups in Clayoquot Sound banned Greenpeace from the region, after the environmental group planned road blockades without first obtaining the approval of Native leaders (see Hamilton 1996). And, in recent years, the Ahousaht band has taken the lead in cooperative ventures with transnational forest companies, forcing environmental groups to rethink how and in what terms they work with First Nations, who often have very different futures in mind than preserving the primeval forest.

In this complex political terrain, it is not difficult to understand the significance of Donna Haraway's (1992, 296) often misunderstood phrase that "nature cannot pre-exist its construction." Far from the idealism that it has been made out to be (Soule and Lease 1995), Haraway states what is obvious: that there can be no one objective rendering of our natural surroundings but many, and that each and every one is infused with relations of power. This is consistent with much poststructuralist thought, which insists on the undecidability of identities such as nature, in part to draw attention to how such identities come to be fixed. The point of emphasizing undecidability is not to arrive at the pluralist position that all statements are equally valid; rather, it turns critical inquiry to the task of examining how different constructions of nature come to matter, and why. It is often the case that statements about nature's "essence" or Native "traditions" must be made strategically, in order to launch a political critique (cf. Spivak 1988a). The danger is that if this move is not made explicit, and its problems carefully attended to, it risks reproducing the game of knowledge as power. Environmental politics must never be a simple matter of speaking nature's "truth."

4. Landscapes of Loss and Mourning
Adventure Travel and the Reterritorialization of Nature and Culture

Rousseau, by his example, teaches him [man] how to elude the intolerable contradictions of civilized life. For if it is true that Nature has expelled man, and that society persists in oppressing him, man can at least reverse the horns of the dilemma to his own advantage, and seek out the society of nature in order to meditate there upon the nature of society.

—Claude Lévi-Strauss, *Jean-Jacques Rousseau*

The possibility of an unalienated subject and an unmediated relation to nature can find expression only within an equally ideological, even utopian, sphere.

—Susan Stewart, *On Longing*

Much of our current critical and political project appears to me as a kind of unrealized mourning in which all of life has become reorganized around something that 'died'.

—Dean MacCannell, *The Tourist*

Ends and Beginnings

Tofino, British Columbia, lies at the end of the road. Here the lone ribbon of asphalt that winds its way west across the mountainous interior of Vancouver Island passes by the town's shops and galleries, turns to descend a short, steep hill, and then comes to an abrupt halt at a government wharf, its uneven pavement replaced by worn wooden planks. Beyond lie the frigid waters, deep fjords, and rugged mountains of Clayoquot Sound.

In various ways, this terminus represents a boundary. As a physical boundary, it separates land from water—cars give way to boats, trucks to barges, the solidity and permanence of the road to the plasticity and buoyancy of the ocean. It appears as a social and racial boundary too, dividing a Euro-Canadian world from Nuu-chah-nulth villages further north. As an economic boundary, it separates an affluent tourist town from villages dependent on fishing and forestry, in which unemployment rates can be as high as 80 percent. And, for many, it represents a temporal boundary, one that divides modernity from what comes before, and that separates a primeval nature from a colonizing modernity that must be its inevitable fate.

Of course, these boundaries are rarely what they first seem. A day spent observing activity on the wharf reveals a very different picture. Far from a place where one world ends and another begins, the two sides of this divided world continuously bleed together. Workers, shoppers, televisions, fish, cedar baskets, groceries, videos and LA Kings hockey jerseys move across this boundary as if it were not there at all, and without ever stopping to declare loyalty to one side or the other. How one understands this wharf depends on one's perspective. For many local residents, Clayoquot Sound is experienced as a single totality, albeit one with a differentiated geography shot through with history, culture, and politics. For them the wharf is not a dividing line but a nodal point in a complex web of social and spatial relations. For visitors, on the other hand, the wharf is often set up—and experienced—as a sharp divide, on which are mapped a series of familiar dualisms (culture/nature, modern/premodern, civilized/primitive, artificial/real, masculine/feminine). By this view, things have an assigned place.

For those under the thrall of these modern mythologies—and who truly eludes them?—it takes little effort to imagine the town of Tofino, with its co-op grocery and smattering of shops, restaurants, and galleries, as civilization's last outpost, beyond which the present age has somehow, magically, not yet arrived. Such double vision would matter little if it were not also a prolegomenon for ways of knowing and moving. Today, visitors flood into Tofino each summer by the tens of thousands to partake in the pleasure of crossing these material-semiotic divides. Most visitors remain near Tofino and its environs, exploring its beaches or perhaps taking a short whale-watching tour, returning to the comforts of RVs, hotels, and restaurants at night. But for many travelers, such timid reimmersions into nature are insufficient. For these so-called adventure travelers, the end of the wharf is less a place from which to peer across into the "before time" of pristine nature than a place from which one *departs*. What appears as the end of the journey to some is for others only its beginning.

This chapter follows this second group of actors, tracing their movements and practices, and through them exploring how nature and culture, discourse and politics, the local and the global, come to be woven together in the rainforests and waters of the so-called wild side of Vancouver Island. At times I will focus on the travelers themselves, tracking their activities and following their gaze. At other points I will step back to examine at greater length some of the cultural logics that may explain their presence in the region and the paths they take.[1] My objective is not to pass judgment on adventure travel. Nor is it to denounce adventure travel as inherently colonizing, a set of practices that "dis-place" and "ruin" the cultures and peoples it embraces. This would be to remain trapped within the same logic of mourning that this chapter explores. It is not insignificant, as Dean MacCannell notes (1989), that much critical theory shares with the traveler a nostalgia for worlds lost. Instead, my aim is to understand adventure travel both as a source of pleasure and as a specific *constellation* of cultural, discursive, and spatial practices that are increasingly central to life on Canada's west coast. As we will see, these practices open new spaces of politics and identity even as they foreclose on others.

In what follows, I offer a reading of adventure travel—and its ideological twin ecotourism—as cultural and spatial practices that emerge within, and as an effect of, ideological formations of modernity that produce subjects who experience the present in terms of loss. As I explore at more length later, this gives adventure travel a paradoxical nature—because what is lost is ultimately an impossible object (origins, purity) that can never be found. Thus, for many it is the search itself that is the source of pleasure (and that comes to be commodified). A point of clarification is necessary here. To posit adventure travel in terms of modernity and loss is to understand it in terms similar to Rosaldo's (1989) notion of "imperial nostalgia." Yet, the nostalgia that I trace is perhaps more pervasive than what Rosaldo describes, for it is not only a response to the physical and cultural dislocations of modernization and imperialism, whereby what has been destroyed (primitive cultures, nature) comes to be eulogized by the very agents of its destruction. Rather, to the extent that modernity exists under the sign of *history,* mourning the loss of what was before is part of what it means to *be* modern. In other words, to the extent that modernity achieves its coherence through a temporal narrative that understands the present as that which transcends, supersedes, and ultimately destroys the past, mourning presents itself as modernity's most pervasive psychical form. Paradoxically, it is in the continuous *failure* to locate the not yet destroyed that we sustain our sense of ourselves as modern, for it provides the very evidence of modernity's destructive force (look, it's already happened here!) and leads to additional rounds of nostalgic yearnings. What I wish to suggest, then, is that adventure travel and ecotourism are best viewed as practices through which subjects both perform and reaffirm the present—and their own identity— as modern. It follows that, far from fulfilling the promise of escaping modernity, adventure travel may be its most pure expression.

Although it is crucial to understand this discursive framework of adventure travel, my foremost concern is to explore the social, political, and ecological consequences of constituting Canada's west coast as a landscape of mourning. To what extent does adventure travel result in a reconfiguration of the region's economic, political, and cultural geographies? How does it produce new spaces for identity and politics? More broadly, how do these cultural and spatial practices contribute to the formation of a unique regional modernity on Canada's west coast? At times I travel some distance from Clayoquot Sound in order to trace how this place "snaps into place" as a site of desire (cf. Stewart 1996), but I will consistently return to these questions in order to trace reconfigurations of place, culture, and politics in the region.

My discussion begins with the movements of a commercial adventure travel expedition as it travels through the waters and landscapes of Clayoquot Sound. Noting its paths and its preoccupations, its incorporations and its blind spots, I argue that adventure travel both imposes and locates an order in the unruly social and ecological spaces of the temperate rainforest. Following this, I embark on an effort to situate this journey—that is, to understand its conditions of possibility in late-modern metropolitan cultures. Rather than end my story here, however, I return to Clayoquot Sound at the end of the chapter to consider at greater length the complex *cultural politics* that arises when metropolitan fantasies of recovery are staged in and through postcolonial landscapes such as Clayoquot Sound. Although at points it may appear that adventure travel is simply the most recent way in which Canada's west coast comes to be "colonized" by external forces, I will suggest something different. Drawing on the practices of a second, more modest, commercial expedition, run by women from Marktosis, a local Nuu-chah-nulth village, I conclude that the *politics* of adventure travel cannot be read off its *rhetoric* in any simple manner, in part because practices of adventure are always historically and geographically situated. In places such as Clayoquot Sound, they occur within what Stuart Hall (1986) has described as "conjunctural" moments, and thus give rise to new and unique social, cultural, and political realities that cannot be known or predicted in advance.

Circumnavigating Vargas

Impressed by the flood of adventure travelers to Clayoquot Sound, I interrupted my study of forest conflicts in the summer of 1994 to study the operations of a relatively new Vancouver-based adventure travel company. The company, with its headquarters set among the galleries, boutiques, artist lofts, outdoor equipment stores, and fruit and vegetable markets of a trendy waterfront redevelopment scheme, offered trips to a variety of "exotic" locations: Haida Gwaii, Belize, Irian Jaya, Baja California, and several others, in addition to Clayoquot Sound. Its specialty was sea-kayak expeditions, and from its small office the company coordinated a global network of clients, guides, equipment, and supplies. Although located in Vancouver, most of the company's clients hailed from elsewhere, responding to advertising campaigns in outdoors and adventure magazines, or articles written by journalists who had accompanied its expeditions. Several times each summer this company led expeditions to Clayoquot Sound and for one six-day period I joined thirteen clients and two guides on a tour that traced an itinerary of adventure through the landscapes of the Sound.[2]

Geographies of Adventure

Our trip began on a Saturday evening, as clients gathered at a small inn on the Tofino waterfront a short distance from the government wharf.[3] As we waited for our guides to arrive, we made introductions and exchanged stories. Some participants had been in the area for several days. Others had only just arrived, tracing transcontinental routes that had taken them from New York, Toronto, or Albuquerque to the small airstrip located just south of the town. After they arrived, the guides provided the group with an overview of the equipment, instructed clients on how to pack a kayak, and gave the group a set of rules and safety procedures to be followed while at sea. This last point deserves emphasis. Although they are packaged as adventure, these expeditions are actually quite safe. As experienced kayakers know, tipping a fully loaded kayak usually takes conscious effort. Indeed, most novices are surprised to discover that what initially seems the most intimidating part of sea-kayaking—the powerful ocean swell—actually poses little threat: it is predictable, and so long as the waves are not breaking, kayaks simply bob up and down like corks in a tub. It is only when it comes time to return to land that the swell comes into play. Most capsizes occur when paddlers misjudge the size and speed of breaking waves, or approach the shore at an oblique angle, resulting in a crash course on wave physics and a change of clothes. Nevertheless, the Pacific

Ocean has its hazards: fog hugs the outer coast during summer months; if one is inattentive, dangerous reefs and partially submerged rocks can appear unexpectedly; squalls can make for choppy waters; and tidal currents can interact with winds to produce dangerous conditions. Thus, despite being "soft" adventure by the standards of many of today's extreme sports, a clear set of safety rules and a well-defined hierarchy of command are rehearsed before clients ever see the water. These rules are reviewed each day, and are part of what enables the trips to signify adventure. At the same time as the presence of risk is reaffirmed through the invocation of rules, however, all efforts are made to diminish its possibility. Two-way radios allow guides to continuously monitor weather reports and to call for information and advice. Guides chart careful and cautious courses through the Sound's islands and channels and clients are warned to stay close to the guides at all times. In short, adventure travel companies may commodify and sell risk, but they take every precaution to ensure that the *appearance* of danger never translates into *actual* peril.[4]

It is the itinerary, though, that most concerns me here (Figure 4.1). The tour departed Sunday morning, riding an incoming tide through a narrow channel to a nearby island containing some of Canada's largest

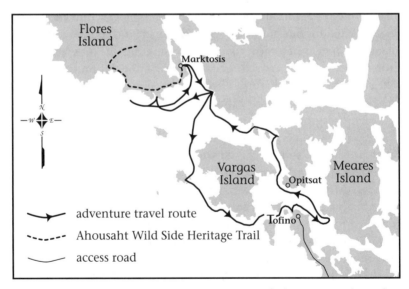

Figure 4.1. Distributing desire across space: routes of adventure travel expedition and Ahousaht Wild Side Heritage Trail. Tofino is the last point accessible by road.

trees.[5] Here clients were taken down a short trail that led to a number of large, ancient trees (several clients struggled heroically to take pictures of these objects whose mass and location deep in the forest defied all efforts to do so). For the remainder of the day, the group paddled along the edge of a waterfowl refuge; skirted Opitsat, a Native village strung out along the shoreline of Meares Island; and worked its way down a long channel to its first night's destination, a small secluded beach out of sight of the lights of Tofino, which lay only a few miles behind us.

Although the itinerary provided a sense of spontaneity (consistent with its narrative of exploration and discovery), the route and campsite were actually carefully chosen to provide views of "nature" and to take us out of view of "civilization," including the former residential school that we passed along the way.[6] Looking out from under the trees lining the back of the beach, clients were able to survey a realm that appeared wholly under the sign of nature: only the lights of distant navigational beacons disturbed the idyllic scene. Here we can pause briefly to make the obvious point that the trip's sense of adventure was partly an effect of the itinerary itself, which traced a selective route through the Sound's waterways and forests in order to remain within adventure travel's system of signification. Two other beaches along the route were actually more suitable camping beaches, especially for a group as large as ours. Yet, for the purpose of "adventure," both had drawbacks. One remained within view of Tofino. The other, although out of sight of the town, had become too popular. On a recent trip, the guides and clients arrived to find that another commercial tour was already encamped. Although the beach had adequate space for everyone, a confrontation ensued between the clients of the two groups: the arrival of the second group, it appears, had threatened the "experience" of the first group. By entering the scene, the second group had inadvertently erased the temporal and spatial distance that the first group had so fervently sought to achieve—and the Sound no longer appeared as an anachronistic space located outside, and prior to, a capitalist modernity. Instead, the landscape switched signs, now appearing fully *within* a global modernity and its circuits of capital and commodities. Nothing drains value from adventure more quickly than *other* adventurers.[7]

The second day saw clients paddle through a series of small, sheltered islands, past the quaint cabins of a number of back-to-the-land squatters, and eventually to the edge of the open ocean. As the group approached its destination, the swell became more notable, and the sound of breaking waves increased. That evening, as clients sat around the evening fire, a nervous excitement was palpable. It was the encounter with nature's *force,* after all, and the challenge to overcome it, that sets adventure travel apart from "mere" tourism. In antimodernist discourse, to experience raw nature unmediated by modern technology is to approach humanity's degree zero—that mythical place governed solely by biology and instinct where humanity had once lived in a direct relation with nature.[8] Thus, the open ocean was both what was most feared and most sought after.

The next four days repeated this scenario in various patterns. Each day brought something new: a spectacular beach; an island with an abandoned Ahousaht village site; wave-lashed headlands; wildlife; high seas; and, of course, wind, rain, and fog. Along the way, the guides pointed out natural features, provided information on flora and fauna, and described aspects of Nuu-chah-nulth culture. "Culturally modified trees" were "discovered" by clients, paddlers found sea caves and blow-holes, and behind one of the camping beaches the group entered an old-growth forest so verdant it absorbed all sound, even a raging sea pounding the shore a short distance away. The only disappointment came on the last day when the group paddled through an area of ocean frequented by gray whales. None appeared.[9] Eventually the group completed a circuit of Vargas Island, landing behind the same Tofino inn from which it had departed six days earlier.

Although mundane in its details, the itinerary deserves careful attention. The length of the trip—six days—was long enough for paddlers to feel that they had experienced a true adventure: muscles ached, skin was salty, hair was matted. No matter how carefully packed in special "dry bags," clothes always felt damp. The smell of campfires pervaded everything. It was long enough that work and the city were temporarily forgotten (although on this trip one paddler—a stockbroker from Nevada—had taped a list of his investments inside his kayak and pulled out a shortwave radio each evening to listen to reports from Wall Street).[10] The trip was also short enough that the discomforts and labor of travel could be aestheticized and endured rather than turned into the drudgery of daily life. Playing at primitivism, after all, can quickly lose its appeal. Perhaps most important, the trip fit well within individual work schedules. One could board a flight in New York on Friday evening or Saturday morning, paddle the circuit, and return in time for a Monday shift while using up only one week of vacation time.

Another important feature of the itinerary was its emphasis on *movement.* With the exception of one campsite where the group spent two nights, the expedition was always on the move. Every effort was made to provide clients with a continuous series of new experiences: wildlife, long, sandy beaches, powerful surf, sea caves, natives, and the rainforest all became part of a carefully chosen chain of sights/sites. To linger too long at any one point would have disrupted one of the most important effects that the itinerary was designed to produce: a sense of exploration and adventure. To retrace one's path would have the same effect, turning the exotic into the same. The ideal journey was therefore circular, continuously producing the "exploration effect" and deferring the sense of return until the very last moment. Likewise, the mode of transport was important. Kayaks are human-powered rather than machine-powered and this not only recalled earlier moments of European travel before the advent of steam and diesel engines, but also allowed for a fantasy of "going native."[11]

Several other aspects of the journey deserve emphasis. Like most tourism, the itinerary was organized around a visual logic (one client shot twenty-four rolls of film over the six days). As we saw in the land-use planning map discussed in chapter 1, this introduced a new commodity form—"scenic landscapes"—to these remote landscapes, one increasingly incorporated into state planning. Thus, it was not only the succession of ever-new experiences that mattered; it was equally important that certain things appeared within the visual field and not others. Many clients expressed strong views about what belonged in the landscape, and guides made every effort to bring these items to the attention of clients. Other elements of the landscape were disparaged and avoided. Most disliked were signs of anthropogenic change, or, for that matter, anything that signified an industrial modernity. Thus, whereas primitive nineteenth-century farmsteads on Vargas Island that had been "reclaimed" by nature were aestheticized, late-twentieth-century clear-cuts were roundly condemned (Figure 4.2). Crucially, not *all* logging was equally disparaged. Evidence of preindustrial forms of logging—such as the marks left in stumps from the springboards that supported hand loggers in the first decades of the twentieth century—were more easily incorporated into adventure travel's nostalgic gaze, because they could be assigned to the category of "heritage." Motorized vessels (including Native fishing vessels), litter, electric lights, and "modern" Natives, in contrast, threatened to disrupt the clients' nostalgic gaze.

When it came to the more explicitly cultural aspects of the landscape, guides made every effort to remain as fully as possible within the domain of "traditional" Native culture, much like the environmental

Figure 4.2. Disturbed viewscapes: clearcuts in Kyoquot Sound. Photograph by author.

groups encountered in chapter 3. At times, clients were invited to participate not only in a fantasy of adventure and territorial exploration, but in a Harrison Ford–like fantasy of archaeological discovery. Guides led clients to locations where signs of *past* Native culture could be discerned within the forest. On one island, for instance, the group was taken on a walk to a midden, then to a clearing in the forest where the outline of an old longhouse was faintly visible amid the lush forest growth (Figure 4.3), and then, finally, to examples of "culturally modified trees" deep in the forest.

What remained unspoken in this sighting of traces and ruins were colonialism's displacements. Nuu-chah-nulth ruins were troped as nature "running its course" rather than the effect of politics and power. This collapsing of Nuu-chah-nulth history *into* nature was underlined by the itinerary of the journey itself. The same walk that brought clients to the midden and longhouse included an old-growth forest, a number of sea caves, and a rugged shoreline pounded by ocean waves. Native culture and pristine nature blurred together in a succession of roughly equivalent sites. This was further reinforced by reading materials sup-

Figure 4.3. Locating "ruins": members of adventure travel expedition in former site of Native longhouse. Photograph by author.

plied by the guides, which provided clients with information on how to properly interpret the natural history and traditional cultures of the Sound (Figure 4.4). These included a collection of the Nuu-chah-nulth myths and stories.[12] Although this set the tour company apart from others, in that the books called attention to the existence and complexity of Nuu-chah-nulth culture, it tended to fix Natives on one side of the great

Figure 4.4. Landscape / text: enframing "nature" and "culture." Photograph by author.

divide between the modern and premodern. Little information was provided about *contemporary* Nuu-chah-nulth life, local politics, or present-day economic and social practices (although if clients had cared to inspect the intrusions of modernity into the Sound—something that they were encouraged to avoid—they would have found these to be part of an *indigenous* modernity in which the categories of "traditional" and "modern" made little sense).[13]

Marktosis

The itinerary of adventure I have traced produced Clayoquot Sound as an "anachronistic space," a space that appeared caught in another time prior to, or outside, (modern) history. As Ann McClintock (1995) notes, this has the effect of producing the modern as the destination to which all cultures must eventually travel (although, arguably, the arrival is always deferred). In this case, the production of Canada's west coast as an anachronistic space is precisely what enables it to become a privileged landscape of mourning. This framing, however, is never achieved without remainder; it is always disturbed by objects that cannot easily be contained within discourses of the pristine and the primitive. This was especially true for the journey I accompanied, which was continuously confronted with the presence of fishing boats, whale-watching tours, and signs of industrial forestry, all of which had the effect of shifting Clayoquot Sound back to the "wrong" side of the great divide between the modern and the premodern. As they are always fated to do, the adventure travelers arrived too late.

Perhaps nowhere was this ambivalence—and anxiety—more pronounced than during the expedition's visit to an Ahousaht village. Consistent with the dominant rhetoric of ecotourism and adventure travel, the owner of the company whose tour I followed subscribed to the notion that alternative tourism should benefit local communities, in contrast to "mass tourism" which he felt brought many more costs than benefits to "host" communities. His solution was to have clients take a meal in Marktosis, the largest Ahousaht community in the Sound. Before I turn to this event, it is worth noting that the ethical principles espoused by adventure travel companies are often belied by economic facts. In this instance, benefits to local Nuu-chah-nulth communities were minimal. All the equipment used on the trip was manufactured in high-tech industries located elsewhere. Virtually all food was imported from Vancouver. Although the region produces a wealth of seafood, the problem of storing and preserving bulk foods (such as fish) precluded purchasing local products. Only bread and some vegetables were purchased from local retailers (none of the vegetables, however, were locally grown). Moreover, while a Tofino warehouse was used to store equipment and goods, and a local inn and restaurant used by the company to house and feed the clients on the first night of the journey, these expenditures remained entirely within the white economy of Tofino, merely underlining the fact that most businesses that cater to tourists in the region are owned and operated by members of the non-Native community, many of whom arrived in the Sound only after the completion of the cross-island highway in the early 1970s.[14]

Aware that adventure travel in Clayoquot Sound benefited non-Natives much more than Natives, the operator arranged for his company's expeditions to take dinner in a Native-run bed-and-breakfast in Marktosis. The guides explained this as a form of charity, and attributed it to the enlightened humanitarianism of the company's owner. This rhetoric, of course, merely obscured the important fact that, as with industrial forestry, the company's profits depended in large measure on the spatial and political dislocation of the Ahousaht from their surrounding territories (which could then be commodified as landscapes of risk).[15] Moreover, although the three hundred dollars paid to the owner of the bed-and-breakfast for the privilege of a "traditional" meal was significant within the community of Marktosis, where unemployment often exceeds 80 percent, it remained an insignificant portion of the total economic activity generated by the tour, and represented less than 4 percent of client fees. Far greater expenditures were made in Tofino, Vancouver, and other white communities, although most of the trip occurred in Ahousaht and Tla-o-qui-aht territories.[16]

Although the rhetoric of "local benefits" belied the manner in which the company profited from colonial displacements, the trip to Marktosis was important for other reasons. It was the sole point in the journey when actually existing Natives—as opposed to mere signs of their past presence—were brought within the expedition's visual and discursive orders. To be sure, it was a rather limited encounter. Upon arriving at the village, clients were met by a pack of dogs, then by a few children who wandered to the beach to see the kayaks. After dragging the kayaks ashore and carefully removing valuables, the group wandered the length of the town's main street in search of the bed-and-breakfast. No interaction occurred with village residents, who appeared indifferent to the group's presence. Instead, the group moved in small groups through the village, stopping to inspect or comment on different elements of this "native-scape." Later, some clients explained that they were disturbed by the extreme poverty; others were troubled by what they saw as conspicuous acts of consumption (satellite dishes received the most comment). They were equally surprised to see that the bed-and-breakfast was housed in an old, weathered trailer, that the "traditional" meal they had been promised included mashed potatoes and root beer, and that a large-screen TV remained on during the entire visit.

What made this visit interesting was *not* its staging of a "cross-cultural" encounter. Little interaction occurred; the women running the bed-and-breakfast remained in the adjoining kitchen, while expedition members sat at a long table and talked among themselves. What interested me was the significant *disjuncture* between this experience of Native life and the romantic, primitivist discourse that structured much of the rest of the trip. This was not the indigenous world that clients had been expecting. The responses of clients varied. Some quickly shifted ideological frames, perhaps prompted by the American Indian Movement (AIM) posters on the walls.[17] Although not directly questioning the primitivist frame deployed in the remainder of the journey, these individuals adopted a more explicitly political stance, interpreting the poverty on the reserve in terms of government paternalism and neglect. Others drew on the familiar terms of "inauthenticity," viewing the satellite dishes, large-screen TVs, political posters, and "non-Native" cuisine as a sign of the *contamination* of Nuu-chah-nulth culture through its contact with modern Canadian society. These things were "out of place." Still others responded through a "culture of poverty" discourse, even repeating well-worn stereotypes of the "lazy Indian." In one instance, this was combined with a developmentalist narrative that not only saw Indians as insufficiently modernized, but attributed this to a cultural lack. As one client explained as the group paddled back to its campsite, the band was sitting on a real-estate bonanza that could be privatized and developed as condominiums for discriminating non-Native buyers, if only the residents would show some foresight and initiative.

The next day the group continued on, traveling through a magical landscape of natural wonder and cultural ruin (more middens, longhouses, and CMTs lay ahead). This does not mean that the experience in Marktosis was forgotten, although it received surprisingly little comment in the days that followed.[18] Still, the side trip to Marktosis is instructive. In the most obvious sense, it reminds us that the conceptual frames that structure alternative travel on Canada's west coast are ideological—they make sense of the landscape in particular ways, incorporating some elements and excluding others. As important, it reminds us that these conceptual frames are far from fixed, but instead continuously disturbed by elements that do not properly fit. This did not only occur in Marktosis; rather, each appearance of modernity, in the guise of technology, industry, "mass culture," or "modern" Natives, threatened to disrupt the temporal narrative that was key to the expedition's nostalgic mode of travel. Forest clearcuts, pleasure boats, and trash (the ultimate signifier of commodity culture) were at once a source of great anxiety and things that required continuous *comment* (as a way of containing the threat they posed to the semiotics of adventure). Individuals felt obliged to narrate what they saw—either in order to expunge certain things from the temporal frame that structured the journey or to place in question the frame itself. The latter is important for it reminds us that travel is often an unsettling affair (cf. Gregory 1994; di Chiro 2000). Although in this case the trip to Marktosis tended to be incorporated by most clients within the ready-made terms of a modernist and ethnocentric vision, this was not true for all. Moreover, there have been recent attempts to introduce a new kind of "reality tourism" to Marktosis that stages the landscape in a much more political frame. I will return to this topic at the end of the chapter.

Geographies of Desire

How does following adventure travelers through Clayoquot Sound add to a discussion of the cultural politics of nature on Canada's west coast? The most obvious point is that while adventure travel and ecotourism hold out the promise of unmediated encounters with nature and other cultures, their practices are governed by a set of discursive norms that are deeply rooted in metropolitan cultures. Tourism involves the consumption of "signs" (Culler 1988). In turn, although these so-called alternative forms of travel are often considered "light" (because they are thought to be less disruptive than "mass" tourism), they are no less infused with relations of power. It is also important to note that adventure is not a quality inherent to a place; it is an effect produced through an elaborate set of discursive conventions and spatial practices. And finally, the itinerary of the trip helps us see some of the ways that productions of nature and culture on Canada's west coast are the result of local and global processes that simultaneously *constitute* and *displace* sites such as Clayoquot Sound within wider social and psychical geographies.

It is this last point that I wish to focus on. One of the arguments of this book is that the temperate rainforest does not name a static set of cultural and ecological relations, but that it is the name for a constellation of elements shaped within global flows of capital, commodities, ideas, and images. It is a space continuously *reterritorialized* as these flows combine to produce new configurations of social nature. This is partly what makes these apparently primeval landscapes so dynamic (and why they are always in the news!). And it suggests why they must be submitted to a conjunctural analysis rather than approached in terms of an underlying essence. But it is just as important to recognize that globalization is not merely about nonlocal forces that touch down in specific sites. This not only problematically situates the "local" as always a residual and reactionary site (see Massey 1994); it also misses the novel social and cultural forms, and political identities, that emerge as globalization works *through* local social and cultural conditions. Thus, for the remainder of this chapter I want to focus on both of these aspects of globalization: the way in which "place" comes to be constituted within local and global dynamics, and the way that this occurs in and through *already existing and differentiated* social formations (cf. Oakes 1999). This should caution us from reading adventure travel in terms of a singular telos (modernization) or a predictable politics (colonization).

Distributing Desire across Space

To adequately understand adventure travel in Clayoquot Sound it is necessary to consider how places like this come to show up as sites to which people seek to travel. What accounts for the presence of the adventure traveler in Clayoquot Sound? What governs his or her appearance? And how does this shed light on the production of space and place within the discursive economies and cultural geographies of a globalizing modernity? To answer these questions, it will be necessary to travel some distance from the waters and forests of the Sound. We can begin with the now commonplace understanding that travel is an elaborate cultural practice that has as much to do with sets of meanings and desires constituted within metropolitan cultures as it does with the site traveled to (Culler 1988). This does not mean that the traveler's experiences are completely scripted in advance. There is, after all, a *materiality* to travel that disturbs the ideological construction of the tourist object. It is also the case, as Derek Gregory (1995) notes, that there are many different ways of traveling to the same place, mediated by gender, race, and class. Thus, at the same time as the discursive conditions that enable and sustain the desire for adventure must be carefully explored as a nexus of power, it is equally important that we not inadvertently take up a deterministic argument that sees the experience of the tourist as always already closed off by ideology.

With these cautions in mind, we can begin by noting that the journey of adventure I just described did not begin when clients launched their kayaks into the waters of Tofino Inlet (or when they left the comforts of home to begin their journeys). This would separate adventure travel from the cultural practices and discourses that enable and sustain it. A full account of the circuits of capital, culture, and fantasy that lead tens of thousands of adventurers to travel to remote places on Canada's west coast is beyond the scope of this chapter: at the very least, it would have to examine the global flows of images, ideas, people, and capital that Arjun Appadurai (1996) argues are central to the construction of imaginary geographies; the communication technologies (i.e., air travel, the Internet) that have so radically reconfigured the social and imaginative spaces of lived worlds for many people living in North America, Europe, and parts of Asia; and the economic relations that have enabled the middle class to be increasingly mobile. Such an analysis would also lead us to the state agencies that regulate the use of land and environment, making some places available to tourism and not others, and to the entrepreneurs who set out to market certain sites as zones of adventure.

My objective here is much more modest. Dennis Porter (1991) argues that at some level all forms of travel cater to desire. Likewise, Mark Seltzer (1992) suggests that travel is a way that desire gets "distributed across space." If so, how are we to understand the desire today for adventure-in-nature among many metropolitan subjects?[19] Why seek immersion, and tests of strength, in remote natural settings? There is no dearth of possible explanations. By some accounts, the quest for adventure is simply hardwired into human nature, part of our genetic makeup (even prompting some researchers to search for the gene that "controls for" adventure).[20] By this view, those who seek adventure have a genetic predisposition to risk, whereas others are more adverse to it. A subset of this approach sees the *modern* phenomenon of adventure as a response to the reduction of danger in everyday life. By this view, adventure is posited as a biological need, but only those in modern industrial societies—especially its managerial and professional classes—are fated to continuously *search* for adventure, in contrast to others (premoderns? illegal immigrants? manual laborers?), for whom adventure remains a part of everyday life. Others understand alternative travel in terms of identity formation. Ian Munt (1994), for instance, explains the rise of "alternative" tourism—and its commodification in niche markets—in terms of the emergence of, and struggles between, new class fragments, figured both as part of broader cultural shifts and as a consequence of the restructuring of capitalism (see Bourdieu 1984; Harvey 1989; Featherstone 1991). These class fragments, Munt (1994, 102) argues, are engaged in a "frenetic struggle . . . in establishing and maintaining social differentiation," something that in so-called posttraditional societies is thought to occur increasingly at the level of taste.[21]

Each approach has its appeal and its adherents, but none is entirely satisfactory. The first reads culture as biology, and thus effaces how the desire for adventure may be an effect of discourse and power. It is unable to adequately explain why adventure travel is so differentiated by class, race, and gender, except to posit that some people have a surplus need for adventure, or, alternately, that adventure is a universal need, but that modern culture isolates people from this experience (some more than others), leading to an adventure deficit. The second approach, in contrast, sees travel as a site for social differentiation. This draws on a long tradition in critical theory that understands tourism as a venue for marking social distinctions.[22] But it too falls prey to problems. On the one hand, it flirts with the notion that identity is something objectively calculated within a matrix of choices—one travels merely for its *signifying function* at home.[23] On the other hand, when it turns to debunking tourism as an effect of niche marketing or as "boutique" tourism, it risks returning to the notion that the consumer is an unwitting dupe of advertising, by which the tourist industry manipulates and manufactures desire.

More recently, travel theorists have suggested other approaches that understand travel as the expression of interlinked social and psychic dynamics, or what Anthony Elliot (1996) refers to as the crisscrossing of fantasy and culture. To understand this, we can begin with the now commonplace notion—usually derived from poststructuralist and psychoanalytic theory—that the subject does not preexist its insertion into social and symbolic orders, but is instead an *effect* of these. This insight is not new (indeed, it has shadowed Enlightenment notions of subjectivity from the start), but it has recently provided travel theorists with a powerful means by which to understand travel as a cultural practice. Such accounts often draw on Jacques Lacan's understanding of the split subject, who can know himself or herself only through an alienated image that comes from outside. By this view, the subject is constituted in terms of a sense of lack, and thus necessarily predisposed toward, and entangled in, the field of the Other (or, more generally, within the "symbolic order"). As Elliot (1996) explains, psychoanalytic approaches help explain how, in order to achieve a sense of plenitude, something "outside" the self is taken "inside," or, in other words, how the Other becomes the terrain of self-formation. It is along these lines that Porter's statement that all forms of travel cater to desire should be understood. By this view, travel involves a fantasy of regression, in the sense of returning to, or recovering, a lost condition of wholeness. In the words of Curtis and Pajaczkowska (1994, 204, 210–11), "travelling . . . is undertaken to restore something that is lacking. . . . the question of what motivates the tourist to set off on a quest for visible experience depends on a much deeper need to return to a pre-social world of imaginary plenitude." They go on to explain that this is precisely why tourism has a *fetishistic* structure, because particular sites/sights and objects are sought out for what they promise the subject in a "repetitious experience of knowing loss and disowning it by substitution" (Pollock 1994, 64).

This is not the place to rehearse the details of psychoanalytic theory, nor to enumerate its problems. Certainly, as a number of writers have noted, psychoanalytic theory has at times been given a transcendental cast that renders it both universal and ahistorical (Gregory 1994). It is not clear that psychoanalytic theory must be framed in this way; indeed, recent work has sought to link psychical dynamics to cultural and historical dynamics (see especially McClintock 1995; Kirby 1996; Butler, Laclau, and Žižek 2000). In other words, while certain assumptions about subject formation are understood as universal (i.e., the relation between the subject and the "symbolic order"), this allows for great variation according to the contingencies of culture and ideology (understood as that which constructs and supports our social reality; see Žižek 1989).

For my purposes, these insights generate two crucial observations. First, while retaining Porter's insistence that the traveling subject is a subject of desire, it should be possible to *historicize* the late-twentieth-century subject-of-adventure in order to analyze how what Griselda Pollock (1994) has usefully called the "territorialization of desire" in travel comes to be expressed geographically in particular ways. This is to suggest that the traveler is neither a bundle of instincts, a knowing, calculating subject, nor a dupe of ideology, but instead an agential subject that emerges at the place where material history and psychoanalysis meet.

Crossing the Great Divide

To explore this crisscrossing of fantasy and culture in metropolitan cultures, let me return again to British Columbia. This time we will not travel to the site of adventure but to Vancouver, where the offices of some of Canada's largest adventure travel companies are located.

Ecosummer Expeditions is among the most prominent of Vancouver's adventure travel companies. Like the company whose itinerary in Clayoquot Sound I described earlier, it is headquartered in the same trendy harbor-front market, alongside the boutiques, galleries, and restaurants that cater to the city's young professionals. Each year it distributes an extensive catalog that describes the company's "journeys of discovery" to various remote locations, from Greenland to Papua New Guinea. Its catalogs are far from unique; similar publications are available from Ecosummer's local competitors and the hundreds of other North American adventure travel companies that advertise in *Outside, Explore, Sierra, Men's Magazine, National Geographic Adventure,* and countless other magazines, or that post displays in outdoor equipment shops such as Recreation Equipment Incorporated (REI), or Canada's Mountain Equipment Co-op (MEC).[24]

Two pages from one of the company's recent catalogs can help us think about the way that desire comes to be "distributed across space." One outlines a journey into the Canadian wilderness; the other, a journey into the distant jungles of Irian Jaya. In the first advertisement (Figure 4.5) we see a group of adventurers departing on a journey through the Bowron Lakes, a series of eleven lakes and connecting rivers in British Columbia's mountainous interior. Several aspects of this advertisement deserve attention. First, like in the nature photographs discussed in chapter 3, the camera is an absent presence, and this absence helps structure the photograph's meanings. For instance, it erases the technological and institutional framework of adventure, allowing the site and the experience of adventure to appear as if they were somehow separate from their enabling conditions. The work of *producing* adventure (in terms of actual labor, but also the work of ideology) is removed from view, and adventure appears not as the expression and extension of modern technological societies—and capitalist relations—that it actually is, but as an escape *from* them. Indeed, the image reproduces the same "great divide" found in Adrian Dorst's photos, and in turn holds out to the viewer the promise of crossing this divide, moving from a world of culture to a world of elemental forces. Conversely, if we bring the camera—and the photographer—back into the visual field, the image loses its effect. All the technological and social relations that enable adventure become visible and adventure comes into view as an extension of modernity, rather than an escape from it.

Equally important, the image constructs the viewing subject as the subject-of-adventure. By situating the viewing subject behind the (absent) camera, looking out into the wilds, the image firmly situates the viewer *in* modern society and asks him or her to ponder the yawning gap between culture and nature, city and country, modernity and its premodern antecedents. Again, this gap is merely an effect of the photograph itself. Indeed, the largest image repeats these same binary oppositions internally. The viewer's gaze is drawn to the bow of the first boat as it breaks the glassy waters. The wake that it produces neatly divides the image—and the world—in two. In front of the wake lies nature; behind it, culture. In front of the adventurer lies the pristine and the untouched; behind, the disturbed and the disrupted. This renders the photograph allegorical: the snow-clad peaks that rise in the distance symbolize the *limits* of humanization, and the canoes, propelled by human power rather than machine power, promise an *immersion* in nature and a recovery and rediscovery of an essential harmony of self, body, and environment.

Figure 4.5. Framing adventure: Bowron Lakes, Canada. Source: Ecosummer (1994). Photograph by Chris Harris.

This wake also defines an anxiety. If the image holds out to the viewer the possibility of crossing over "into" nature—that impossible place of nonculture—then the wake haunts the adventure traveler as his or her troubling double. As we all know, things "follow in the wake"; it is impossible to move forward without bringing them along. Herein lies a central paradox of ecotourism and adventure travel that is rehearsed endlessly in adventure travel magazines like an originary trauma: at the same time that adventure travel promises to leave culture behind, it never does.[25] Like trying to jump out of one's shadow, the adventure traveler is always already "in the wake": his appearance disturbs the very myths that sustain the journey, which therefore must be reiterated again and again. In this light, the smaller second photo that depicts a direct encounter with wildlife takes on added significance, for it not only presents a specific detail of what might be experienced on the trip, but it stands as a guarantee that the adventurer actually *does* come into contact with "nature itself" (as opposed to the inauthentic or staged nature of urban zoos and theme parks).[26] Thus, it is not just the case that adventure travel traffics in the real, but that nature itself necessarily follows a logic of supplementarity: one always needs another sign of its presence.

At first glance, the second advertisement (Figure 4.6) promises an adventure of a very different kind. Here readers are invited to imagine themselves on a trek into the "lost world" of Irian Jaya, where they will find remote cloud forests and people who "are on the cusp of the stone age and have given up cannibalism only in the last 20 years." The text and photos on this page do not evoke pristine nature, but instead "primitive" pasts and "natural" cultures where recently discovered tribes emerge from the mists of time. Again, the everyday is bracketed and the exotic highlighted. The primary photograph depicts a Dani chief adorned with signs of savagery—nakedness, exotic jewelry, body piercing, penis gourd, and face paint. No modern technologies or Western clothing are allowed into the frame. The effect, as Rojek (1993) explains, is that the scene appears liminal, as if magically outside the present. Here is the ultimate adventure, taking travelers deep into the mythological space of the premodern Other.[27] The accompanying text reinforces this view. The journey is described as "rigorous" and "demanding," nostalgically invoking a time when travel was *travail*. "Our exploratories," the catalog intones, "are physically strenuous . . . suited only for those in excellent condition and prepared for the unknowns of adventure." Prospective participants are asked to contact the company in writing, summarizing their experience.

Figure 4.6. Framing adventure: Irian Jaya, Papua New Guinea. Source: Ecosummer (1994).

Much can be said about these images. Clearly they are both gendered and sexualized. Ann McClintock (1995) suggests that an "erotics of ravishment" has long organized (male) European travel. In European lore, she argues, non-European sites became "porno-tropics" for the European imagination, sites over which Europe projected its forbidden desires and fears. That the individual portrayed is male does not contradict so much as complicate this observation. In this image, the sex of the person portrayed is less important than that the body is *eroticized*. Indeed, the image remains fully within normative depictions of non-Western male sexuality that shuttle between a fear of unbridled sexuality (signified through the penis gourd) and the feminization and/or impotence of the non-Western male (signified through the naked, yet adorned, body). Thus, despite the portrayal of a male Dani chief, the advertisement is consistent with a masculinist discourse of adventure, something echoed in the text, where Irian Jaya is presented as an "unknown land," its territories as yet "uncontacted." The adventurer is invited to penetrate into the veiled, secret interior of the island, which is portrayed as passive in the face of the aggressive thrust of modernity (to which it "may soon fall prey"), and before the "strenuous effort" of the adventurer, who is seen beating his competitors to the prize (a rape fantasy thinly veiled as a paternalistic concern over uncontrolled contact).[28] It is also worth noting that Ecosummer's discourses of adventure do not merely conserve an already existing terrain of gender and race (which would make them mere reflections of ideology rather than a site for politics), they actively construct the traveler *as* a gendered (male) and racialized (white) subject.[29]

Although the differences between these journeys can be misleading, they are in important respects very similar. Following the logic of commodity culture, each trip—like brands of soft drinks or makes of automobiles—turns on its differentiation from others. Yet, crucially, what differentiates the two examples I have given is less that one discovers "primitive culture" while the other promises "nature itself," than it is the *degree* of rigor and adventure found in each and thus the distance that each marks out from highly technologized metropolitan cultures (indeed, the "culture" of the second is—through a discourse of primitivism—conflated with nature, allowing both trips to appear as journeys into worlds that are primarily *biophysical* rather than *cultural*). The first trip transports travelers into the wilds, presenting the promise of communion with a primal order that exists before and beyond the modern metropolis. The second trip takes the trekker further into the place of the West's darkest mythologies—the tropical rainforest—a world ruled by biological drives and instincts, not by cultural codes and institutions. Thus, the second trip, described as the "leading edge of adventure travel today," stands as the guarantee of "authenticity" for all other Eco-summer trips, establishing a continuum of adventure that runs through the company's entire catalog.[30] Yet, both turn on the construction of "anachronistic space," the sense that these places are temporally different from the modern world that the traveler inhabits.

These are fascinating images, but it is important that we avoid reading them as the source of the desire for adventure. The more important point is that these advertisements make sense to readers; that is, they are effective not because they dupe consumers into believing that this is what they want, but because they tap into a discursive terrain that is shared by advertiser and consumer alike. We do not need recourse to false consciousness or the culture industry to make sense of this. Indeed, such notions hinder more than they help, because they assume that desire actually has a proper object, rather than being infinitely substitutable. Our task is instead to consider why these advertisements—and this type of travel—should have such widespread appeal.

Landscapes of Mourning and Recovery

Advertisements like Ecosummer's must therefore be read symptomatically, as giving expression to deeply seated cultural discourses that shape the "territorialization" of desire.[31] Drawing on the advertisements, then, I wish to develop three arguments. First, that adventure travel turns on, and gives spatial expression to, a discourse of *modernity-as-loss*. Second, and related to this, that adventure travel appears as a form of *mourning* that is at the same time a source of pleasure. And third, that this gives adventure travel a fetishistic character, in that this mourning fixes on certain things or places (such as Clayoquot Sound). As we will see, this is deeply paradoxical, for the construction and consumption of landscapes of mourning have the perverse effect of affirming our modernity even as we seek to escape it (MacCannell 1976; see also Kaplan 1996). In this sense, constructions of Clayoquot Sound as landscapes of return and recovery serve important ideological functions, not in the sense of hiding from us elemental truths, but in the sense of "suturing" a social reality (modernity) through forms of spatial consciousness.[32]

Being Modern: Loss and Mourning

As evident in the two advertisements, adventure travel pivots on a sense of historical rupture, and on the double coding of modernity as both progress and decline. Both assumptions can be found with regularity in contemporary discourses of adventure travel and ecotourism. Consider, for instance, the words of one promoter: "Active travelers prefer to *dig down* through the *thin crust* of modern civilization to their destination's ancient *core*; unexcited by nature that has been *brought to heel,* they want it without fences, *untamed* and *pristine*" (Bill 1994, A2; emphasis added). Here all the elements of a discourse of modernity as decline are conveniently gathered together. The author's geological metaphor divides the social world into two distinct domains: surface and core. The surface—modern civilization—is but a "thin crust." It is the realm of the fleeting and inauthentic, the merely historical in contrast to the eternal and the unchanging. "Modern civilization," then, is understood as both a departure (a moment in which something else has been left behind) and a veil (that which obscures the "real" and which masks origins and essences). In turn, "active travel" is portrayed as a practice of recovery, an effort to get beneath the present in order to recover that which has been lost. In all of this nature is a privileged trope. Not only is modernity the realm of nature's violent subjugation—here we find echoes of the Frankfurt School, which saw both external *and* internal nature brought to heel—but modernity itself is coded as unnatural.

The notion that modernity distances humanity from its source is, of course, not new. Richard Bernstein (1992) has argued that at least since the French Revolution modernity has been understood within Western culture and philosophy in two opposed yet complementary ways. On the one hand, it has been equated with enlightenment and progress; on the other, with denial, repression, and domination. Although opposite, they are similarly teleological, one a story of the future perfection of nature, the other a story of the ineluctable loss of human nature (self, freedom, and authenticity) and ongoing destruction of external nature. Thus, the great divide between the modern and the premodern that lies behind this double vision is itself freighted with *moral* implications. Such divides, Derrida (1976, 114) argues, "[have] most often the sole function of constituting the other [culture or nature] as a model of original and natural goodness." On the other side of this great divide, non-Europeans and premoderns appear as an index to a hidden and good nature, or the zero degree against which one can track the ongoing degradation of the modern West. As Derrida goes on to note, such notions were already central to the social and political thought of Rousseau, who distinguished between a modern civilized existence and a state of communal grace from which moderns were cut off.

Today, this notion of modernity-as-decline echoes widely. It is found in critiques of consumer culture, made as early as the 1970s by social critics such as Christopher Lasch (1979) and conservatives like Daniel Bell (1976). For these writers, modern civilization is narcissistic, individualistic, and inauthentic. In such societies, the self is dissipated. Nor is the left immune from similar tropes today. Critics of postmodern culture, for instance, have combined the same sort of mandarin pronouncements of deterioration and inauthenticity, arguing that the real has been emptied into "simulation," a world of signs unmoored from their referents and where the commodification of experience has "hollowed out" meaning from everyday life (Zukin 1991; Sorkin 1992). Anarcho-primitivists trump all, suggesting that humanity's fate was sealed when agriculture first took root. We could find countless other examples. The point is not to amplify constructions of modernity-as-decline; a related, and equally potent, discourse of modernity-as-progress runs parallel at every point. Rather, it is to note the almost obsessive rehearsal of a rupture that lies behind notions of decline *and* progress, and in turn to see how the modern is constituted through the compulsive repetition of this founding break. When companies such as Ecosummer advertise their excursions as trips to places "off the beaten track" or as yet "undiscovered," they appeal directly to this continuously reiterated sense of modernity as an epoch of loss.

This may help put a finer point on our understanding of the specific practices and geographies of adventure travel. It is tempting to understand Ecosummer's journeys of discovery in terms of nostalgia—a longing for that which has been, or is fated to be, destroyed by the incursion of modernity (or elsewhere by European imperialism; see Rosaldo 1989). This approach has proved immensely productive for thinking about the desire among many Western travelers to witness that which is about to disappear, and it helps us understand why rainforests and primitives have occupied such a privileged place in Western imaginations for more than a century, not only in popular culture but in anthropology too. Rosaldo argues that nostalgia of this sort operates as an apologetics for domination, because the act of eulogizing that which has been destroyed erases personal and collective accountability (see also Kaplan 1996). But these insights can perhaps be pushed further. As I hinted at the outset, the notion of nostalgia does not quite get at the very pervasiveness of mourning in modernity. Rosaldo (1989, 69) defines imperial nostalgia as a "mourning for what one has destroyed." In other words, nostalgia emerges as an *effect* of historical displacements caused by imperialism or projects of modernization. But this is to relate nostalgia most closely to *modernization,* that is, as a belated response to forms of creative destruction unleashed by modernity. Without rejecting this, I wish to suggest something slightly different: that mourning is an irreducible element of *being* modern, and that this is tied less to the actual destruction of the premodern than to the sense of *temporality* that defines and pervades modernity.

To understand this, we need to attend to the way in which the modern is inaugurated as a condition at the same time as "history" becomes a privileged term in Western epistemologies (Foucault 1970). Thus, to be modern is to be imbued with a historical consciousness; it is to always situate the present in relation to a past that has been displaced and superseded, such that the present is, by definition, constituted through loss. To say this differently, modernity is the epoch of mourning simply by virtue of its historical self-reflexivity, by virtue of its own self-definition as a time that exists through the loss and/or transcendence of that which was *prior.* To be modern, then, is to live in a permanent condition of loss, and, to the extent that the present remains the epoch of history (there are signs, perhaps, that this period is coming to an end?), we remain within an age of mourning.

What I am suggesting is that mourning is a more persistent condition than what is usually meant by nostalgia, which is often discussed as a side effect of modernization. To this we can add some additional observations. The first follows almost by definition. To be modern is to mourn; we perform ourselves *as* modern by mourning. By this definition, then, adventure travel and ecotourism do not escape from modernity; rather, through their practices they inaugurate or reaffirm the modern. Indeed, to the extent that modernity exists in part as a habit of thought (Latour 1993), it must be continuously reiterated in this performative mode for it to retain its ideological hold. Second, because this is an unrealized mourning in the sense that it remains unconscious, it comes to be expressed as a *fetishism,* whereby the "truth" of mourning is converted into, or attaches itself to, a thing.[33] This is clearly evident in the fetishization of nature, natives, and ruins in the advertisements and practices of adventure travel discussed earlier, but, in a sense, this "thing" is endlessly substitutable.[34]

In passing, it is worth noting that the same notion of modernity-as-loss that can be traced at the level of society is repeated at the level of the self. We are familiar with this pessimistic account of the individual in modernity from the Frankfurt School, especially in the years following

World War II. For these writers—Theodor Adorno, Max Horkheimer, and Herbert Marcuse most prominently—modernity was characterized by forms of instrumental reason that increasingly repressed individuals and alienated them both from external nature and their own inner nature. To be sure, this is a very one-sided reading of the Frankfurt School, and it is important to remember that they remained heirs to Enlightenment aspirations and the belief that modern forms of reason also held possibilities for greater autonomy (and equality).[35] Still, the prevalence of the former reading is perhaps indicative of the pervasive nature of discourses of modernity-as-decline. Bernstein (1992, 42) captures the reading well. For Adorno, he writes, "the domination and control over nature inexorably turns into the domination of men over men (and indeed men over women) and culminates in sadistic-masochistic *self-repression* and *self-mutilation*. . . . The hidden 'logic' of Enlightenment reason is violently repressive; it is totalitarian." Marcuse (1955), similarly, mapped the dissolution of the individual subject within the psychic and social fields of modernity. Human subjects, for Marcuse, were continually subsumed and manipulated by a rationality imposed on them from the outside, and in capitalist modernity this took increasingly restrictive forms (cf. Elliot 1992).[36] That this turns on a barely veiled notion of human nature is clear. From this perspective, what was needed was the recovery of the core of selfhood—the degree zero of the self—that still existed, timelessly residing in the unconscious.[37]

At first blush it may appear that we have traveled some distance from Canada's west coast and the itineraries of adventure that have been mapped over its territories. Yet, perhaps we have not traveled so far at all. Arguably, adventure travel and its associated forms—ecotourism, "active" travel, and so on—spatialize this discourse of modernity-as-loss.[38] How else do we explain the appeal of the sentiments expressed in the recent tourism campaigns by Canada's Yukon Territory, which picture a group of backpackers on a windswept ridge looking down into a pristine wilderness valley: "Somewhere down there is your soul" (Figure 4.7)?[39] Or, as noted in an article on mountain sports, "the interior adventure is as thrilling, if not more so, as the physical feat" (Foehr 1995, 36).

Theorists of travel have rarely noted the parallels between critical theory and tourism, but they have noted the degree to which travel is scripted as such a recovery. Traveling is not just spatial displacement. In the words of Curtis and Pajaczkowska (1994), it is understood and expressed as a "restorative process," a way of "reaffirming self-identity," of locating a "unity of self," a sort of archaeological process that uncovers an essence lost in the maelstrom of modernity and its repressive techni-

Figure 4.7. Adventure travel and the journey into the "inner self." Source: Yukon Tourism Bureau.

cal rationality. What makes these narratives so powerful is that they foster the belief that there exists something behind or prior to our everyday experiences. In this sense, Foucault's (1980) dictum that our discourses on the self give rise to the illusion that there exists something *other than* bodies, organs, somatic localizations, functions, anatamo-physiological systems, sensations and pleasures, and so on is particularly relevant. What I have argued is that contemporary discourses on modernity-as-decline, and the loss of the self in modernity, rest on the assumption that a real, prelinguistic or precultural realm of unalienated and essential cultural and psychic relations exists. It is precisely the truth effects of these discourses that license projects of recovery and journeys of rediscovery. To say this differently, although adventure travel offers the promise of more "authentic" encounters, and stands in opposition to mere tourism with its trade in stereotypes of the Other, adventure travel remains *within* the discursive economy of European modernities, and in its projection of its metropolitan anxieties to distant sites, stages a spatial encounter where the "non-West" becomes (yet again) the terrain for the remaking of Western subjects.

Nature as Fetish

This brings me to a final observation that has been implicit in everything to this point. If adventure travel and ecotourism are a kind of longing that at once mourns that which has passed and reaffirms our modernity, then, as we already saw in the two Ecosummer advertisements, nature becomes the privileged site for modern subjects to exorcise/exercise their modernity.

More important for my purposes, this brings us full circle to Clayoquot Sound, which now comes into view not as a landscape peripheral to modernity, but instead one that, along with other similar landscapes of mourning, is central to modernity's continual reaffirmation of its difference from the past. It is a landscape that, by being placed outside modernity, comes to define modernity. Indeed, it is precisely this inside/outside duality that is constitutive of modernity that the "active" traveler traverses with a mixture of anxiety and pleasure. It is this double vision that lies behind, and authorizes, quests for "the Arcadian prelude to industrialization," and that gives pleasure to "outrunning 'Time's winged chariot' and the forces of modernity" (Curtis and Pajaczkowska 1994, 191, 202).

Two additional points require emphasis. First, if travel is fetishistic, a "repetitive experience of knowing loss and disowning it by substitution" (Pollock 1994, 64), then perhaps no other substitute substitutes so well as nature. Posited as a sphere separate from culture, yet also as culture's primal source, it holds out promise for projects of return and recovery. Indeed, the primitive is in many ways simply a subcategory of the natural. William Cronon (1995, 80) captures this turn to nature well in his discussion of the North American wilderness ethic, which, he argues, stages the journey into nature as a move toward the sacred and toward a world of essences. North Americans, he writes,

[appeal] explicitly or implicitly, to wilderness as a standard against which to measure the failings of our human world. Wilderness is the natural, unfallen antithesis of an unnatural civilization that has lost its soul. It is a place of freedom in which we can recover the true selves we have lost to the corrupting influences of our artificial lives. Most of all, it is the ultimate landscape of authenticity . . . it is the place where we can see the world as it really is, and so know ourselves as we really are—or ought to be.

This is hardly new, nor is it only an American preoccupation. In his *Triste-Tropiques,* for instance, Lévi-Strauss revealed a similar antimodernism when he turned to the forest as a space of alterity. At one level, he expressed this in terms similar to Benjamin's melancholic emphasis on transience in nature. The forest, he wrote, "keeps man at a distance and hurriedly covers up his tracks . . . [its horizons] soon close in on a limited world, creating an isolation as complete as that of the desert wastes" (quoted in Arshi, Kirstein, Naqui, and Pankow 1994, 340). On the other hand, he saw the *encounter* with the forest as getting back to the core of

the self, as a stripping away of modernity and its ocularity and the recovery of that which was lost in the "fall": "a few dozen square yards of
forest are enough to abolish the external world; one universe gives way to
another which is far less flattering to the eye but where hearing and smell,
faculties closer to the soul than sight, come into their own" (341).[40] Set
against the tropical rainforest, European forests appeared to Lévi-Strauss
overly humanized and demythologized, separated from the rainforest by
"a gap so great that it is difficult to find words to describe it" (ibid.).[41]
Again, these tropes are reiterated in one of the first Ecosummer (1990)
catalogs: "to stand in a cathedral of giant cedar and spruce . . . is to get
into contact with what has spiritually sustained our species since we
began."

Second, it is important to recognize that discourses of modernity-as-
loss give rise to geographical projects. Adventure travel is not only about
mourning loss; it translates mourning into a movement through space.
Griselda Pollock (1994, 64) captures this well. The search for an originary, primal moment, she writes,

> requires a territory on which this temporal ellipsis can occur. It re-
> quires a spatial encounter with imaginary figures whose difference
> must be construed and then marked in order that the sense of loss,
> lack and discontinuity characteristic of metropolitan modernity can
> be simultaneously experienced and suspended by a momentary vi-
> sion of a mythic place apparently outside time, a "before-now" place,
> a garden before the fall—into modernity.

We are perhaps closer now to an understanding of the routes traced
by adventure travelers through the coastal forests and seascapes of
Clayoquot Sound. In tracing paths through the Sound, they subtly inscribe a Rousseauian notion that premodern societies—and ultimately
nature—are somehow closer to the origins of humanity. While some
have suggested that forms of alternative travel are an expression of postmodern sensibilities, in the sense that they reveal a certain reflexivity
(Beezer 1993; Munt 1994), then this is a reflexivity that often looks
back to, and searches for, a period prior to the advent of the modern.[42]
In so doing, it reprises some very *modern* themes.

Reterritorializing Clayoquot

To say that ecotourism and adventure travel are a form of mourning is
not the same as saying that they bring no enjoyment. Indeed, exactly the
opposite is true. Not only are there many simple pleasures involved (on

the order of sensation), but there is pleasure in mourning itself. One way to understand this is that it is precisely mourning that stabilizes our identity as moderns; it provides constancy rather than crisis. Here we find a central paradox: to the extent that loss is what marks us as modern, repeating this loss—here, through a spatial itinerary—provides homeostasis.[43] In important ways, then, ecotourism and adventure travel work to *conserve* the modern bourgeois subject, not to disrupt or traumatize it. In psychoanalytic terms, adventure travel brings pleasure because it is on the side of the symbolic (or the signifier), rather than the real. It is also, loosely paraphrasing Cronon (1995), a means by which the subject manages the crisis of modern ecology, externalizing nature and engaging in acts of recovery, all the while disavowing the (toxic) natures produced in the practice of everyday life.

These are provocative notions that I cannot take up here. In the remainder of the chapter, I want to travel back to Clayoquot Sound in order to return to a set of questions that I raised earlier about adventure travel as a spatial practice that reterritorializes nature, culture, economy, and politics in consequential ways. I have already noted that proponents of adventure travel and ecotourism promote these industries as low-impact forms of travel, because they leave sites "untouched." I want to challenge this assumption, not in order to dismiss adventure travel as merely the same as other forms of tourism, but instead to displace the untouched/contaminated duality entirely. There is no untouched Clayoquot that is in danger of being defiled by external forces (such as the traveler). As I have made clear, Clayoquot Sound is best seen as a constellation of actors and forces that is continuously in flux. The question that faces us, then, is not how does adventure travel "corrupt" its site, but instead how does it work to reconfigure an already dynamic and differentiated social and ecological landscape?

It is necessary first to call into question the assumption that adventure travel and ecotourism are innocent forms of travel that leave the sites visited intact. This view attains its force through the assumption that both forms of travel are primarily visual, and that vision is passive. Witnessing nature—or pristine cultures—is thought to have no lasting effect (how else can Ecosummer worry over its object soon "falling prey" to an expanding world, yet not problematize the presence of the adventure traveler?). Against this it must be stressed that in the apparently passive gaze of the adventure traveler can be found the active *production* of landscape as an ordered totality. Nor is this ordering random or arbitrary; as we have seen, it is governed by a discourse of modernity-as-loss.

Viewed in this light, it becomes possible to see adventure travel as bringing about that which it so insistently seeks to avoid. Despite its supposed distance from an intrusive mass tourism, in its demarcation of the premodern and identification of this space with essential and timeless forms, the critique of Western modernity as colonizing that lies at its core doubles back as colonialism's most paradigmatic form, a sort of environmental orientalism that images modernity's Other as fixed and immutable.

A similar argument can be made through recourse to John Urry's (1990) claim that the tourist gaze is highly disciplined, seeking out specific sites and views. Not just anything "fits" within the visual logics of mourning. Adventure travel and ecotourism enframe nature and culture, and they do so in particular ways. This may not have a direct impact on local culture and politics, but it subtly insinuates itself into the landscape through the language of value, and the form of the commodity. For the landscapes of Clayoquot Sound to retain their value as commodified sites for acts of recovery, activities that threaten the landscapes' ability to signify a time "before the fall" must be restricted. Indeed, many of the conflicts surrounding Clayoquot Sound today can be explained in terms of competing cultures of nature and their *visual* rather than explicitly ecological logics. Because the gaze of adventure travel corresponds with what Urry describes as the romantic gaze—concerned primarily with the pristine and the picturesque—there is great pressure to freeze these landscapes in the past as primal nature. To return to a theme introduced in chapter 1, this is no less a production of late-capitalist natures than are the clearcuts of industrial forestry, or the genetically modified foods of Monsanto, although the differences between them matter, and each must be analyzed on its own terms. Whereas industrial forestry abstracts one commodity (timber) from its ecological and cultural context, adventure travel abstracts another ("viewscapes") from their ongoing historical construction and places them in the mythic time of the premodern.

That adventure travel is not merely passive in relation to the historical construction of place, nature, and identity was made abundantly evident in the Clayoquot Sound Land Use Decision (see Figure 1.1). This act of state planning did not simply divide the landscape into regions open to industrial forestry and regions that were to be protected for ecological reasons, but introduced a whole range of other land uses that were designed specifically to conserve this place as a landscape of mourning. This took the form of protecting visual and cultural resources, such as the "scenic corridors" found throughout the Sound (what is conserved, of course, is nothing less than the scenography of modernity itself). Oddly, these corridors received very little attention in the weeks and months that followed the release of the plan, except in debates over what level of forestry would be allowed to occur within them. In their calculation of forests preserved from clearcutting, state forestry officials and forest industry spokespersons included these areas, whereas environmentalists classified these as regions still open to forestry. Crucially, neither questioned the cultural, or political-economic, logic of the gaze that ordered the landscape *as* scenic.

There is no innocence to be found in adventure travel and its kin. By classifying some elements of the forest as visually sensitive, the gaze of the modern bourgeois subject is projected onto nature as one of its constituent parts. This has a range of effects. For instance, in these areas only alternative harvesting methods are to be used, shifting what falls within and what falls outside the "economically viable" forest.[44] Indeed, some have argued that the landscape is more *valuable* as a tourist landscape than as a resource landscape, and this shift in the region's economy has been touted by political elites in Vancouver and Victoria—as well as certain sectors of local communities—as an economic boon to forest-based communities caught in the decline of industrial forestry (cf. Ecotrust Canada 1997).

The benefits and costs of tourism are hotly debated in Clayoquot Sound. It is not at all clear that tourism—and especially alternative travel—provides the economic salvation that its boosters promise. Wages in the tourism sector are notoriously low, and workers are poorly organized. Conversely, forestry has historically provided well-paid, highly skilled jobs for local (white) male residents. During the long boom in the forest industry after World War II, the nearest city to Tofino—Port Alberni—had the highest per-capita incomes in the province. There is no doubt that with the shift to tourism Tofino has benefited, but provincially this impact has been markedly uneven. For every Tofino there are endless other villages whose surrounding landscapes cannot be so easily incorporated and commodified as sites outside or prior to modernity. Moreover, as I showed in the case of the company whose activities I followed, many of the economic benefits of adventure travel return to the metropolitan centers where these companies are located. Also overlooked in boosters' accounts is that the economic development of Clayoquot Sound is uneven. Tofino's boom as a tourist destination has not been matched by economic development in the various Nuu-chah-nulth villages scattered around the Sound.[45] Most of these towns remain *outside* the spatial and economic circuits of adventure capital. Few adventure travel companies employ Natives. Even apparently enlightened adventure travel companies, such as the Vancouver company whose excursion I accompanied, are organized out of distant offices, employ guides from the city (almost all are Euro-Canadians), import food from outside the regions traveled to, and organize their tours in such a way that clients can arrive, witness, and depart without spending any additional time in the region. In short, these trips are organized such that adventurers can quite literally "drop in" to the nonmodern for brief visits. In this sense, adventure travel might be described as a form of "flexible tourism." With little or no physical investment in local communities, adventure travel companies—and individuals—can simply redirect their resources at will. When Clayoquot Sound no longer appears adequately exotic, or its landscapes fail to signify modernity's "outside," it is a relatively simple task to shift activities to another site.[46] Indeed, at the time that I accompanied the commercial tour of the Sound, several companies were looking for alternative sites on account of Clayoquot Sound's being too popular.

The political economy and semiotics of adventure can also not be disentangled from the region's histories of colonialism. Indeed, it is not insignificant that just as local First Nations begin to obtain more control over, and benefits from, forestry on their traditional territories (often through hard-fought legal battles), the high-paying jobs that have for so long been denied individuals in these communities are jeopardized. In BC, the adventure travel industry has found its strongest allies among preservationists. Their common agenda is often explicit. "The race is on," declared an early Ecosummer brochure, "the last of the ancient rainforests are being logged. . . . Wilderness travel is one of the surest ways of recruiting new troops for the war" (Ecosummer 1990). The company whose tour I accompanied made annual contributions to the Friends of Clayoquot Sound (a local preservationist group) and distributed Sierra Club brochures to its clients. At first glance this seems admirable, bringing "ethics" into business, but it is not necessarily viewed positively by First Nations. As a local Native woman explained, this conjoining of tourism and preservation was something to fight rather than encourage: "Current events force us to remain ever vigilant that growing pressure from tourism and environmental groups will not lead to park expansion in our lands and waters" (Charleson 1992, A9).

Here we get to the crux of the matter. Within the postcolonial landscapes of Canada's west coast, the incorporation of nature within a discourse of modernity-as-loss is far from innocent. Although offering certain opportunities to Native residents who succeed in inhabiting its terms, it is not at all clear that this staging of the (anti)modern represents the best path of modernization for Clayoquot Sound's Native communities. With its rugged topography, it is difficult to find areas in Clayoquot Sound that are *not* "visually sensitive." In short, privileging the gaze of the bourgeois modern subject has serious consequences; it involves massive shifts in investment, employment, and wealth production that remain unannounced in the apolitical language of "scenery" and "vistas." It also has significant ecological consequences. This is not just because the pattern of resource use will be different in an economic regime based on tourism (thus producing a different nature), but also because adventure travel reorders nature through a *visual* logic, not an *ecological* one. My point is not necessarily that the latter should be privileged over the former, but that we need to acknowledge that as a landscape of mourning, those areas that are to remain open to logging, and those that are not, are increasingly determined in relation to metropolitan fantasies of loss and recovery and not necessarily by concepts drawn from conservation biology or other allied disciplines.[47]

Inhabiting the Nonmodern: The Women of Ahousaht

My objective has not been to dismiss adventure travel so much as to call attention to some of the cultural, economic, and political effects of remaking west coast landscapes as sites of recovery and renewal. Far from being a "light" form of travel, adventure travel can be remarkably heavy. In this sense, mourning *is* as weighty as it sounds: for it demands that places such as Clayoquot Sound—and its people—carry the burden of modernity's loss.

The politics of adventure, however, is not determined in advance. Clearly, to live in Clayoquot Sound is to be continuously entangled in the mythopoetic space of others, to find one's home a site of fantasies and desires that are not one's own, and to experience in powerful ways the continuous decentering of place and nature in the dreamworlds of distant actors. Yet, the discursive and spatial practices of mourning do not merely enclose place and nature on Canada's west coast in the eulogizing embrace of nostalgia, they also give rise to new spaces for identity and politics. To ignore this would be to tell the story of adventure travel solely as the most recent example of Western cultural imperialism, visited (again) on unsuspecting Natives. Although First Nations in Clayoquot Sound today find their landscapes displaced within global circuits of capital, commodities, and identity, they are not passive victims of global forces. Nor were they in the past (see Clayton 1999). To explore this, I return to the same village our band of adventure travelers visited as part of their commercial tour.

Walking on the Wild Side

In the summer of 1996, two years after the tour group I accompanied dined at a bed-and-breakfast in Marktosis, twenty youths cleared an eleven-kilometer trail through the surrounding rainforest, linking the village with the sandy beaches and headlands on Flores Island's exposed Pacific Coast (see Figure 4.1).[48] More than three kilometers of the "Ahousaht Wild Side Heritage Trail" were built as a boardwalk in order to protect ecologically sensitive sections, and at intervals along the trail artfully crafted cedar signs were posted, giving hikers background about specific places, Ahousaht cultural practices, historical events, or stories associated with specific locations (Figure 4.8).

The trail is a modest affair. Accessible only by a long (twenty-kilometer) boat ride from Tofino, its remote location means that only the most intrepid will find their way to its unmarked trailhead at the end of a dusty Marktosis road. Although the scale of the trail makes it appear the antithesis to large-scale corporate tourism, it is important to remember that the very possibility of such a trail—with its invocation of wildness and aboriginal heritage, and with its ready-made, albeit modest, clientele anxiously searching for such sites and experiences—situates this trail within, rather than outside, the discursive space of anti-modern modernisms. But by now we should be skeptical of any claims of exteriority, and indeed, to explore such a project on these terms would be to miss the point entirely. What concerns me here are the ways that in this project, place, nature, and identity come to be *reworked* in new and unique ways.

Local accounts trace the rebuilding of the trail to 1993, when a group of women in the village of Marktosis decided to enter the adventure travel/ecotourism industry, despite suspicions toward tourists among many Ahousaht. The concept behind the "Walk the Wild Side" project, as it was then known, was relatively simple. With each passing year in the late 1980s and early 1990s, an increasing number of "unregulated" adventure travelers had arrived in Clayoquot Sound.[49] In the early 1990s, for instance, the Ahousaht estimated that as many as eight thousand tourists visited Flores Island, many camping on the island's spectacular outer beaches.[50] The Ahousaht argued that they received no benefits from these visitors (almost none ventured to

Figure 4.8. The forest as cultural landscape: historical marker from Ahousaht Wild Side Heritage Trail.

Marktosis), yet bore all the costs of pollution and environmental degradation.

Most immediately, then, the "Wild Side" project had some simple economic and political objectives: to redirect some of the economic benefits of ecotourism and adventure travel into the local economy and to regain some control over the use of Ahousaht traditional territories. Thus, together with the newly formed Nuu-chah-nulth Business Association's booking office in Tofino, and with support from the Vancouver-based Western Canada Wilderness Committee, the women promoted a package tour to visitors to Tofino.[51] For a set price, clients were transported by a Native-owned sea bus to the village of Marktosis. Here they were taken on a guided tour of the village, a walk through the ancient forests and across the white-sand beaches that ring the island, and then back to the "Arts of Paawac" center, where arts and crafts made by local Ahousaht artists were displayed for sale. Clients were then ferried back to Tofino. Run as a nonprofit organization, the economic benefits derived from the project remained almost entirely within the Native economy.[52] Yet, this only touches on one aspect of the project, for the selling of adventure and ecotourism was not an end in itself, but connected to other political struggles.

Articulating Sovereignty

Although a modest effort, the "Wild Side" project became a way for the Ahousaht to extend claims of sovereignty over their traditional territories. To understand this, we need to remember that the sovereign territoriality of the Canadian state is something achieved, rather than given, and that it must be continuously reasserted through a set of practices (cartography, law, etc.). In Clayoquot Sound, Canadian sovereignty has operated in part through the cartographic and legal fiction of "Crown land" and the cartographic incarceration of First Nations on tiny reserves (see chapter 2). These sovereign claims must be continuously reasserted through state practices in order for them to retain their force, but it is precisely this logic of performativity that renders these claims both arbitrary and unstable. What interests me is the manner in which the Ahousaht Wild Side Heritage Trail disrupts the territorial logic of state sovereignty.

We can see this occurring in several ways. For example, the trail visits a series of beaches that lie *outside* the narrowly delineated spaces of Ahousaht reserve lands, including several beaches on Crown land that are assumed to be public and frequented by adventure travelers. By this simple transgression, the Women of Ahousaht foreground the contested nature of these spaces.[53] This is more than the disarticulation of the nation, however; it also rearticulates the Ahousaht as both a *cultural* and a *territorial* collectivity. The guides extend this claim by relating the nature of Flores Island to Ahousaht culture, and vice versa, thereby remarking on the many ways that these landscapes have been used by the Ahousaht in the past and continue to be used in the present. Middens and signs of Native tree use are identified, plants and animals still used in Ahousaht economic and cultural practices discussed, and Ahousaht names for sites provided. When I first visited the project, the Women of Ahousaht had prepared a small guidebook that they gave to each client. This described Nuu-chah-nulth culture, explained the tribe's territorial practices, and gave details about its social organization, its systems of ownership and property (*ḥaḥuulḥi* [proprietary rights] and *tupaati* [hereditary privilege]), food gathering, history, and archaeology. Significantly, the guidebook reversed the primitivist ecology of many environmental groups and adventure travel companies that begin with nature and incorporate indigenous peoples within the region's natural history. In this case, it was only after sociopolitical and historical information about the Ahousaht was provided that the guidebook turned to the region's natural history (cf. Dorst and Young 1990). In July 1997, this booklet was replaced by a more extensive guidebook published by Stanley Sam Sr., an Ahousaht elder (Sam 1997) with assistance from the WCWC and the Washington-based Natural Resources Defense Council (NRDC). Although clearly a venue for the Wilderness Committee to pursue its agenda to keep Flores free from industrial forestry (an agenda shared by the NRDC and its key spokesperson, Robert Kennedy Jr., who wrote the introduction to the book), the volume extended Ahousaht cultural, political, and territorial claims in a similar fashion.[54] In short, where adventure travelers saw nature, the Ahousaht insisted first on culture.

These may be small gestures, but they speak of everyday practices of (re)signification through which alternative futures are imagined and articulated. The project also allowed the Ahousaht to achieve some measure of control over the movement of people through their village, and to manage the impressions they gathered. Previously, visitors to Flores Island learned what little they knew about the Ahousaht from non-Native sources (books, Tofino residents, word of mouth). Most avoided Ahousaht sites altogether, except those that appeared as ruins. Those travelers who did disembark in Marktosis often wandered through the village "viewing" Ahousaht life without any contact with the people. As one of the women on the project explained, residents could not control where visitors went or what they saw, and thus had no means of shaping their impressions of the village and Ahousaht life.[55] By taking visitors on guided tours, the Ahousaht found that they could claim a level of control over how visitors viewed and understood their home. As the group explained on its Web page, the walk sought to provide clients with the enrichment "that comes from seeing another way of life, rather than looking for the 'beach paradise' of the tourist posters."[56]

These sorts of "reality tours" are increasingly popular in the United States, and have been used by local activists to raise awareness of political issues about which most travelers would be unaware (di Chiro 2000). In this case, the tour rearticulates what is significant in the landscape, and encourages clients to see it through new eyes. Clients on the "Wild Side" tours were shown much more than forests, beaches, and signs of land use. The tour also took in the village's new school (and the considerable artwork that adorns the building), new developments in the town (including the "Arts of Paawac" gallery and the commercial center that it is in), and other sites that are particularly meaningful to the community. More than this, by gathering tourists together and guiding them through town, the Ahousaht in a sense reframed the tourist gaze. As tourists move down the streets, it is *they* who come under the watchful eye of village residents, rather than the reverse. The "Wild Side" tour is therefore not simply about adventure and ecology, it is also about building a public image for the Ahousaht as a *modern* Native community with historical claims to ownership of the neighboring lands and forests.

The Cultural Politics of Mobility

The project was also tied in to other politics, including what I describe here as a politics of mobility. According to one of the women who initiated the project, the most pressing reason for developing the "Walk" was to save the Native-owned sea bus—the *Spirit of Marktosis*—that ran twice daily between the village and Tofino.[57] At first glance saving a sea bus might appear a relatively inconsequential goal, but on closer inspection it can be seen as far more central, for it is articulated with issues that cut to the core of Ahousaht life and identity.

For several years in the early 1990s, the sea bus servicing the town had operated at a loss, and there was concern that the service might be shut down. The twenty-kilometer journey from Flores Island to Tofino includes a crossing of Russell Channel, a stretch of water exposed to the open ocean and often subject to heavy seas in winter.[58] For village residents who did not own boats or owned unsafe vessels, and for those whose age or health rendered them unable to safely operate them, the end of service brought the specter of isolation. This was especially true for women in the community, many of whom depended on the ferry as the only means of transportation around the Sound. This had further ramifications. Some felt that the loss of the ferry would increase pressure on community members to move away from the village, reversing a trend in the 1990s that had seen off-reserve residents returning to the village. As is the case in many Native villages on British Columbia's coast, residents of the village (especially young men) regularly leave to find employment in such cities as Port Alberni, Victoria, and Vancouver. Today, only half of Ahousaht band members live in Marktosis, with the rest scattered over Vancouver Island and the mainland. For these individuals, the sea bus provided a crucial link that tied an Ahousaht diaspora with its home. In turn, this allowed for the sorts of movements between and across social, political, and economic spaces that were necessary for the Ahousaht to participate in the global economy yet retain a place-based identity.[59] This is of great importance, for, like most Native groups, Ahousaht identity is closely tied to the land, and in particular to the band's traditional territories on Flores Island. By enticing tourists to take the "Wild Side" tour, the community was able to take a small step toward sustaining both the sea bus and a uniquely *Ahousaht* identity in a world where centrifugal forces threatened at every moment to disarticulate community, place, and identity.

Reconfiguring Social Relations

The "Walk" also became articulated with a range of local political issues surrounding gender, forestry, employment, youth, and local control over law and justice. Although I will touch on only a few of these, they help illustrate a more general point: in the complex assemblages of nature, culture, and politics that compose Clayoquot Sound, the cultural politics of adventure turns out to be remarkably complex.

As with many isolated resource-based villages, Marktosis tends to have a strict gendered division of labor, with men who are employed working on fishing boats or in the forest industry, and women charged with child care and domestic labor. Gender politics therefore infused the "Wild Side" project. As explained by members of the Women of Ahousaht, women in the community found themselves far more isolat-

ed than men, probably even more isolated than women in mainland resource towns. This was related in part to the lack of road access. The end
of the sea-bus service, therefore, would have had a far greater impact on
the lives of women than men, who regularly plied the waters between
the two settlements in their own boats. As important, women were
faced with almost no work opportunities within the village, and thus
had to seek work outside Marktosis or be dependent on a male wage
earner. The development of the "Wild Side" project, together with the
"Arts of Paawac" center, promised much-needed employment and training opportunities for village women.[60]

By tapping into the growing ecotourism industry, Ahousaht women
were able to develop employment opportunities that were not tied directly to either forestry or fishing, two industries that have historically
been reluctant to employ them. This had some unexpected consequences. For one, it allowed dissident voices to question whether the
Ahousaht *should* participate in the forest industry when, or if, land
claims were settled. In other words, adventure travel provided for villagers the possibility of imagining alternative social and economic futures. Like many other First Nations, the Ahousaht are deeply divided
on forestry issues. As mentioned earlier, the promise of the "new tourism economy" is often wildly exaggerated, and justifiably distrusted. Yet
it is also true that Native communities, faced with chronic underdevelop-
ment, find large-scale, unsustainable forestry an almost irresistible
temptation. The "Walk" has also been articulated with other social and
political spaces. Elderly residents benefited from maintaining the transportation link with Tofino. In turn, some suggested that the project
could be used to wrest back control over the definition and administration of justice from the state by providing band members convicted of
minor offenses with opportunities to serve "community time" as volun-
teers on the project. Likewise, the building of the trail provided the
community with an opportunity to employ youth from the village, and
to provide opportunities for cooperative work with non-Native youth
from white communities.[61]

Conclusion: Refiguring Global Flows

It is important not to overstate the potential of the "Wild Side" project
and others like it. Indeed, the future of the project is presently in doubt, a
victim of economics and local politics. The arguments in this chapter go
beyond the success or failure of individual projects, to explore the ways
in which landscapes such as Clayoquot Sound are intensely historical,

discursive, and political, even though they are often assigned to sites "outside" history, language, and politics. My concern is not simply to debunk ideologies that see these landscapes as nature itself, but also to understand the processes that shape these places. Through the example of adventure travel and ecotourism we can recognize the ways in which specific environmental imaginaries—deeply embedded within, and constitutive of, modernity—come to shape the production of space and place. Understanding Clayoquot Sound as a landscape of mourning helps us see not only how closely interconnected are the remote and the near, the primitive and the modern, the natural and the artificial, but also the political work that these categories and distinctions do. Each exists as a necessary condition of the other, and, perhaps more important, these become part and parcel of ever-changing productions of space, place, and nature. Indeed, whereas historical materialists have long noted that the production of nature is related to *economic* processes (such as uneven development [Smith 1990]), it is increasingly clear that social nature is a product of myriad cultural discourses and practices.

This also helps us understand the *overdetermination* of social natures. I mean something more than Althusser's (1969) sense of multiple dynamics conspiring to cause a particular outcome. Rather, following Laclau and Mouffe (1985), I wish to retain the earlier psychoanalytic use of the term, which situated overdetermination within the symbolic order (and thus having no "ultimate literality") in order to suggest that social nature emerges at the intersection of a plurality of nonidentical discursive and spatial practices that achieve only a precarious and contingent fixity. Accordingly, there is no pre-given path for a green, anticolonial politics. The social natures produced in Clayoquot Sound are always unstable, always sites of contestation, always subject to multiple, irreducible processes. It is precisely for this reason that we must distinguish between adventure travel as *colonizing* in its incorporation of other natures and cultures into a particular modernity (something that I traced in the first sections of this chapter), and as only one part of a much more complex *conjunctural* moment where what emerges are new and previously unimaginable spaces of identity and politics. Like all cultural practices, adventure travel is an ambivalent site of politics and power, one that at once opens possible futures and forecloses on others. In places such as Clayoquot Sound, adventure travel is much more than an innocent movement through a timeless landscape; it is also somewhat less than an enframing of nature and culture that is totalizing and inevitable in its form. Rather, it is one among many flows of people, ideas, images, and capital that are implicated in the ongoing reterritorializations of space, place, and landscape on Canada's west coast. Understanding their logics, tracing their effects, and articulating within them possibilities for alternative futures is the task of criticism.

Canada's temperate rainforests are complex landscapes that have no essential or pregiven meaning. Our understanding of them is rooted in, and routed through, specific events and practices. Not all of these practices are as physically demanding as adventure travel or require one to test one's mettle in the face of nature's "forces." Some, like those of the landscape artists I discuss in the next chapter, involve other ways of perceiving, experiencing, and representing and are shaped by different temporal rhythms and spatial processes. What they share is movement, translation, and transformation: the breaking and relinking of sign-chains, the ongoing remaking of nature, and the reworking of colonialisms, past and present.

5. BC Seeing / Seeing BC

Vision and Visuality on Canada's West Coast

The Northwest Coast will always remain in our minds as Carr saw it.

—Ruth Appelhof, *The Expressionist Landscape*

If it could be said that there is an observer . . . it is only as an *effect* of a heterogenous network of discursive, social, technological and institutional relations. There is no observer prior to this continually shifting field.

—Jonathan Crary, "Modernizing Vision"

Exhibiting Landscape

In 1997 the Vancouver Art Gallery (VAG) hosted a small exhibition of two landscape painters: Emily Carr and Lawrence Paul Yuxweluptun. Carr (1871–1945), the daughter of English immigrants, is an iconic figure in Canadian culture, revered for her images of west coast Indians and verdant rainforests painted during the first half of the twentieth century (see Figures 5.1, 5.2). Yuxweluptun (1957–), a Coast Salish artist, gained prominence in the 1990s for his striking re-visioning of BC's natural and cultural landscapes (see Figure 5.3). Although the exhibit, curated by Andrew Hunter, received only modest attention, the juxtaposition of these artists raised important political and theoretical questions about vision and visuality on Canada's west coast in the aftermath of colonialism. On the one hand, it questioned the central place accorded Carr's work in Canadian culture and challenged assumptions about its political neutrality. The decision to place Yuxweluptun's *Thou Shall Not Steal* (three painted wooden panels emblazoned with the words NATIVE LAND CLAIMS) amid Carr's celebrated images of Native villages and verdant rainforests alerted viewers that the visual arts—and not only politics and law—can be complicit with colonial power.[1] On the other hand, it raised anew a set of questions about the incorporation of Native art within Canadian cultural institutions. The show was

Figure 5.1. Emily Carr, *Kitwancool* (1928). Oil on canvas. Glenbow Collection, Calgary, Canada. Reprinted by permission.

one among many exhibits of contemporary First Nations artists staged in Canada during the 1990s, a decade that witnessed the rediscovery of Indian art by curators, collectors, and critics alike.[2] By the mid-1990s, the "indigene" was fashionable again in the art world, even more so if indigeneity was articulated in a way that brought European and non-European aesthetic traditions and cultural identities into tension, displaying hybrid forms that troubled distinctions between Europe and

Figure 5.2. Emily Carr, *Cedar* (1942). Oil on canvas, Vancouver Art Gallery, Emily Carr Trust VAG 42.3.28. Photograph by Trevor Mills. Reprinted by permission of Vancouver Art Gallery.

its others, or between the categories of art and artifact. Yuxweluptun's work—with its self-conscious mixing of Northwest Coast Native motifs with European modernisms—fitted this slot better than most, and was added to the National Gallery's permanent collection in the first half of the decade.

Such incorporations are always double-edged. Although they are a sign of increased visibility for aboriginal peoples, they also contain echoes of earlier incorporations, such as those that occurred in the 1920s and 1930s when Native peoples and artifacts were romanticized and incorporated as the nation's heritage and origin, while the dispossession of Native lands continued unabated (cf. Rushing 1995; Watson 1995). It would be overly cynical to dismiss these recent exhibitions entirely, for many important differences exist. Today Native artists are often involved in planning the exhibits and frequently write commentaries for them. Moreover, the staging of Native cultural production as an integral component of the contemporary multicultural nation is a marked departure from earlier incorporations that positioned them as residual elements of the nations' premodernity. Indeed, as we will see in the case of Yuxweluptun, incorporations of the modern indigene within Canadian culture are ambivalent, since they disturb narratives of the

Figure 5.3. Lawrence Paul Yuxweluptun, *Scorched Earth, Clear-cut Logging on Native Sovereign Lands, Shaman Coming to Fix* (1991). Acrylic on canvas; no. 36950 National Gallery of Canada, Ottawa. Reproduced with permission of artist.

unitary postcolonial nation, even as they are brought within the state's institutional frames.

As important as these issues are, this is not the place to adjudicate questions of appropriation and incorporation. Rather, my interest in the works of Carr and Yuxweluptun—and in Hunter's exhibit—lies in what they can tell us about the cultural politics of nature on Canada's west coast.[3] Specifically, my interest centers on the way in which Canadians' *environmental imaginaries* of the coast have been constructed and contested. Juxtaposing the work of these artists helps us not only to relate nature, culture, and visuality, but to see these as fractured, overlapping fields that carry in them complex and often ambivalent relations of power. It also helps us to think about what we might call the temporal rhythms of today's cultures of nature—the way that forms of vision and visuality from the past continue to shape the present. As I will show, to the extent that Carr's work continues to hold a central place in the archive of images through which Canadians know the west coast, the colonialist visuality that informed her seeing surreptitiously irrupts as a persistent trace. Visuality on Canada's west coast is not a synchronic, unified totality that evolves through time; rather, it carries multiple, irreducible histories within it, elements from the past that take on new life and new forms in the present. For geographers interested in the cultural politics of nature today, attending to these temporal rhythms is critical.

Why these painters in particular? Despite a wealth of BC landscape artists, it is almost impossible to discuss visuality in BC today without reference to Emily Carr. In the words of one writer, Carr is "intrinsic" to the Canadian imagination (Walker 1996). Her art is displayed routinely in the nation's galleries and museums. It is reproduced endlessly on postcards and calendars. Photographers try to emulate her vision, poets find inspiration in her artistic struggles, and playwrights bring her life to the stage. Dog-eared posters of Carr's work hang on the office walls of local environmental organizations, realtors, and travel agents. Even scholarly books on BC by historians and economists with no stated interest in art or aesthetics place Carr's work on their covers. For many Canadians, Carr *is* BC; her ubiquitous presence needs no explanation, no supplementary text.

Borrowing from Homi Bhabha (1994), I suggest in this chapter that Carr's work continues to play an important pedagogical role in Canadian culture, educating Canadians to see west coast natures according to a particular visual logic. It is not insignificant that of the many galleries at the VAG—itself located at what Stan Douglas (1991) has described as the "proprioceptive center" of the city—only the Emily Carr gallery is permanent. All other exhibits come and go, a testament to the fleeting nature of artistic trends. This leads to a series of questions. What does it mean to privilege Carr's work today, almost sixty years after her death? And what does this tell us about the temporalities of the postcolonial? When Carr traveled the inlets and passages of the coast in the early decades of the twentieth century, what did she see? What social, spatial, and ideological relations enabled and constrained her gaze? And how are these relations *carried forward* into the present? To answer this last question, it will be necessary to inquire into Carr's popularity. What accounts for the metonymic relation between "Carr" and the "west coast"? How did her vision come to stand in for the west coast as a whole?

Yuxweluptun's work stands in counterpoint. Like Carr, his work depicts west coast scenes. He also has gained critical and popular acclaim, although nothing approaching Carr's iconic status. Like Carr, Yuxweluptun too seeks to bend vision in new ways. But here similarities end, for whereas Carr in her time sought to move landscape art away from the merely picturesque, it is now Carr's work—and the sort of seeing her work represents—that Yuxweluptun seeks to call into question. Indeed, if Carr has come to exemplify "BC seeing," then Yuxweluptun can be seen to paint at its ideological and political limits. This chapter focuses primarily on Carr, in order to explore how west coast environmental imaginaries have been set in place. I will return to Yuxweluptun's work at the end, not as a corrective to Carr but as a reminder that nature, culture, and visuality in postcolonial BC are fractured fields, shot through with power and politics.

Framing Carr: "Artist of the West Coast"

National artists are made, not born. This simple statement reminds us that we need to look elsewhere than the artist and her work in order to understand Carr's ritualistic invocation in the present. Indeed, Carr's central place in Canadian culture is arguably as much a historical accident as a product of her artistic excellence, owing in part to her work's fitting well into an ideological slot available within an anxious Canadian nationalism in the years immediately preceding World War II. Although she had been painting for more than thirty years, she was relatively unknown until the late 1920s and 1930s, when she was discovered by critics, and came to be discussed as a western Canadian counterpart to the

Group of Seven, a loosely organized circle of artists in eastern Canada. These artists were celebrated for forging a "Canadian School" of landscape painting at a time when nation-states such as Canada were seeking to establish unique cultural traditions.[4] In Canada, this meant finding cultural forms that were different from those in Europe, but also from those in the United States. What was thought to set the Group of Seven artists apart was their immersion in the Canadian landscape, from which they were said to obtain their inspiration and their uniquely Canadian style.[5]

From 1927 onward, Carr's work was viewed in similar terms, although perhaps more than that of any Group of Seven painter, her work came to be associated with a specific *place*—British Columbia. Even before her death in 1945, she was described as the artist who responded most intensely and openly to the west coast, capturing its unique character, and distilling its landscapes into its two defining elements: a monumental, albeit vanishing Native culture, and a powerful, primeval nature. In this section, I focus on two key aspects of Carr's critical reception: her construction as an "organic" artist, and the division of her life and work into early, middle, and mature periods. As we will see, coupled together, these narratives have largely determined how Carr has been viewed, and which of her works has been most frequently exhibited and reproduced.

Carr as Organic Artist

Although it is difficult today to imagine the coast apart from Carr's prolific work, this association was not assured. Nor was it simply the logical outcome of *how* she represented the region's landscapes. Rather, it was the product of critics and historians who consistently framed Carr's art in terms of an organic relationship between the artist and her surroundings. For much of the past six decades, this view has dominated Canadian popular culture.

The curator and historian Doris Shadbolt has perhaps done the most to popularize this view. Her influential book-length study and catalog of Carr's work, for instance, begins with a simple heading: "Artist of the Canadian West Coast." The double meaning of the "of" is not, I suspect, entirely accidental. In Shadbolt's eyes, Carr was at once an artist who *portrayed* west coast landscapes, and an artist who was *of* the west coast. For Shadbolt, the two had a necessary relation: one could truly be an artist of west coast landscapes only if one was of the coast. Immediately following this heading a passage establishes what is now the most widely accepted understanding of Carr and her art:

Emily Carr was born in Victoria on Vancouver Island 13 December 1871, died there 2 March 1945, and lived most of her life within a few blocks of the house where she was born in the James Bay district of that city. Her genius throve in the island's isolation from mainland British Columbia and in the province's isolation from Canada and the world. The two great themes of her work derived from the most characteristic features of that region—a unique and vanishing Indian culture, and a powerful coastal nature. It is logical to think of Carr and the Canadian West Coast at the same time, for her paintings and her writings bear the indelible imprint of her long attachments to the place where she was born and where she chose to remain. (Shadbolt 1979, 11)

Here seeing is conjoined with proximity to form a powerful rhetoric of unmediated vision. For Shadbolt, Carr's art was organic, a product of a long and sustained encounter with west coast landscapes.

Shadbolt is not solely responsible for this interpretation. Critics in the 1930s and 1940s had already made this connection. In a *Saturday Night* review Mortimor Lamb (1933, 3) described Carr as a west coast painter "free from all influences than those created by her own responsiveness to a scenic environment itself elemental, austere and yet scenically romantic."[6] In the same magazine, Ira Dilworth (1941, 26) added to Carr's mystique, describing her as an artist who fought her way to the objects themselves. Carr, he explained, traveled "with a courage that we at the present day can scarcely understand," traveling into "the most solitary parts of the British Columbia coast," overcoming "poor transportation" and "often extremely dangerous stretches of sea." Nor was Carr immune to understanding her art in this way, especially after she was championed by eastern Canadian critics in these terms following the inclusion of her work in a 1927 National Gallery show titled *Exhibition of Canadian West Coast Art: Native and Modern*. In her autobiographical writings, Carr (1946, 1966) implied that her response to these places was somehow different from—and superior to—that of other Canadian artists who also sought to depict west coast scenes. Yet, in her opening paragraph, Shadbolt goes beyond what Carr ever claimed for herself, tightly binding this seeing to notions of rootedness, and an acquaintance that only comes to the artist with time, experience, and, perhaps most important, *isolation*. The implications of Shadbolt's narrative are clear: if one looks long and hard enough, one begins to see completely; if one remains "in place," one's vision remains unpolluted by outside influences—the filters of culture and ideology fall away, and things become present in their true form.

Today, we have learned to be skeptical of such claims. Notions of a pure space outside global circuits of capital, commodities, and information seem rather quaint, even naive (Appadurai 1996; Gupta and Ferguson 1997). So do suggestions that vision is unmediated (Crary 1988; Haraway 1991). Indeed, it now goes without saying that critical analysis must interrogate the *situatedness* of vision in order to draw out the complex conditions that inform not only what *is* seen, but what *can* be seen (Rajchman 1988), or what Derek Gregory (1994) refers to as "spaces of visibility." Accordingly, some of Carr's biographers have placed much greater emphasis on the social, psychological and political conditions that may have shaped her life and enabled her work (see Tippett 1979; Blanchard 1987; Moray 1993). It is now well known, for instance, that despite living much of her life in Victoria, Carr was thoroughly enmeshed in a web of extralocal social, political, and aesthetic relations that linked her "isolated" world in Victoria with artists in eastern Canada, New York, Paris, and elsewhere (even before her so-called discovery by eastern critics). She was also in close contact throughout her career—through books, friends, and numerous trips—with developments in Western culture and aesthetics. Downplayed in rhetorics of unmediated vision were the many years Carr spent studying abroad (San Francisco, 1890–93; London, 1899–1903; Paris, 1910–12), her acquaintance with ethnologists, her contacts with members of the Group of Seven, her appreciation of Fauvism, Expressionism, and Cubism among other streams of Postimpressionist modernisms, and her trips to places such as Toronto and New York, where she had contact with prominent artists such as Lawren Harris and, in passing, Georgia O'Keeffe. Nor does such rhetoric allow for a full acknowledgment of what it meant that she read Walt Whitman religiously, or that for a period in the early 1930s she struggled to come to terms with, and incorporate into her art, the spiritual vision of theosophy.

Far from unmediated by contact with a wider world, Carr's art (and journals) show a wealth of references to the work and writings of others. Her work was decidedly intertextual, in the sense that it obtained its meanings through its relation to other works. This intertextuality troubles the received image of Carr as an organic artist, because it places her work in a complex web of relations. This was evident even to Shadbolt, who at points interrupts her narrative of "isolation" to acknowledge Carr's international connections. Yet, in the final analysis, she sustains her reading of the purity of Carr's vision by appealing to a well-worn discourse of art as an expression of an inner journey:

> Emily Carr was an individual who did not sit comfortably in close company. Like any artist, she drew on whatever art sources were available to her, but she was never truly in step with any group, movement, or trend. She was inspired by the strong fresh look and the vigorous nationalist sentiment of the Group of Seven. She did not, however, adopt their painting conventions and she was fundamentally different from them in the degree to which she finally made her work the vehicle of feeling, and in the strong mystical tendency she shared with [Lawren] Harris alone. . . . Because of the quality of Carr's painting and its originality . . . affinities are bound to suggest themselves and to hint at a larger relationship with other artists whose strength, like hers, grew out of isolation, particularly an *inner isolation* (1979, 194–95; emphasis added).[7]

Again, Shadbolt was neither the first to deploy these terms nor the last. In 1939, Eric Newton opined that Carr "belonged to no school" and that her inspiration was "derived from within herself" (Newton 1939, 344–45). A 1990 National Gallery exhibit repeated this message, and with its permanent Carr gallery the VAG continues to encourage visitors to understand Carr in this way, as if, in contrast to other artists whose work must always already be seen as embodied, historical, and partial, Carr's work transcends the particular.

Dividing Carr

To this narrative of unmediated vision has been added another, which is equally as important for my story. Curators, critics, and art historians have regularly framed Carr's life in terms of a progressive chronological and artistic development that proceeds forward from her early accomplishments to her mature years (those that followed after her trip to eastern Canada in 1927). Although more recent biographers have raised questions about Carr's isolation, they have essentially followed Shadbolt's division of Carr's life into three periods. The first (1907–13) corresponds with what has come to be considered Carr's documentary period, during which she produced more than two hundred sketches and paintings of west coast totem poles. The second (1914–27) corresponds to a period of relative inactivity following Carr's unsuccessful attempt

to sell her "ethnographic record" of Native villages to the newly established provincial museum.[8] The third (1928–45) is identified as Carr's
"mature" period, during which, Shadbolt argues, Carr discovered a
"deeper vision" that took her to the essential spirituality and formal aesthetic properties found in Native artifacts and west coast landscapes:

> It was not until she was in her late fifties that she came to understand
> the Indian's carvings as the expression of his experience in a primor-
> dial environment and was led in turn to a bolder perception of the
> forests and coastal landscape. When she finally was able to identify
> the primal energy she found in nature with the spiritual energy she
> was seeking as the manifestation of God, the elements in her art be-
> came integrated at a higher level. . . . At this point her art reached its
> *full range of effectiveness.* (Shadbolt 1979, 195; emphasis added)

As Gerta Moray (1993) explains, this developmental story has effectively devalued work Carr did before her mid-fifties, and elevated the work
she produced in the last two decades of her life. In Shadbolt's 1979 biography and catalog, for instance, fully four-fifths of the text and plates
focus on the period *after* 1927, even though Carr had previously produced hundreds of canvases and sketches. Likewise, because only work
from the latter period is cloaked with the mantle of artistic excellence,
it is such paintings as *Kitwancool* (Figure 5.1) and *Cedar* (Figure 5.2)
that have come to represent the most sophisticated expressions of "BC
seeing."

Revisiting Carr

This developmental narrative matters in part because it shapes the
archive of west coast images that Canadians have ready at hand. Carr
may be closely associated with the west coast, but only certain of her
works circulate widely, while others are rarely seen. To return to an earlier question, this partly determines *how* the past circulates in the present,
giving shape to the uneven and differentiated rhythms of postcolonality
(cf. Hall 1996).

This uneven reception may be important in other ways too. Gerta
Moray (1993) has argued that in her early years Carr brought to bear on
BC landscapes a decidedly anticolonial optic, one that was much dimin-
ished in her popular later work. I focus on this here at some length, not
only because it forces us to consider the implications of "dividing" Carr,
but because it necessarily leads us back to one of my initial questions
concerning the *manner* in which Carr framed BC landscapes.

Appropriations or Interventions? Debating Carr's Legacy

Moray's argument is in part a response to a brief yet sharp critique of Carr's work written by the Haida/Tsimshian critic Marcia Crosby (1991). In it, Crosby examined Carr's work alongside a number of other non-aboriginal Canadian writers and artists who had "appreciated and supported indigenous art forms and culture at a time when no one else did" (270). Far from celebrating this appreciation and support, however, Crosby showed its paradoxical nature, writing that such work "ha[d] more to do with the observers' own values" (ibid.) than with the situation of individuals or First Nations peoples, or the character of First Nations sites visited and depicted. In Crosby's view—drawing heavily on Edward Said (1978) and Robert Berkhofer (1978)—the Indians that white Canadian artists sought to represent were always, and could only ever be, imaginary Indians, because these artists were never able to step outside their own cultural frames.

In Carr's case, Crosby argued that her writings and paintings were notable for their "produced authenticity," portraying only a pre-contact or uncontaminated Native life. As much as she paid tribute to the Indians she loved in her journals and short stories, Crosby suggested that Carr's Indians were nostalgic figures rather than the people who took her in gasoline-powered boats to the abandoned villages that she painted. She finds the same to be true of Carr's paintings and sketches of Native artifacts, which "intimate that the authentic Indians who made them existed only in the past, and that all the changes that occurred afterwards provide evidence of racial contamination, and cultural and moral deterioration" (276). Such portrayals devalued present-day Native cultures in the guise of celebrating their past.

It is worth noting that Crosby's interventions occurred concurrently with two important political and intellectual developments in Canada: the (re)assertion of First Nations land rights, which gained momentum in the 1980s and 1990s; and the culture wars that brought into view the relation between representation and power (Dawkins 1986; Ryan 1992) and cultural appropriation (Todd 1990; McMaster and Martin 1992; Fulford 1993; for discussion of these debates, see Coombe 1998). Writing into these controversies, Crosby sought to map the connections between cultural production and (neo)colonialism. By showing how apparently sympathetic portrayals of Native peoples were not free from colonialist power and knowledge relations, Crosby radically undermined claims for a disinterested literary or artistic expression, or, for that matter, for the possibility of a "love" for Indians independent from the cultural discourses that shaped how Native peoples could be viewed and represented. For Crosby (in contrast to Shadbolt), representation was always already implicated in cultural and political dynamics that exceeded the stated intentions of artists. To "re-present" Natives (or nature) was therefore to enclose them within the frames of reference of the artist, claiming Natives aesthetically, and, by extension, politically.[9]

Although Crosby prodded artists and critics in BC to view art and power as simultaneous, determining whether an artist's work was "colonialist" may be far more complicated. So argues Gerta Moray (1993) in a study that situates Carr's early work within the fragmented (post)colonial cultural and political fields of early twentieth-century British Columbia. I will follow Moray's argument at some length, in part because she provides a compelling case, but also because by attending to its shortcomings I will develop my own critique of the celebrated rainforest images that Carr produced in the last years of her life.

I focus here on two aspects of Moray's argument. First, responding to critics such as Crosby who reduce Carr's political consciousness to her social location, Moray suggests that not all representational practices by non-Natives were equally implicated in colonialism. To support her argument, Moray provides a careful reading of the social worlds that Carr inhabited, suggesting not only that settler society in early British Columbia was significantly more heterogeneous in terms of ethnic, class, and religious differences than these critiques allow, but also in terms of competing political opinions. Painting within this context, Moray argues, Carr's paintings cannot be reduced to a singular and seamless "colonial discourse." Second, Moray argues that this was especially true in the case of Carr's early years, before her discovery by eastern critics, when her effort to document Native villages may have deliberately challenged the racist views held by the British gentry, missionaries, and state administrators. Moray notes, moreover, that it is precisely work from this early period that has been largely overlooked by critics and historians who, consistent with their developmental narrative, privilege only Carr's later work.

It is worth noting that the differences between Crosby and Moray are symptomatic of a shift in the early 1990s from the study of colonial discourse (often presented in Said's totalizing terms) to an analysis of colonialism as a fractured and contested field in which the operation of power is found to be more ambivalent and uncertain (Bhabha 1994; Gandhi 1998). Moray, for instance, places great stress on the existence of multiple and overlapping social, cultural, and political fields, and notes the presence of vastly different ideological positions regarding the "Indian question" during Carr's lifetime, explaining that these could often be held simultaneously and in contradictory ways by individuals. Significant contrasts, for instance, could be found in attitudes toward Indians held by white settlers, Anglican and Catholic missionaries, Hudson Bay Company employees and their families, and the British gentry that dominated political and economic activity in the province (see also Harris 1997; Christophers 1998). Carr moved within and between these groups, and would have been aware of political struggles over land and culture. It is well known that Carr preferred to see herself as colonial or Canadian rather than as part of the British gentry to which her family belonged, and her writings often harshly criticized the efforts of missionaries, even as she relied heavily on them for her travels and contacts. Likewise, Carr strained against the protocols of Victorian femininity. Her career as a painter departed significantly from those of other white female artists who lived and worked in the region but who remained more closely bound to notions of the picturesque and to subject material drawn from gardens or town life.[10]

This fractured field suggests to Moray the possibility that oppositional positions were available *within* non-Native society. As Moray puts it, since "images are generated within a *field* of political and psychological interests, not all such imagery is equally closed, unsympathetic or reductive" (1993, 9). Rather than write totalizing critiques based on identity, she argues, Carr's work must be studied for its "local levels of engagement and meaning" and evaluated in terms of "the local circumstances of her work's making" (14, 16). Moray's point has considerable merit. Many of Carr's contemporaries considered Native peoples to be culturally inferior, and destined to wither away in the presence of European civilization. In important ways, Carr's work contested this narrative, emphasizing instead the *vigor* and *scale* of Native life, and evaluating settler society in far less favorable terms. On this basis, Moray concludes that Carr's paintings not only ventured into subject matter that most women painters (and many men too) were either unwilling or unable to depict, but were also intended to stand as a vindication of Native traditions against accounts by missionaries and administrators who condemned what they saw as the Indian's dark, heathen past. "Through her choice of images, her style of representation and her textual exploration of her works," Moray writes, "Carr took quite specific issue with certain views of Native culture being propagated at the time by missionaries, government agents, anthropologists and the press" (107).

We can now see the major weakness of Crosby's argument. Locked in a totalizing logic, she forecloses on the possibility that representational practices by non-Natives could actually contest colonial relations. Deborah Poole (1997, 7), in a different colonial context, argues that to understand the role of images in the construction and maintenance of cultural and political hegemonies "it is necessary to abandon that theoretical discourse which sees 'the gaze'—and hence the act of seeing—as a singular or one-sided instrument of domination and control." It follows, then, that the massive archive of images produced on the west coast in the first decades of the twentieth century—including Carr's own—carries in it histories and perspectives that are irreducible to a single ideology, both as a whole, and as individual visual artifacts. It also follows that it should be possible to locate differences *between* images, and ambivalences *within* images, such that the relation between colonialism and visuality becomes seen as somewhat less stable or direct, and thus also to imagine alternative visualities not so wedded to a colonialist imaginary.[11]

That vision and visuality was not a closed field in early twentieth-century British Columbia can be illustrated by comparing four images of Blunden Harbour, a Kwakiutl village located near Alert Bay and a site that Carr depicted in one of her most celebrated paintings. All were produced by non-Natives in the first decades of the twentieth century. Charles F. Newcombe's photograph, taken in 1901, positions the observer with a view of the length of the boardwalk running between the village and the water (Figure 5.4).[12] From this perspective, the eye is drawn to two aspects of the photo: the carved figures that dominate the scene, and the activity occurring along the boardwalk between these figures. This image lends itself to various interpretations. On the one hand, the oblique angle of the photograph emphasizes the carved figures over the activity of villagers, exoticizing the scene in ways that were common at the time. Individual houses merge to form a common background, removing signs of heterogeneity and presenting Native culture as a cohesive, singular unit, consistent with the prevailing views in anthropology that saw "primitive" cultures as unified and undifferentiated.

Yet, different interpretations are, and were, available. Newcombe's photo was part of a massive effort to document and collect the culture and artifacts of the west coast in the first decades of the twentieth century, a project that included anthropologists such as Franz Boas (with whom Newcombe was in contact). Thus, like many of Newcombe's photographs, the image is *ethnographic* in quality; it seeks to record

Figure 5.4. Blunden Harbour. Photograph by Charles F. Newcombe (1901). Canadian Museum of Civilization, image number 99682. Reprinted with permission.

everyday life (even if striving for an ethnographic present). The inclusion of Native activity, for instance, refuses to let the totems be abstracted entirely into the space of aesthetics or mythology, as was typically done by non-Native artists in the first decades of the century (see also Rushing 1995). A blanket is left casually on the stairs. Pots and bowls are placed randomly on the boardwalk. Logs float in the water. Small knots of people engage in various activities, some quite clearly commercial. A child addresses an adult while another watches. Eventually, the eye grows tired of the static sculptures and shifts to the life that swirls intensely among them. In other words, these monumental forms—although clearly the focus of the image—remain intimately tied to the daily life of village inhabitants.

Is this, then, a simple case where Native artifacts were decontextualized and relocated within the aesthetic space of European primitivism, such that they became little more than exotic elements in narratives that had less to do with everyday Native life and more to do with the concerns of metropolitan cultures? Or, in light of the prevailing view in the early 1900s that Native cultures were dying off, does the image bring to light a vibrant Native culture whose present is fully *modern* but still *continuous* with its past? And does it matter what Newcombe intended? How was his image received by others, or, for that matter, by viewers today?[13]

Other images may provide clues. As Moray herself notes, this image had a varied history. It reappeared, for instance, in the memoirs of the Anglican missionary Thomas Crosby, published in 1914, thirteen years after the photo was first taken. Here it appears juxtaposed with a second photo of the "new Bella Bella" (Figure 5.5). Regardless of Newcombe's initial intentions, the image is now asked to bear the weight of other meanings. While the "new" Bella Bella, with its individual English-style milled-wood homes and linear street plan, is set up as an example of a "Christian" village, Newcombe's Blunden Harbour, with its totemic figures, is portrayed as a "heathen" village, in need of the gospel. Here we see the immense difficulty of fixing the meaning of visual images. At one level, Newcombe's image may have reminded viewers that a vibrant Native presence continued to exist along the coast, despite attempts by missionaries and the state to extinguish Native traditions and assimilate Native peoples. Yet, at the same time, Newcombe's perspective and choice of objects could lend support to a different political agenda, incorporated into a discourse of progress that imagined the tranformations of savages into Christians (and Canadian citizens) and that in turn authorized the legal and political efforts of church and state to drive a wedge between a savage past and a modern future. Newcombe's perspective was not innocent: it enframed Blunden Harbour through a particular optic and supported certain interpretations, albeit always more than one.

The partiality of Newcombe's perspective, as well as the views of Anglican missionaries, becomes apparent when placed against yet another view of Blunden Harbour, this time taken by members of Edward Curtis's film crew in 1915 (Figure 5.6).[14] Here Blunden Harbour is framed from a point at sea, a perspective that brings very different elements into view. For one, it provides depth of field in sharp contrast to Newcombe's oblique angle, which emphasized only surface. It also permits the differentiation of individual buildings while at the same time diminishing the importance of any exotic artifacts, which, while still present, are often indistinguishable from their background.[15] Moreover, the photograph makes no effort to exclude the milled-wood buildings set among the cedar-plank houses, undermining the modern/primitive

NEW BELLA BELLA—A CHRISTIAN VILLAGE.

A HEATHEN VILLAGE—AN APPEAL FOR THE GOSPEL.

Figure 5.5. A Christian village—a heathen village. Source: Crosby (1914).

Figure 5.6. Blunden Harbour. Photograph attributed to Edmund A. Schwinke (circa 1914). Milwaukee Public Museum, negative number 3512. Reprinted with permission.

divide so crucial to colonialist discourse, and which Curtis, in his massive twenty-volume *North American Indian,* did so much to reinforce.[16] Significantly, the image also disrupts the Anglican discourse of the assimilated Indian. Whereas the images in Thomas Crosby's memoirs establish savagery and a European modernity as oppositions and thus incompatible, here we see an image from much the same period that reveals a hybrid, regional modernity that exceeded the binary terms provided by colonial and missionary discourse.

This brings us, finally, to a painting by Emily Carr dating from around 1930 (Figure 5.7). It is based on Newcombe's photograph, but the differences are stark. Carr not only monumentalizes the Native artifacts, but abstracts them entirely from their cultural and social surroundings. The boardwalk and canoes are emptied. Gone is the hustle and bustle of daily life. No people are present, no activity occurs. What little differentiation between buildings that could still be found in Newcombe's photo has been removed entirely. All that remains are the last vestiges of a passing race fading before the inevitable advance of a foreign and incommensurable modernity, presented as aesthetic and spiritual resources for the modern nation whose teleology—and territoriality—is merely assumed. Ironically, one of the most telling critiques of the work

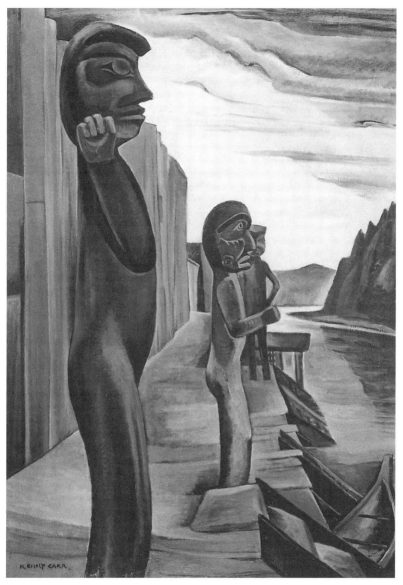

Figure 5.7. Emily Carr, *Blunden Harbour* (circa 1930). Carr's painting is based on Newcombe's photograph (Figure 5.4). Note the removal of all signs of contemporary Native life. Oil on canvas, no. 4285, National Gallery of Canada, Ottawa; purchased in 1937. Reprinted with permission.

is found in an obscure memoir by Jim Spilsbury (1990), an early aviator along the BC coast. Spilsbury notes that whenever *he* flew into this village, including several times around the same period that Carr painted her canvas, hundreds of people crowded the boardwalk in what was a vibrant node of economic and cultural exchange.

My point is not to dismiss out of hand Moray's assertion that Carr's work contested prevailing colonialist views, and thus to side with Crosby. After all, Moray would likely object that Carr's painting was done during her later period, after her aesthetic concerns as a modernist painter, and her exposure to primitivism and theosophy, had reoriented her initial anticolonial impulse and led her to treat Native artifacts only in terms of their formal aesthetic qualities. (I will explore this argument at more length momentarily.) Rather, my point is to show that one cannot speak of a singular and undifferentiated *European* gaze, nor posit a simple opposition between colonialist and anticolonialist representational practices. Vision and visuality are multiple and unstable, and both the production and the consumption of images must be placed within complex social and political contexts.

This is the point that Moray wishes to make for Carr's early "documentary" paintings, which, she argues, posed a serious challenge to the discursive and political practices that underwrote colonial relations in the province. In the next section, I examine this argument, both to explore its validity and to question whether it makes sense to speak—as Moray and Shadbolt do—in terms of a *rupture* between this work and Carr's later work, including her famous rainforest images (which are often seen in terms of a second rupture between her Indian and forest themes). Moray simply inverts Shadbolt's valuation of Carr's work. Where Shadbolt finds an undeveloped aesthetic, Moray finds a political agenda. Where Shadbolt later finds a fully developed aesthetic, Moray finds a strangely depoliticized gaze (see also Linsley 1991). Both, however, retain the broad outlines of a developmentalist narrative that divides her life into stages. The remainder of this chapter suggests a different reading of Carr, one that questions the language of rupture in order to think about continuities, and one that questions whether rendering Native people visible is always an anticolonialist gesture. By exploring Carr's early documentary project, I will suggest that her early work was perhaps less anticolonial than Moray supposes. Further, I will suggest that in important ways this work laid the groundwork for the remarkable decontextualization of Native artifacts that occurred in her later work, and her eventual shift from the Indian village to nature itself, in which Native artifacts disappeared entirely and a new focus on nature—as *even more* original or primal—came to the fore.

Observing Nature and Natives: Space, Culture, and Visuality

Between 1907 and 1912, Carr engaged in an extensive program of recording totem poles in village sites up and down the BC coast, eventually producing more than two hundred canvases. There was a sense of urgency to Carr's task. Along most of the coast, the practice of erecting such poles had declined, and in the two decades prior to her project a large number had been removed and relocated to museums and fairs (see Cole 1985; Blackman 1992).[17] Many others showed signs of deterioration and decay. Carr would later write that in the face of these rapid changes, her project was to record what remained before they too suffered a similar fate.

Moray argues that Carr's work in this period was framed by scientific and political interests more than aesthetic ones (precisely why Shadbolt disparages them). By comparing Carr's village paintings with photographs of the villages taken during the same period, for instance, she demonstrates how Carr strove for accuracy and detail. From points outside each village, Carr painted panoramas designed to place each pole in relation to the others (Figure 5.8), and in relation to the village as a whole. From vantage points inside the villages, Carr painted detailed

Figure 5.8. Emily Carr, *Tanoo* (1913). Oil on canvas. Province of British Columbia Archives PDP02145. Reprinted with permission.

studies of individual poles (Figure 5.9). Thus, Carr produced a detailed *geography* of all of each village's totem poles, and an equally detailed *iconography* of each pole. The contention that Carr understood her project more in scientific terms than artistic ones finds further support in her attempts to have the provincial government purchase the paintings for its new provincial museum as a permanent record, and also in the public lecture that she gave on the event of their initial display in Vancouver in 1913.

For Moray, the documentary nature of Carr's project is evidence of her political intent. To be sure, Carr's exhaustive record would have reminded white settlers that the coast was never an empty land without prior residents, upsetting "the formal and abstract division of the land into exploitable parts" (Linsley 1991, 231). Further, there are important differences between Carr's work and that of other women artists at the time who, by avoiding Indian content altogether, reinforced settler society as normative. And there is some basis for arguments that a political impulse in this early work was blunted over time. Carr's pre- and post-1927 work diverges in important ways. In comparison to her later work, Carr's documentary project gives far more attention to the *details* of individual poles and their village contexts (compare Figure 5.1 with Figure 5.9). People were more often present (although limited in terms of how they appeared). After 1927, scenes similar to those represented in her early work became increasingly stylized in accordance with modernist conventions, most notably Expressionism and Cubism, and also in terms of the primitivism that pervaded both.

I am less convinced, however, that these add up to an anticolonial optic. Part of the problem lies in the willingness of biographers and critics to take Carr's statements about her art at face value as the *truth* about her paintings. Some thirty years after completing this work, for instance, Carr explained that her project had been to "picture totem poles in their own village settings, as complete a collection as I could" (Carr 1946, 211). This statement, along with comments from a 1913 lecture in which she discussed in some detail the mythology of each pole, and other comments disparaging missionaries (among others), has been invoked as evidence that Carr sought to *vindicate* Native traditions in the face of settler hostility.[18] According to Moray, for instance, it was only much later, after her reception by eastern Canadian critics as a western counterpart to the Group of Seven, that Carr would shift registers, appropriating Native art and artifacts as resources for her own aesthetic and spiritual journeys.[19]

Although there can be little doubt that Carr's documentary project

Figure 5.9. Emily Carr, *Skidegate* (1912). Oil on card on board, Vancouver Art Gallery, Emily Carr Trust VAG 42.3.46. Photograph by Trevor Mills. Reprinted with permission.

broke with certain colonial norms, I think it is important to question recent efforts to defend her from critiques such as Crosby's. Invariably,
Carr's defenders appeal to her *writings* in order to support their assertion
that she was sympathetic to the predicament of Native peoples. Douglas
Cole (2000, 152), for instance, turns to Carr's autobiographical *Klee
Wyck,* in which she described many of her travels to Native communities. Here he finds a portrait "of a people living in a transient present,
caught in conflicts between old and new, between traditional and colonized modern, who are making their way as best they can." Carr's reflections, Cole writes, were of everyday life, not an imaginary authenticity.
Where these efforts go awry, I think, is in their attempt to interpret
Carr's *visual* work through her *written* work. This is to make two related
mistakes. First, it recuperates a notion of intentionality whereby Carr's
statements about her work (often written decades later) come to stand
in *for* her work. Second, it posits Carr as a unified subject, whereby her
writings and her paintings are interpreted as the product of a single, unified vision. Intuitively this seems reasonable (if she wrote sympathetical-
ly of Native life, would this not be reflected in her paintings?), but it
fails to recognize that *writing* and *painting* are different genres governed
by very different codes. These two common moves allow Carr's statements to become a substitute for a thorough examination of the *visual
cultures* in which she moved and worked. In their own ways, Crosby,
Moray, and Cole each fail to recognize that Carr the writer and Carr the
visual artist need not have been the same. Curiously, Carr's interpreters
say little about what sort of objects Carr's gaze fixed upon. Why should
these poles have so captured Carr's attention? Why not *other* objects and
figures? Why poles and not people? This was a time of great economic
and cultural transition for Natives. Fishing canneries, sawmills, and
other resource industries were increasingly central to the social and economic life of the province, and many Natives took part. Yet no evidence
of this appears on Carr's canvases. One might ask why labor is so invisible, especially if it involved modern technologies. Or why the sites
Carr portrayed were so often vacant.[20] Put more pointedly, why should
recording *totem poles*—even with an emphasis on accuracy—represent a
defense of First Nations? What sort of vindication is this?

In the face of Carr's documentary intentions, we need to ask a different question. What did it mean to be an *observer* of peoples and landscapes on the west coast in the first decades of the twentieth century?
How was the desire to "see" channeled along certain lines? What social,
technological, discursive, and institutional relations shaped the visual

fields through which Carr moved? How was it that some things—such as totem poles—showed up as objects of interest and not others? In her memoirs, Carr (1966, 315) makes the curious statement that "the wild places and primitive people claimed me." But what does it mean to be "claimed" by wild places and primitive people? How does an object claim its observer? Despite the celebrated image of Carr as a solitary woman artist traveling in a remote and hostile environment, the focus by critics and biographers on her eccentricity, and recent efforts to locate an anticolonial optic in her work, there are good reasons to see her vision of nature and culture on the west coast as remarkably conventional, perhaps even more so than those of other, less celebrated travelers.[21]

Perceiving the Discrete Indiscriminately: Emily Carr and the Tourist Gaze

Let me look at this documentary project in more detail. In *Growing Pains,* Carr (1946) traced the origins of her documentary project to a trip through the Inside Passage to Alaska that she took with her sister Alice in 1907. Traveling the popular tourist route, Carr encountered for the first time many of the Native artifacts that were to become her all-encompassing passion. This would lead to additional trips over the next decades, often including side trips off the beaten path. Typically, discussion of these trips has been cited as evidence that Carr, more than any other artist, had an intimate relation to the land and captured the essence of the coast, since they entailed considerable effort and hardship and allegedly took her away from metropolitan culture to the domain of the objects themselves (see Dilworth 1941; Shadbolt 1979).

There can be little doubt that Carr's journeys were out of the ordinary (although not unprecedented for a white woman). Yet, such accounts studiously ignore the very conditions that enabled and shaped possibilities of seeing for white settlers and travelers in BC at the beginning of the twentieth century.[22] If the observer is an *effect* of material and discursive practices, rather than an intentional actor, it becomes important to ask how not only the objects seen, but seeing itself, have a material history (Rajchman 1988). It is also important to bring the *body* of the observer into view, because vision must always be anchored in material practices. In his brilliant *Techniques of the Observer,* for instance, Jonathan Crary (1990) documents the disciplining of modern observers in the nineteenth century, and the visual technologies and scientific practices that were central to this. Crary's key insight—that seeing is an effect of heterogeneous relations—can be extended to the BC coast at the end of the nineteenth century in order to consider the technological, social, discursive, and spatial relations within which subjects were constituted as observers. Crucially, any account of seeing in BC in this period must take on the question of *movement*. Here we might follow Nicholas Green (1990), who, in his superb study of nineteenth-century Paris, traces the role new transportation technologies played in the reconfiguration of vision and landscape in both the city and its environs. The increased appeal to nature and the search for the picturesque, he explains, were not merely a reaction to industrial modernity, but were enabled by new forms of mobility that allowed the bourgeoisie to experience city and country as distinct domains. Indeed, one could argue that this mobility—and constructions of "city" and "country" that they allowed—was not merely something available to the bourgeois subject, but constitutive of emerging bourgeois subjectivities.

Similar dynamics were present in British Columbia during the first half of Carr's life. Cole Harris (1997) has sketched the outlines of a system of transportation and communication that emerged in the province during the first decades of the twentieth century. Drawing on the Canadian economic geographer Harold Innis, as well as Gilles Deleuze and Félix Guattari, he links these emerging networks with the spatial extension of imperial and colonial power. His point is simple yet critical: technological and spatial transformations at the time were not culturally or politically innocent; they *repositioned* people and things. For instance, the time–space compression that came with these networks allowed for capital to colonize space in British Columbia in new ways. From wagon roads to railroads, postal service to telegraphs, paths opened for people, capital, and commodities that were not previously available, enabling new social, economic, and cultural relations and drawing formerly distant places together (cf. Harvey 1989). In short, these transformations *reterritorialized* BC and ushered in new spaces of identity, culture, and politics.

It is worth pausing here to emphasize a key point: there is not one history of space, one of society, another of technology, and yet another of vision; these must be considered simultaneously. This insight has consequence for how we understand Carr's project. To take one example, in the last decades of the nineteenth century, steamship service radically reordered both the geographies and the optics of social life in the region. Service began on the south coast as early as 1836, and in the north by the 1860s, but the volume of travel along coastal routes accelerated most rapidly during and after the 1880s.[23] By the 1880s, regions outside the orbit of Vancouver and Victoria increasingly found themselves within the economic, political, and cultural influence of these cities, and through them connected to east coast centers of culture and commerce. This had a number of important effects. To begin with, it helped produce the "west coast" as a distinct region both in terms of a set of material relations, and also in terms of emerging identities and imaginative geographies.[24] As well, increasing linkages along the coast had the paradoxical effect of strengthening distinctions *between* the city and its surrounding country, even as both were increasingly coupled within a unified space-economy. To say this differently, in the years before Carr's trips, the countryside in BC had become an important resource for its emerging metropolitan cultures (see Williams 1973; Jasen 1995). We should not be surprised, then, to find that for bourgeois travelers, the Inside Passage had become a rite of passage.

The extension of regular and rapid service along the coast altered Native–white relations too. In earlier periods, travel by Europeans had been dependent on establishing close, if unstable, relations with Natives. By the early 1900s, with a modernizing system of transportation and the spatial incarceration of Natives within a system of reserves, the movement of Europeans was increasingly distanced from direct encounters with Native peoples. No longer as reliant on their knowledge and labor, the emerging space-economies of late-nineteenth-century British Columbia often bypassed their villages. As Harris (1997, 183) explains, "Improvement in transportation and communication enabled the world economy to use BC's space not through Native intermediaries, as during the fur trade, but by distributing Western technologies, labour and settlers across the land." For my purposes, it is important to recognize how this allowed Native villages and people to be viewed in a new light. No longer essential to networks of social and economic exchange within which the recognition of coevality was of paramount importance, Indian villages could now be viewed as anachronistic elements of a landscape that was passively viewed without direct contact.[25]

This brings us to a key point: the very way that Europeans looked at landscape underwent significant transformations during the period. In his classic study of railway travel in the nineteenth century, Wolfgang Schivelbusch (1986) argued that new transportation technologies gave rise to "panoramic travel." Before this period, travelers belonged to the same space as perceived objects (what he calls the space of landscape).

New transportation technologies resulted in travelers seeing objects through the apparatus that moved them through the world (geographical space). Modern transportation technologies, he wrote, "choreographed the landscape. The motion . . . shrank space, and thus displayed in immediate succession objects and pieces of scenery that in their original spatiality belonged to separate realms. The traveler . . . acquired a novel ability . . . to perceive the discrete . . . indiscriminately" (60). This involved a degree of detachment and reorganized the hierarchy of senses. The relation of the traveler to the landscape was now predominantly *visual,* a development accentuated by an array of popular and widely available visual devices such as binoculars and cameras that accented the visual appropriation of sights.

Although much slower than trains, steamship service at the turn of the century also remade the BC coast as a succession of sites/sights. This is evident in travel brochures from the period (Figure 5.10). Travel along the coast was now self-contained in what we might call "scenery machines" that moved along the region's passages and inlets. Detached from local intermediaries, and insulated from the hardship of travel, movement was rationalized into a regular itinerary of fixed stops and predictable arrivals. From the comfort of the ship deck one could view nature—and "savages"—as a passing spectacle. The effect of such distancing, of course, was to evacuate specific meanings assigned to places and incorporate them into a mythological space that had little to do with the lived details of the sites and much more to do with a discourse of European modernity. It is not insignificant, I think, that during this period Native villages were far more frequently appraised aesthetically than were non-Native settlements (which did not lend themselves to such detachment) and much more prone to being situated *outside* history in contrast to white communities, in which history was thought to unfold.[26]

By the 1870s, the route to Alaska had grown rapidly in popularity among tourists, many from the United States. Among the first tourists was John Muir, the noted conservationist, who traveled the passage in 1879 and wrote "half-booster, half-scholarly" articles in the *San Francisco Evening Bulletin.*[27] In 1890 alone, almost two decades before Carr headed north, more than five thousand excursionists moved along the route. So extensive was this tourist trade that the itineraries of steamships were altered in order that the most spectacular scenery en route would be passed during daylight hours (Turner 1977).

It is important to situate Carr's movements within these dynamics.

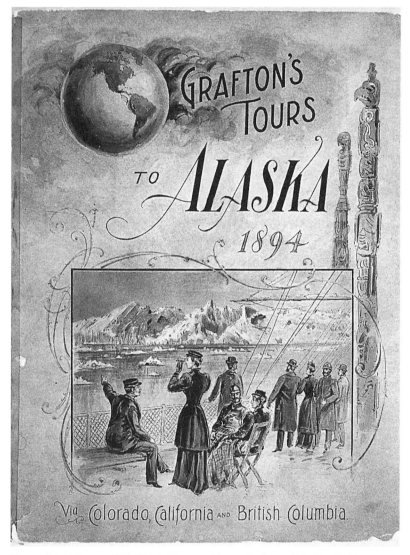

Figure 5.10. Cover of Grafton's Tours guide to Alaska and the Inside Passage.
Source: Grafton (1894).

At the time she and her sister explored the region, as many as thirty-
eight steamships were operating along the coast, with departures from
Vancouver occurring almost daily. This was far from uncharted territory.
Indeed, measured on a per-capita basis, travel along the coast by residents of Vancouver and Victoria was likely far more common in Carr's

time than it is today (adding a measure of irony to the rediscovery of these sites by adventure travelers today). By the 1890s, numerous guidebooks and accounts of the Inside Passage were in circulation. These guidebooks outlined important sites/sights, scripting in advance an iconography of landscape that viewers would recognize as they passed through waters that were in important respects already known in advance.[28] As Badlam's *The Wonders of Alaska*—already by 1891 in its third edition—explained to readers, they could "follow these pages and be fully informed of all the principal points of interest along the Inland Sea, with its innumerable islands, the great resources of this wonderful country, its Native villages, the grandeur of its scenery, the traditions of the Indians, the success of its mission schools and the extension of its civilization" (1891, iv). Nor did these guides spare detail. Grafton's 1894 guide ran to eighty-seven pages, its cover distilling the Inside Passage into two "typical" scenes—glaciers and totem poles—anticipating by several decades Carr's doubled vision of "pristine nature" and "Native culture." The same company explained that its tours offered "an excellent way in which to journey and see strange lands under auspices which will enable [travelers] to learn and see more than if they should go alone. Where to go and what to see has all been arranged. . . . [Travelers] *only have to keep their eyes open and look*" (8; emphasis added).[29]

Clearly, this was a well-disciplined gaze that found certain objects and scenes to be of interest and others not. What travelers saw along the Inside Passage had as much to do with a series of discourses at home, and the organization of the view from on board, as it did with the land and the people themselves. This does not mean that visuality on the coast was singular. Carr, for instance, went to great lengths to distance herself from tourists. In her autobiographical writings, she was cruelly disparaging of her fellow travelers, complaining about their fabricated experience and decrying their destructive impact on Native villages. Her description of Hazelton, BC, was typical:

> Hazelton was a tough little mining town. It had three hotels, rough and turbulent as the rivers. . . . The coming of rail eased travel and gentled Hazelton's hotels. Tourists came. The G.T.P [Grand Trunk Pacific Railway] ran its line close to several of the Indian villages and the tourists looked curiously to see how our Aborigines lived. They did not see real Indian life, the original villages, or the grand old totem poles because, flattered and boastful, the Indians tore down their crude but grandly simple old community houses, built white

man's houses—shoddy, cheap frame buildings. They turned the totem poles that had faced the river, welcoming visitors who had come by canoe up Skeena: they turned them to face the railroad by which visiting tourists now came. They loaded garish commercial paint over the mellow sincerely carved old poles till all their meaning and beauty were lost under gaudy, bragging show off. . . . The Indians hurriedly made baskets and carving too—careless, shoddy things to catch the tourist's eye—which brought in a few dollars but lowered the standard of their handicraft. . . . Everything was changed, cheapened. (Carr 1953, 60–61)

It is tempting to take this as evidence of a countervision. Yet, as much as such fault lines within this visual regime are important, they do not add up to a position of exteriority. Beyond the fact that her statements were often written long after the events she described, and thus may suggest motives retroactively, it is important to remember that criticism of "mere tourism" is often framed in the very terms that structure the tourist experience in the first place: authenticity. Indeed, criticism of tourism was at the time as common as tourism itself, and as W. Jackson Rushing (1995) has shown for the United States during the period, Carr's disparaging attitudes were shared by many other artists who sought to capture the "real" Indian.

For the most part, Carr traveled the same routes as tourists, and her apparent departures were necessarily shaped *in relation to* these routes.[30] One stop on Carr's initial journey to Alaska is illustrative. The town of Sitka was an important station along the popular route Carr followed in 1907. In the years immediately preceding Carr's trip, a "totem-pole walk" had been established for tourists in the forest behind the town. This consisted of poles that had been removed from Native villages along the coast. Indeed, many of these poles were as well traveled as the visitors who came to see them. Prior to their display at Sitka they had been exhibited at the Louisiana Purchase Exposition in Saint Louis, and then again at the Lewis and Clark Exposition in Portland (Knapp 1980; Cole 1985). The irony—that these poles were "repatriated" not by local Natives, but by the white population of Sitka, and then only to be set up *yet again* as objects for tourists—was not lost on Carr. In a later collection of stories, Carr expressed profound disdain for this walk.

When in delicious remembering my mind runs back to Sitka, what does it see? Not the . . . Totem-Pole Walk winding through trees beside a stream named Indian River, named for the pleasing of tourists but really the farthest possible distance from the village of Indian shanties—a walk ornamented at advantageous spots with out-of-setting totem poles, transplanted from the rightful place in front of an Indian chief's house in his home village, poles now loaded with commercial paint to make curiosity for see-it-all tourists, termed by them "grotesque," "monstrous," "heathenish"! . . . Tourists came back from their twenty-minute stopover at Sitka and lectured on Indian totem poles and Indian *un*culture, having seen a few old Indian squaws who spread shawls on the end planks of the wharf and sold curios. . . . Civilization cheapened her Indian. (1953, 80, 82)

Whether Carr held these opinions at the time is unclear. We know that Carr proceeded to paint the scene, suggesting that she had not yet developed her virulent antitourist rhetoric. What I wish to stress, however, is the close affinity between the tourist who travels to Alaska to see the poles (in an experience of "produced authenticity") and the artist who seeks to return them to their "rightful place." What appears as the negation of the tourist gaze is perhaps merely its most pure form. Significantly, it was not the attention paid by the tourist to totem poles (rather than the details of modern Native economies and culture) that Carr decried, but rather the displacement of these poles from their "rightful place" and the application of commercial paint.

We are now in a position to see why understanding Carr's desire to return the artifact to its original site as an explicitly anticolonial project is problematic. Despite what has been read into Carr's statements, her visual interest lay not in the village settings but in the *cultural artifact,* reproducing the same abstraction and commodification of the object that she had experienced at Sitka. Conceiving of her work as a documentary project, Carr went to great lengths to be exhaustive and representative, locating the poles in their "correct" places and faithfully reproducing their detail. Although these paintings do depict some of the details of houses and villages, they remain secondary to the documentation of the artifact, much the way dioramas were used in the same period to authenticate museum displays (Haraway 1989). When people are present, they are often little more than placeholders in a background scene; we learn almost nothing about their activity. The village setting functions merely to inform the viewer that Carr, in contrast to the tourist, trafficked in the "real." In short, as much as she hated the inauthenticity of the tourist spectacle, her love of Indians was channeled along remarkably similar lines. Like the tourist, Carr found that these "awkward objects" had attached to them a "sort of fascination" (Ballou 1890, 228).

Salvage Anxieties and Primitivist Fantasies

Carr (1966, 185) would later write that "a picture equals a movement in space." Although she was referring to her desire to express movement (such that the whole of her canvas would be articulated together and the eye would "swing through the canvas with a continuous movement and . . . not find jerky stops"), the phrase perhaps applies equally to her own movements: one cannot understand Carr's vision apart from the routes she traveled.

Tourists were not the only travelers who followed these paths. By the end of the nineteenth century the Pacific Northwest had become the site of intense activity by anthropologists and collectors. Already in the 1880s, Franz Boas was studying the Kwakiutl settlements that twenty years later provided Carr with a wealth of visual objects. Numerous others also moved along the coast: members of the Geological Survey of Canada such as George Dawson; the Krause Expedition for the Bremen Geographical Society; J. A. Jacobsen for the Berlin Museum; Dr. Sheldon Jackson in Sitka; the Jessup Expedition for the American Museum of Natural History; collectors working independently for the Smithsonian, or for museums in Princeton, Chicago, Philadelphia, and New York; missionaries such as Crosby and Hall; local collector C. F. Newcombe; and many other traders, colonists, artists, and photographers (Cole 1985, Blackman 1992; Jacknis 1992). Many of these collectors and artists would have shared boats with Carr, used the same white contacts in the nearest non-Native settlements, and employed the same Native guides and informants.

Here I wish briefly to trace another related element of Carr's "BC seeing." As already seen in her preoccupation with authenticity, Carr's desire to "picture totem poles in their original setting" resonates with the rhetorics of "salvage," a point readily conceded by Moray and many other critics. Carr's "Lecture on Totems" (1913), delivered at the end of her documentary project, lays this out starkly:

> Soon, soon, the old villages will be a thing of the past. Even in the north it is the same. Settlers are pouring into the new country; silently, gradually the old is slipping away. We hardly know where or how long till we wake to the fact that they used to be and now are not. . . . [The] object in making this collection of totem pole pictures has been to deposit there wonderful relics of a passing people in their own original setting. The identical spots where they were carved and placed by the Indians in honour of their chiefs.

Where Native culture was "slipping away," Carr, like her contemporaries, imagined creating a permanent record in which the anthropologist (or artist) was viewed as the safekeeper of authentic Indian culture. As evident in this passage, Carr's acquaintance with, and careful attention to, Indians was not centered on the ongoing life and adaptation of a people negotiating a modernizing world in which tradition, culture, and identity were being continuously reworked (much as they were for Europeans too). Carr found herself claimed by "the primitive people" and "wild places"; her gaze rested on details of poles that were "relics of a passing race."

It is difficult to see this as anything other than the imperial nostalgia that it was. True, Carr's work may have registered the continuing presence of Native people in the province, but it tended to frame them only as anachronistic relics, and thus may have done as much to reinforce the widely held assumption that Indians had no future. There is little sense of the massive social, economic, and technological changes that were occurring along the coast. Brass bands and gasoline streetlights had no place in Carr's imaginative landscape of Native relics (cf. Crosby 1991). A great deal remained outside the bounds of her "organic" vision. Ironically, Carr's defenders have appealed to the region's *modernization* to justify the absence of people in Carr's images, because the villages that Carr visited were often abandoned by able-bodied workers seeking employment in the province's growing resource sector. Yet, Carr seemed strangely unable to reflect upon this as an essential element of what it meant to "be Indian" during her documentary period.

Carr's salvage anxieties were further inflected by nationalism and regional sentiment. Indians were often viewed by Canadians as distinctive elements of a national heritage; the desire to preserve relics of Native culture was akin to preserving ruins in Europe. Perhaps nowhere was this more apparent than in western Canada. Already in 1892, officials in the provincial government had expressed concern over the loss of what many considered the province's heritage:

> With regard to Indian curios, it is much to be regretted that so many are being exported from the Province. The funds at the disposal of the museum are quite inadequate to meet the high prices demanded by the dealers, so that it is to be hoped all who can will assist in securing rare specimens, and that persons visiting the outlying districts will avail themselves of their greater facilities for bartering with the Indians, and add to the, at present, somewhat meager collection now in the museum. (Quoted in Chorley-Smith 1989, 142)

Despite such concerns, the extraction of artifacts accelerated further toward the end of the nineteenth century. In 1900, Franz Boas, then assistant curator of the Ethnology Division of the American Museum of Natural History, claimed that the museum's Jessup North Pacific Expedition had in its first three years secured 6,600 ethnographic artifacts and nearly two thousand specimens of physical anthropology (Rushing 1995). In 1905, the province responded to this crisis by establishing what was to become the Royal British Columbia Museum, with the intent of keeping at least part of the region's heritage at home.[31] At the time of Carr's journeys, however, its collection was considered incomplete and hopelessly inferior to the extensive collections of Northwest Coast Native artifacts held by museums outside the province (Newcombe 1909).

Like her contemporaries, Carr was alarmed by the export of Native artifacts to locations outside Canada. Influenced by public interest in the museum, the activity of ethnographers, and the export of artifacts, Carr conceived of her documentary project as a lasting record and hoped to interest the province in purchasing the collection.[32] Again, her "Lecture on Totems" is instructive. In her concluding remarks, she connects salvage ethnography with the search for a national prehistory:

> I glory in our wonderful west and I hope to leave behind me some of the relics of its first primitive greatness. These things should be to us Canadians what the ancient Briton's relics are to the English. Only a few more years and they will be gone forever into silent nothingness and I would gather my collection together before they are forever past. (Carr 1913, 53)[33]

Again, although this provides a view into a prevailing current of colonial culture, it is important not to collapse Carr's paintings into her writing, but instead to ask what *visual* form such imperial nostalgia took. We can see this by returning to Carr's work, but also by comparing it with that of others. As already observed, Carr's record was focused primarily on artifacts, not people, although, as Moray correctly notes, people were more present in these images than in Carr's later work. It is equally important to recognize that Carr's vision was widely shared. This shared perspective is clearly evident in the remarkable similarities in viewpoint and subject matter between Carr's painting of a stairway and longhouse in 1912 (Figure 5.11) and a photograph of the same scene by Edward Curtis taken around 1914 (Figure 5.12). Curtis is today known widely for his remarkable twenty-volume photographic/ethnographic catalog of the

"North American Indian" in which he mapped a geography of Indian
types and tribal characteristics (both physical and cultural). There is no
evidence that Carr and Curtis ever met, or that either had seen the other's
work. But in many ways this is beside the point, for what is remarkable is

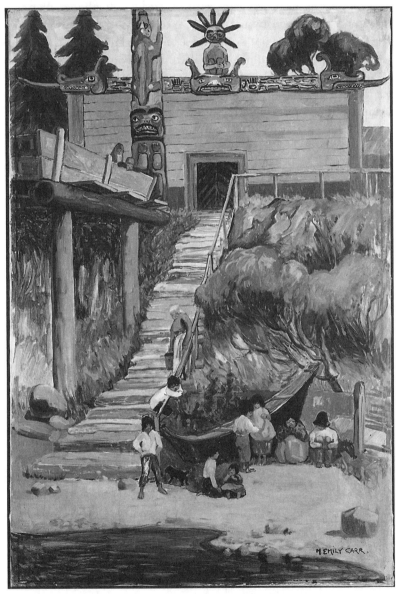

Figure 5.11. Emily Carr, *Memalilaqua, Knight Inlet* (circa 1912). Oil on canvas,
no. 5043, National Gallery of Canada, Ottawa; purchased in 1950. Reprinted
with permission.

the degree to which both found the same objects ready at hand (both materially and ideologically) and the manner in which both apprehended them from almost exactly the same viewing point.[34]

Like Carr, Curtis was concerned to document Indians in their precontact condition (Holm and Quimby 1980; Jackson 1992). The striking similarities between their projects becomes evident in Curtis's (1907, 1) "General Introduction":

Figure 5.12. Tenaktak House, Harbledown Island. Photograph by Edward Curtis (1914). Reproduced from the Collections of the Library of Congress.

Months of arduous labour have been spent in accumulating the data necessary to form a *comprehensive and permanent* record of all the important tribes of the United States and Alaska that *still retain* to a considerable degree their primitive customs and traditions. The value of such a work, in great measure, will lie in the breadth of its treatment, in its wealth of illumination, and in the fact that it represents the result of personal study of a people who are *rapidly losing the traces of their aboriginal character and who are destined to become assimilated with the "superior race."* (Emphasis added)[35]

Perhaps here we can see that Moray's painstaking reconstruction of Carr's viewing sites, by which she establishes the documentary nature of Carr's work in contrast to her later aestheticizing of these artifacts, may actually tell a far different story than the one she intends it to. In her study, Moray mapped the place from which Carr sketched each pole in order to show the care Carr took to accurately represent each totem and the ensemble of totems in their original settings. As Moray explains: "these maps, together with an extensive study of ethnographic photographs in museum archives, and of secondary literature on the sites, have been used to reconstruct what Carr would have seen on her visits and her activity on each site" (1993, 6). Yet, what Carr would have seen was not merely a product of her presence (and diligence) but of her construction as an *observer.* Moray simply reproduces Carr's ideological positioning as a detached observer with a narrow focus on "relics" who at the same time remained quite uninterested in the wider cultural and economic geographies of the region. Hence, these maps, and the historical photographs from which they are based, do not so much attest to an objective scientific interest, and a deliberate anticolonial political intent, as tell us about a visuality grounded firmly in the discursive and spatial organization of the region's postcolonial cultures. Carr's viewing sites were chosen specifically to capture totems. Her viewing position, often on beaches leading to or fronting the villages, reproduced the sovereign gaze of the European traveler surveying the village from afar rather than attending to village interiors. As is today common knowledge, Edward Curtis also went to great lengths to frame Native scenes (Figure 5.13). Carr's perspective was little different (Figure 5.14).

Fortunately, Carr's paintings and Curtis's photographs are not the only images from this period that have survived. On his trips, Curtis was accompanied by an assistant, Edmund Schwinke, who recorded Curtis recording his Indians. Figure 5.15, shot by Schwinke during a break in Curtis's filming, shows a very different world from that which

Figure 5.13. Reconstructing pre-contact Native societies: Edward Curtis in Blunden Harbour, circa 1915. Photograph by Edmund A. Schwinke. Courtesy of the Burke Museum of Natural History and Culture; reprinted with permission.

Figure 5.14. Emily Carr painting in Tanoo, 1913. Province of British Columbia Archives F00254. Reprinted with permission.

was allowed to appear in either Carr's or Curtis's work. Not only do we see Curtis (center left) playing to the camera, swinging on part of a set built to simulate "authentic" culture, but the Native actors hired to portray their culture are fully aware of the staged nature of the project as a whole. Clowning for the camera, an actor to the left of Curtis holds a decapitated trophy head while others have crowned a second head with a spruce-root hat and adorned it with eyeglasses. Far from traditional or the last vestiges of a dying people overcome by a destructive and incomprehensible modernity, Native peoples—like those in this image—found themselves negotiating their own paths of modernization, paths that *included* Carr, Curtis, and the many others who traveled the coast, but paths that were rarely permitted to enter the frame of either artist's official record.

Going Deeper: From Primitive Artifacts to Primal Nature

What relation does Carr's early work have with her later work, and in particular her celebrated forest paintings? Most accounts of Carr's life and work posit a break between these periods, as well as a shift of emphasis partway through the latter period when the rainforest replaced totem poles as her primary object of interest. I wish to call into question

Figure 5.15. Edward Curtis and Indian actors clowning for camera, Blunden Harbour (Curtis is swinging on the railing). Photograph by Edmund A. Schwinke (circa 1914). Courtesy of the Burke Museum of Natural History and Culture; reprinted with permission.

this doubled narrative of rupture by focusing on a number of important continuities. As noted earlier, my intention is not to resurrect the fiction of the authorial subject whose life can be read as a single, unbroken trajectory from beginning to end. But to bend the stick too far in the other direction, such that temporality and discontinuity become the same, is, I think, equally problematic. It is not insignificant that for the remainder of her career Carr retraced the outlines of this original journey, although the specific details of her itinerary changed. Nor is it a minor point that she returned at times to material collected during her documentary project, reworking the canvases and images in different ways as her artistic sensibilities and aesthetic influences changed. This occurred as early as 1912, while Carr was in France, and before her documentary project was complete.

For Carr's interpreters, her trip to France is usually seen as an omen of future changes.[36] Moray (1993, 240), for instance, writes that Carr's years in France "would cause a major shift in her conception of what was significant about Native culture," and that after it her work would shift away from naturalism to a more individual, subjective vision influenced by various modernist movements. Yet, a different point could also be made: that this new aesthetic had as its precondition the "artifactual vision" of Carr's so-called early period in which totem poles and other

Native artifacts had already been significantly decontextualized. Indeed, given her initial concern with authenticity, it is curious that Carr appears to have had no serious misgivings about moving from a documentary objectivity to a more self-consciously subjective vision, to the point of painting over existing works. But perhaps this should not come as a surprise. The Postimpressionist movements that emerged in Europe during the first decade of the twentieth century—Expressionism, Fauvism, and early forms of Modern Primitivism—shared at least one assumption: that the move *away* from naturalism was in reality a move toward a less *normalized* vision and that this, somewhat paradoxically, allowed the apprehension of a more fundamental form and spirit. In the first decades of the twentieth century, perhaps more than at any other time in the history of Western art, artistic expression was seen as an attempt to express a reality that lay behind mere appearances, a search for informing principles that transcended culture, politics, and history altogether (Clifford 1988).[37] Primitivism, in particular, sought after the elemental. In other words, Carr's movement away from description—already evident in France—was likely done in the name of a purer vision.[38] Naturalism only reproduced surface to the disregard of depth; modernism, in its various forms, assumed greater depths beneath surface appearances.

Moray suggests that this more active engagement with modernist aesthetics introduced a tension into Carr's documentary project, but maintains that prior to the end of the project in 1913, she "successfully combined these apparently conflicting goals [ethnography and modernism] into a heightened form of documentary painting" (1993, 412). Yet, if we question the character of this documentary painting to begin with, another story comes into view. European modernisms—with their search for more elemental forms and their appeal to "primitive" sources—may simply have heightened a propensity for displacement and abstraction that was already present in Carr's early work. Only now, rather than seeking to return the tourist object to its original place—a peculiar project on its own—Carr studied the artifacts themselves for formalized rules of form and composition and as the expression of deeper spiritual forces.

This displacement and abstraction of the Native artifact would be accentuated a decade later. After her attempt to sell her documentary project to the provincial museum failed in 1913, Carr entered a less productive period, during which she ran a boarding house in Victoria. This so-called middle period ended in 1927, with Carr's inclusion in a National Gallery exhibition titled *Exhibition of Canadian West Coast Art—Native and Modern.* The curator of the National Gallery, Eric Brown, had been introduced to Carr's work in the early 1920s and in 1927 asked Carr to submit works to the exhibition. These were displayed alongside Native artifacts—masks, painted hats and oars, ceremonial robes—as the further refinement of a uniquely west coast, albeit primitive, aesthetic. As numerous critics have noted, the exhibition not only enrolled Native art as a resource for cultural nationalism, it established Carr as a mediator between this primitive past and the nation's modern present.[39] Scott Watson (1995, 62) summarizes this succinctly:

> In order to create an authentic, modern art, reasoned the early modernists, Canadian art must have a rooted relationship to place, it must assimilate the Native arts of what was Canada before Europeans came—Canada's "timeless" past. In turn, those arts must be contextualized as high art within a modern discourse about art; their purely aesthetic qualities must come to the fore. Native art was accordingly stripped of local meaning and placed within the horizon of universal expression and timeless form.

Similar dynamics prevailed south of the border (Rushing 1995). Although eastern Canadian art critics were clearly responsible for placing Carr in the role of mediator, it is important to note that they did not magically conjure up this image of Carr, nor was Carr's work an accidental nominee for the position: her depiction of totem poles and house posts fitted the cultural nationalists' agenda well.

For my purposes, this show is significant in two respects. It provides insight into the kind of interpretations Carr's early works could bear, and lends support to the argument that, despite Carr's stated intentions (which were, at best, contradictory), her early depictions of Native artifacts were markedly decontextualized, even if they returned the objects to their village settings. The show was also a moment of revitalization in Carr's career as a painter. From this point on, Carr would receive considerable acclaim, and would have much more frequent contact with eastern artists and cultural elites. This in turn would shape her work in important ways. Lawren Harris acquainted Carr with theosophy and with the transcendentalism and vitalism of Walt Whitman. Other Group of Seven painters provided inspiration or competition.[40] After viewing A. Y. Jackson's Skeena River paintings, for instance, Carr began to realize the many ways BC landscapes could be approached through a far more aestheticized and stylized vision, and began to accept her designation as BC's member of the Group of Seven:

> I loved his things, particularly some snow things of Quebec and three canvasses up Skeena River. I felt a little as if beaten at my own game. His Indian pictures have something mine lacks—rhythm, poetry. Mine are so downright. . . . I worked for history and cold fact. Next time I paint Indians I'm going off on a tangent tear. There is something bigger than fact: the underlying spirit, all it stands for, the mood, the vastness, the wildness, the Western breath of "go-to-the-devil-if-you-don't-like-it", the eternal big spaciousness of it. Oh the West. I'm of it and I love it. (1966, 5)

The totem poles and other artifacts that preoccupied Carr in her earlier work remain present—at least until the early 1930s—but they are now seen in new ways. Influenced by Whitman's vitalism, Carr brings a new attention to movement, space, and spirituality. The totems lose almost all detail as they become monumentalized and metaphorical (see Figure 5.1).

Between 1928, when Carr again traveled north to many of the sites that she had visited earlier, and 1932 when she shifted her gaze to the forest, she no longer conceived her work as a "record." As Moray, Shadbolt, and others have noted, she instead found in the totems expressions of what she believed to be the insights and values of Native culture, and also resituated these artifacts as symbols of the *spirit* of the Canadian west. Shadbolt's summary of this period brilliantly—albeit unwittingly—captures the significance of this shift, while reinforcing the conventional view that these works were the culmination of Carr's *intimate knowledge* of Native art: "Her understanding of Indian art is not in fact reflected in her work until after 1927, when she strips the poles of *excessive detail,* removes them from *distracting settings* and concentrates on their *sculptural strength* and *expressive energy*" (1979, 30; emphasis added). In these last Indian paintings, villages lose all specificity, often simply incorporated into the enclosing forests, such that their totems are surrounded only by a threatening wilderness, stripped of any cultural context whatsoever (see Figure 5.16). Increasingly, Carr drew the totem and the forest into the same horizon of meaning. Native cultures—reduced to the expression of a vague primeval spirit—appear to emerge from nature itself. Shadbolt (1975, 33) codifies this reading of Carr's work, but her canvases bear it well:

[Carr had] long since known and felt the art of the Indian. . . . there comes a moment of insight when she sees its profound and inevitable relationship to the environment in which it developed. *The totem poles literally come out of the giant trees.* But as brooding silent watching presences they involve a more profound generic relation to the hostile world of the rainforest. (Emphasis added)

Shadbolt claims that this was the moment when Carr recognized the essential and eternal bond between Native and nature such that the artifact now stood metonymically for a more original *natural* identity, common to all humanity, but most fully expressed by primitive peoples. Thus, Shadbolt argues, Carr recognized in her last Indian works and subsequent forest paintings that "the Indian carved into his pole the same silences, mysteries, orifices and excrescence as nature has in hers" (1975, 36). Native life mirrored nature, and the totem poles, Shadbolt implies, were merely signposts leading Carr deeper into the primeval forest, revealing a domain "vast and rich enough to provide [Carr] with pictorial metaphors for all the expression she wished to express."

Shadbolt's view no doubt reflects what Neil Smith (1996) describes as the almost "instinctive Romanticism" that pervades North American culture. But similar views were prevalent in Carr's time too. Maurice Barbeau, an acquaintance of Carr and curator of the National Museum of Man, argued that Indian art found its source in nature itself (see Nemiroff 1992). Indeed, although unreflexive about the terms of her analysis, Shadbolt is not far off the mark. Already in Carr's work from the early 1930s, totem poles are depicted as an element internal to nature. The irony hardly needs mention: in her later works, Carr placed the totem back in the forest, reproducing the same decontextualization that she so stridently decried in Sitka. At times they are simply pictured at the moment of their reclamation by nature. In these paintings, totems lose their verticality, overwhelmed by growth. They are, finally, vanquished, as much by nature as by the cultural violence of colonialism. Eventually, Indian themes disappear entirely from Carr's work, replaced by the primeval forest.[41]

That Carr comes to the forest *through* the figure of the Native, and ultimately its erasure, does not mean that the attention her forest paintings have received is undeserved. Shadbolt (1979) finds her forest images to be "dramatic," "vital," "poetic," and "mystical," adjectives repeated often by critics and laypersons alike. Others find in her works the expression of social protest, reading them through the lens of late-twentieth-century

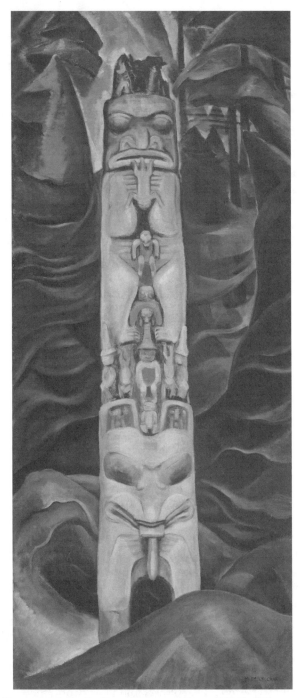

Figure 5.16. Emily Carr, *Totem and Forest* (1931). Oil on canvas, Vancouver Art Gallery, Emily Carr Trust VAG 42.3.1. Photograph by Trevor Mills. Reprinted with permission.

environmentalism (Linsley 1991, 236). We have scant evidence that Carr intended her forest paintings to be statements of environmental protest—the language of ecosystem ecology, after all, was not available at the time—nor do her writings support such a view.[42] But this does not mean that her paintings do not allow such interpretations. Indeed, there is merit in the view that Carr's rainforest paintings disrupted the objectifying gaze of extractive capital. A number of critics have noted that Carr's forest is highly eroticized (much like Georgia O'Keeffe's landscapes). Robert Linsley (1991) usefully reads this eroticism in two ways. On the one hand, the paintings reveal that Carr took "an obvious pleasure in the sensuous plasticity of paint" (230). But more than this, he suggests that "her sexualized landscapes could be read as a literal loving of nature, and as such an expression of an antitechnocratic, life-affirming critical position that still has the capacity to offend bourgeois proprieties. . . . Carr's work opens up . . . the dream of reconciliation, of closing off that distance from the exploitative relationship to nature" (231, 232). By this view, it is partly through their eroticization that Carr's forests are given value *apart* from their commodification as timber, and that they contain the potential to "shock" the viewer into seeing the forest differently.

The relationship between landscape, vision, gender, and sexuality has been well rehearsed by feminist scholars (see Kolodny 1975; Pratt 1992; Rose 1993; McClintock 1995). For some, the gendering—and sexualization—of nature is implicated in its domination and destruction within patriarchal and capitalist social formations (Merchant 1980). Although fiercely debated within feminist scholarship, it has also been asserted that women are somehow closer to nature (through either biology, ideology, or livelihood).[43] Carr's work has occasionally been approached through this lens too, as expressing a relation to nature that only women have privileged access to. Other interpretations, in contrast, find in her forest paintings the expression of a repressed sexuality.[44] Shadbolt (1979, 140), for instance locates here a "sublimated eros" and "imagery of strong sexual connotation": hollows and openings in the woods, phallic poles, stumps and tree trunks, openings and enclosures that "vibrate" with light and movement, trunks that "thrust" upward into the sky, an earth that "fecundates," and so on.

For Shadbolt, this becomes yet further evidence that these forest paintings reveal Carr "painting out of her deepest self," expressing "primary content." Rather than explaining Carr's sexualized imagery—and thus her forest imagery—as an expression of a more *essential* Carr (consistent with Shadbolt's narrative of Carr's development from an early to a mature artist whereby her art eventually reached its "full range of effectiveness"),

I wish to emphasize the ambivalence of Carr's gendered and sexualized imagery, and raise questions about what Shadbolt's analysis leaves out. That Carr's imagery is both gendered and sexualized is surely true, and in a social and economic context where extractive capital and patriarchal relations were hegemonic, this was undeniably provocative. Although the interiors of her forests appear dark, mysterious, and veiled—familiar tropes of feminized nature—hers is not a nature that passively awaits its unveiling before a penetrating and possessive masculinist gaze. Few of Carr's forest images invite possession by the viewer. In examples such as *Cedar* (Figure 5.2), they are viewed from inside, from a vantage point that precludes grasping them in their totality. More often than not, her forests appear impenetrable, to the point where they threaten to engulf the viewer. Indeed, the images can effect a paranoid anxiety in viewers, undermining their ability to master space.

Viewed in light of the ecological—and social—catastrophe of much industrial forestry, it is perhaps justified to find in Carr's images a utopian statement of hope. Yet, such interpretations fail to attend to other important aspects of these works. Specifically, they ignore the histories of seeing that brought Carr to the point where the forest could stand for primal energies and forces, and, moreover, could *stand apart* from its cultural surrounds. Lawren Harris is often credited with prodding Carr to end her fascination with totems and move to the more *elemental* forces of wild nature of which the totems were only an expression. In one of the most quoted letters in Canadian art history Harris wrote to Carr that

> the totem pole is a work of art in its own right and it is very difficult to use it in another form of art. But how about seeking an *equivalent for it* in the exotic landscape of the Island and coast, making your own form and forms within the greater form. Create new things from the landscape." (Quoted in Shadbolt 1979, 76; emphasis added)

That Lawren Harris and Emily Carr could find in nature an equivalent for the totem pole suggests that, for both, nature belonged to the same discursive and visual economy as did Indians: one could be substituted for the other. Here it is important to recall my earlier discussion of Carr and the tourist/ethnological gaze. Once Native culture is staged as a natural culture—the artifacts no longer part of the life of a modernizing people, but symbols instead of the Indian's unity with nature—it is a simple matter to conflate the two. By purifying Native culture—emptying it of any foreign or contaminating modern elements—the nature of which it is a mere element itself appears pure. Native culture (past) and nature (present) become the same, the former simply an expression of the latter's temporality. There is no need, therefore, to mark a contemporary Native presence in a natural landscape.

What must be stressed is that Carr's forest paintings—and not only her images of Native artifacts—retain aspects of a colonialist visuality. This differs from other accounts that understand Carr's shift to the forest as the result either of an inner journey (whereby her journey into nature mirrors her increasing maturity as an artist) or of external influences (her immersion in Walt Whitman, her introduction to theosophy, the prodding of Harris, and so on) that caused her to abruptly change course. To be sure, these influences were important. Yet, they did not lead Carr to abandon a love of Natives for a love of nature. The latter was already anticipated in the former. The effect of these accounts has been that Carr's forest paintings have been treated in very different ways from her Indian works, and often discussed as if unrelated to them. With the rhetoric of *rupture* written twice over in the accounts of critics, historians, and biographers, it is easy to lose sight of those elements in her first Indian paintings that remain as irreducible traces in her celebrated rainforest images. If we lose sight of these traces, we lose sight of the relationship between our present-day "environmental imaginaries" and the colonial histories of seeing that remain embedded in them.

Painting Land Claims: The Salvation Art of Yuxweluptun

Carr's forest paintings are among the most celebrated images of west coast nature in Canadian art and culture. It is not, I suspect, a matter of mere coincidence that we find echoes of Carr's vision in contemporary forest images. Recall for a moment the two photographs discussed in chapter 3 (Figures 3.5 and 3.3): a single totem lost in the forest; the forest's mysterious interior. Or consider the ecotourist's search for ruins in the primeval forest. Is this simply what is there to be seen, or is it the echo of a colonialist visuality that continues to reverberate in the postcolonial present?

I want to end this chapter by returning to the exhibition with which I began. Since the mid-1990s, the work of Yuxweluptun (Lawrence Paul) has received considerable critical attention, in part because his canvases and sketches so thoroughly disrupt conventional ways of seeing on Canada's west coast. Paintings such as *Scorched Earth, Clear-cut Logging on Native Sovereign Lands, Shaman Coming to Fix* (Figure 5.3), for instance, are undeniable challenges to the legacy bequeathed not only by west coast artists who preceded Yuxweluptun, but by a broader colonialist visuality that has framed the region for so many Canadians. I will return to this painting momentarily.

First, let me add a note of caution. Juxtaposing Yuxweluptun and Carr, as Andrew Hunter did in his Vancouver Art Gallery exhibition in 1997, and as I do here, carries certain dangers. For instance, it risks setting up Yuxweluptun's "BC seeing" as a corrective to Carr's, as if it were somehow less ideological in character. Such a view is not intended, and runs against the grain of my argument. If we understand visuality as always already *historical,* then we must understand Yuxweluptun's work also in terms of its enabling conditions (of which one is, undoubtedly, the *experience* of colonialism and its displacements). It is also important that to understand Yuxweluptun's rise to prominence as being about more than the artistic qualities of his work: after years of Native road blockades in support of land rights, attacks on the neutrality of art in Canadian art criticism, and concerns over Eurocentrism in Canadian culture, the stage was set in the 1990s for reappraisals of the nation's most cherished images and for a search for new ways of seeing.[45] Perhaps more problematic is that Hunter's exhibit risks reducing Yuxweluptun's art to a reaction to colonialism, paradoxically recentering European colonialism as that which most defines the experience of contemporary First Nations peoples. This would be a grave mistake. Although attending to post-coloniality means recognizing the ways in which places and cultures have been "worked over" by colonialism, it does not mean that the experience of the colonial subject is reducible to colonialism alone. Indeed, one of the troubling aspects of postcolonial studies has been its tendency to privilege subjects and practices that best fit its theoretical interests in hybridity and transculturalism, while other practices that do not so readily reflect the effects of colonial encounters are left unexamined. Certainly, one of the appeals of Yuxweluptun's work—along with other Native and American Indian artists such as Carl Beam, Teresa Marshall, James Luna, and Hachivi Edgar Heap of Birds—is that it fits the postcolonial slot rather well (much as Carr's work fitted an anxious nationalism in the 1920s). This is not to discredit the work of these artists; rather, it is to keep in mind that there are *other* ways of "being Native" that tend to escape critical commentary.[46]

A final, related problem with comparing Yuxweluptun and Carr is the risk of reducing visuality on the coast to a simple Native/white binary. This not only sets up a dubious opposition, but erases important differences *among* Native and Euro-Canadian artists (not to mention other artists who, in a thoroughly multicultural province, do not fit this binary frame).[47] Yuxweluptun's work can be made representative of Native art only with considerable violence to the diversity of artistic production among First Nations artists, repeating Carr's tendency to collapse different Native groups into a single, undifferentiated Indianness. "Being Native" is not a fixed site; it consists of heterogeneous and shifting locations that must be understood in their relationality and in their singularity. Likewise, it would be a grave mistake to understand Yuxweluptun's art as purely *oppositional.* This would merely recapitulate an old nostalgia for a place of pure difference, as if it were possible to speak, write, or paint from a location that was somehow external to the social, cultural, political, and technological relations and dislocations of postcolonial global capitalism. Yuxweluptun is, after all, a graduate of the Emily Carr College of Art and Design, and among its most illustrious. Yuxweluptun makes no such claims of exteriority for his art, even as he insists on its political intent and on the unique perspective that he has as a First Nations person who lives in the long shadow of colonialism. Rather than oppositional, I understand Yuxweluptun's work as *differential,* in the sense that it disrupts or refracts a colonialist visuality, not in the name of returning to a mythic place prior to the toxic neo/post/colonial environments that he so brilliantly depicts, but instead from within these heterogeneous and uneven social and visual fields, in order to take the present as a point of departure for imagining more socially just productions of nature.

With these caveats in mind, let me turn to Yuxweluptun's work.[48] Like Carr, Yuxweluptun seeks to depict the "truth" of west coast landscapes. But, unlike Carr, who sought to escape history in order to capture the region's underlying essence, Yuxweluptun (1995, 1) understands his "BC seeing" as irrevocably historical: "if I am to talk about my work, I will have to start with the past." This statement offers several readings. In one respect, it points to biography and thus to Yuxweluptun's unique experience as a First Nations person. Born of an Okanagan mother and a Coast Salish father, Yuxweluptun grew up an urban Indian, living in Kamloops and in the Vancouver suburb of Richmond.[49] Both of his parents were heavily involved in Native politics, not only locally, but intraprovincially, with the result that Yuxweluptun was not only exposed to the complexities of Native–white relations in the province, but also to the diverse experiences of different First Nations communities across the province.[50] Thus, in contrast to the sedentarist metaphysics (Malkki 1997) that has so dominated Western representations of non-Western Others (which assume that the identity of the

Other is bound to place in contrast to the mobility of the European subject), Yuxweluptun's life has been lived *between* places, and *between* cultures (see also Gupta and Ferguson 1997). This mobility is mirrored in his artistic training. Avoiding schools that taught Native artists traditional methods and techniques, Yuxweluptun attended the Emily Carr College of Art and Design from 1978 until 1983.[51]

But the matter of history can also be understood more broadly. Yuxweluptun's statement alerts us to the fact that the reality he paints is *historical*: he paints the legacy of colonialist practices both in the past and in the present. To understand the present—and his experience of "being Native"—he has to attend to the histories of colonial governance that incarcerated First Nations on postage-stamp–sized reserves, attempted to disrupt Native cultural practices, and mapped a web of disciplinary practices across the spaces and bodies of aboriginal peoples. Every brushstroke in Yuxweluptun's work is therefore a measured and reflexive encounter with the changing configurations of Native lands and cultural identities as they are lived and experienced today—differentially—by First Nations peoples across the province. Thus, when he speaks of his work, he speaks not of color, line, or style, but of genocide, despotism, and concentration camps; not of an inner vision or of surrealism as a modernist aesthetic but of a life that was *made surreal* by the administrative apparatus of a colonialist state; not of primeval nature but of a toxicological colonialism (Yuxweluptun 1995).

What is of interest to me, then, is the manner in which Yuxweluptun asks us to imagine other paths through the complicated terrain of vision and landscape in postcolonial British Columbia. Evident immediately, for instance, is his refusal to separate art from politics. In interviews, he has referred to his work as "salvation art" (see Townsend-Gault 1995). Salvation, in this instance, is both multivalent and paradoxical. It plays on a long history of efforts by missionaries to drive a wedge between Native people (especially children) and their cultural and religious traditions. It also appropriates the teleology of salvation history, although with a final destination somewhat different from the one conventionally imagined. It also, not unlike Emily Carr, points to an enduring *spiritual* relation to the land among First Nations peoples, albeit a spirituality that does not collapse natives into nature, but instead exists as a resource for efforts to heal an environment exploited by industrial capitalism.

At the heart of Yuxweluptun's art, though, is the matter of land. As Charlotte Townsend-Gault (1995) has summarized in a pithy statement, Yuxweluptun does not paint land*scapes,* but land *claims.*[52] Key in this effort has been the rearticulation of nature as a cultural rather than purely physical artifact, and thus a domain crisscrossed by already-existing political claims. Here the distance between Carr and Yuxweluptun is most pronounced. If Carr purified coastal landscapes into two elements—a decaying Native culture and a brooding, spiritualized, all-encompassing nature—Yuxweluptun does the opposite. Nature in his work does not overwhelm a vanquished culture, nor is Native culture collapsed *into* nature in such a way as to erase the violence of colonial dispossession. Rather, so-called natural landscapes are assembled out of elements of Northwest Coast Native culture.

Yuxweluptun's *Scorched Earth* (Figure 5.3) is illustrative of these multivalent aesthetic and political gestures. We can read it, for instance, as an ecological critique of the violence of industrial forestry. The tree stumps in the foreground, tracing a line backward to denuded slopes, and the brown squares of clearcuts in the distant hills speak of the violence wreaked on the forest. Indeed, both the land and the sun weep in the face of this travesty, as does a figure resting on one of the distant hills. But this image articulates much more, for Yuxweluptun does not simply paint a "green" critique, but asserts Native *rights* to a land being destroyed (Townsend-Gault 1995). The various components of the landscape—trees, mountains, the sun, even the tree stumps—are constructed out of Coast Salish motifs. *There is no nature apart from these.* Whereas for Carr Native culture was merely a fleeting expression of nature, for Yuxweluptun nature is an artifact of culture.

We see this again in *Clayoquot* (Figure 5.17), where clearcuts are depicted as gaping wounds in a cultural landscape, the square spaces of the rationalized forest gouged into hills formed from Coast Salish ovoids.[53] Even the tree stumps are constructed from Coast Salish ovoids in a clear claim of sovereignty over territories already worked over by industrial colonialism. The presence of the shaman in *Scorched Earth* stakes out a similar position: any future healing of this wounded earth will hinge on the recognition of Native land rights and Native cultural and environmental traditions.

There are additional complexities to Yuxweluptun's work that deserve brief mention. Earlier I noted that Carr envisioned her early documentary project as an attempt to establish a permanent record. Yuxweluptun claims to do something entirely different: "my work is *to record*" (quoted in Townsend-Gault 1995, 12; emphasis added). The shift is subtle, but the implications are immense. Whereas Carr's desire to produce a record assumed a static culture threatened by the forces of modernity, Yuxweluptun sets out to record ongoing struggles over identity and land that exceed the traditional/modern dichotomy. There is an important lesson in this for any progressive, antiracist environmentalism. For Yuxweluptun, Coast Salish culture is not something fixed and therefore under threat from a colonizing modernity. Modernity does not arrive from the outside like a stealthy intruder insinuating itself into the

Figure 5.17. Lawrence Paul Yuxweluptun, *Clayoquot* (1993). Reproduced with permission of artist.

cultural traditions of the Other. Rather, the traditional is always already dynamic, a shifting constellation of ideas, forms, and beliefs that are resources for building alternative modernities at once continuous with, but not imprisoned by, the past.

This is also, I think, the way in which Yuxweluptun's so-called borrowings from surrealism must be approached. Much has been made of the artist's debt to Salvador Dalí, but the matter is more complex. To be sure, Yuxweluptun finds in surrealism a method that expresses the fantastic nature of life as a First Nations person; operating under the sign of reason, European colonialism bent the coordinates and lines of Native culture, distorting the real in very material ways. As he puts it, "my reality was surreal" (quoted in Townsend-Gualt 1995).[54] Again, this can be pushed further. To speak of his use of surrealism as a "debt" is to forget the *prior* borrowings that lie at the heart of European modernisms (see Townsend-Gault 1995, 1999; Watson 1995). If we reinsert into the story the circulation of Northwest Coast Native artifacts in European culture during the late 1800s and early 1900s, we can see European modernisms as enabled at least in part by the jolt that these objects gave to conventional ways of seeing. Indeed, it is only through an act of selective memory that modernist aesthetics has been made out to be a development *internal* to Europe. To speak of Yuxweluptun's surrealism as something borrowed merely reinforces a story centered on the extension of European modernity across a global stage in which the only actor is the West writ large. Indeed, such an assumption can be surreptitiously smuggled back in through the much-hyped trope of hybridity. Yuxweluptun makes a much more radical *proprietary claim* to modernism itself, repatriating what had been earlier appropriated.[55]

Crucially, though, surrealism does not return to the west coast without remainder; it has been reworked in the meantime. To reclaim modernism by way of Dalí is in itself a provocative move, even more so given fierce battles over what counts as, and who can speak for, Native tradition among First Nations and non-Natives alike. This issue infuses Yuxweluptun's work, much as it did Carr's. His efforts to destabilize the tradition/modernity distinction have led him to inscribe Coast Salish motifs onto hot dogs and automobiles, "queering" conventional notions of Native culture. Again, this is serious play. It engages demands from white society that Natives be traditional in order to preserve their status as Native, but it also places in question efforts to recover cultural traditions among First Nations. His solution is to recode change as intrinsic to Native cultures, and thus a sign of vibrancy rather than decay: "I work from the native perspective that all shapes and any elements can be changed to anything to present a totally native philosophy. It allows me to express my feelings freely and show you a different view" (Yuxweluptun 1995, 1).[56]

This holds an important lesson for any antiracist environmentalism that seeks to move beyond the instinctive romanticism that pervades North American culture. Not only do many environmentalists posit an equivalence between Natives and nature, they also tend to foreground Native activists who properly fit this image. Yet cultural traditions—including environmental practices—are fiercely contested *within* Native communities. If, within art criticism, figures such as Yuxweluptun have come to the fore, in environmental circles they remain firmly on the margins. The lesson here is that environmental imaginaries among First Nations are as historical and contested as among non-Native Canadians.

Ultimately, Yuxweluptun provides us with a differential optic, one that displaces rather than inverts colonialism's binaries. The Native reality that he paints is traditional but not static, modern but still Native. West coast landscapes are ecological, but cultural all the same. His "salvation art" does not seek an idyllic past, it points to an open future, one no longer constrained by a national(ist) landscape imaginary that relegates First Nations to the margins by placing them "in" nature. His is only one voice among many; his articulation of culture and nature is as unique today as Carr's was more than a half century ago. And, like hers, his "BC seeing" is as embedded in specific historical conditions. Indeed, from the perspective of a radical environmentalism, the value of his work may turn out to be not only its specific recoding of the landscape, but its timely reminder that First Nations speak in many different tongues. Examined more carefully, the "natural" Native—a figure so cherished in Canadian culture—refracts into multiple positions. To collapse this plurality into one merely repeats an earlier colonial violence.

Within the fiercely contested spaces of Canada's west coast, learning to see differently—and to see difference—is critical for new imaginings of nature, culture, and nation. Artists such as Yuxweluptun will no doubt play a key role in this task. Seeing differently, however, does not mean rejecting artists such as Carr. It demands taking her "seeing" seriously. Rather than mystifying Carr's work as transcendental, and then placing it unexamined at the very center of Canadians' imagined geographies of the west coast, her work must be understood as a potent site where what fits within a west coast modernity is at stake. For a progressive postcolonial environmentalism, rethinking these cherished images is an urgent task.

6. Picturing the Forest Crisis

Immutable Mobiles, Contested Ecologies, and the Politics of Preservation

Canada's forests are disappearing.

—Greenpeace, *Cutting Down Canada*

We need . . . to look at the way in which someone convinces someone else to take up a statement, to pass it along, to make it more of a fact.

—Bruno Latour, "Visualization and Cognition"

Into the Clearing

In the late 1930s, Emily Carr produced a series of paintings set in the landscapes surrounding Victoria. By this time, large tracts of timber in the region had been felled, leaving gaps in the forest. Painting in these newly opened spaces, Carr turned her attention away from the forest's interior to the region's expansive and expressive skies. Today, these images of forest clearcuts are often interpreted in terms of a wounded landscape, and Carr is seen as among the first to call attention to the devastation wrought by intensive forestry. Whether Carr understood her project in these terms is not clear. Although her journals and memoirs occasionally refer to forestry as a kind of violence, such statements are few and scattered. She more frequently spoke of this period as one in which she was seeking to capture *space* and *movement.* For this, forest clearings were ideal. Indeed, if these paintings were meant to be environmental statements, there is little indication that they were taken up as such by the public. This should come as little surprise. In a province built on the exploitation of natural resources, clearcuts and smokestacks represented progress. At the time, images of industry were as likely as images of nature to appear on postcards. Further, the forest resources of the province were considered limitless. Only foresters thought otherwise, and it was primarily among them that a new vocabulary was emerging—"forest inventory," "sustainable yield," "mean annual increments"—that would eventually inform Justice Sloan's watershed report, and would be in part responsible for the new social and spatiotemporal logics of forest management that were set in place at midcentury (cf. Mulholland 1937; Sloan 1945). Nor was there a language readily available by which to understand forestry as destruction. For Carr's images to become *environmental* statements required that they be wedded to different concepts and metaphors—"webs of life," "habitat," "biodiversity"—and these became available only with the wide dissemination of ideas from ecology in the 1970s.

The changing fortunes of Carr's paintings of forest clearings call attention to wider shifts in the discursive terrain of BC forest politics. That Carr's canvases are taken today as political, and not just aesthetic, statements is in part because they now appear to illustrate a larger truth about the forest's "disappearance." This chapter considers how this statement— "the forest is disappearing"—has been constructed, taken up, passed along, and made a nexus of international political mobilization. As we will see, this simple statement opens up a Pandora's box in which nature, culture, science, and politics are tangled together in a tight knot.[1] It also takes us to other, even thornier questions about whether this is the most pressing story, or whether there are other stories about the fate of the forest that should be told instead. To explore these questions, I will again begin with an artifact from BC's "war in the woods."

The Disappearing Forest

Perhaps no other image has had as great an impact on BC forest politics as one that was produced and circulated by two environmental groups— the Wilderness Society and the Sierra Club of Western Canada—in the early 1990s (Figure 6.1).[2] The image consists of what appear to be two satellite images. The first, located on the right, is dominated by dark colors (forest green in the original), and shows the extent of the "ancient temperate rainforest" on Vancouver Island in 1954. On the left, a similar image shows the remaining ancient forests in 1990: a fragmented, discontinuous patchwork, surrounded and dissected by "modified landscapes" (pale yellow in the original). Other shades cover small sections of each image, and a legend explains that these refer to bog forests, ice, urban areas, and so on. Placed together, the two images tell an alarming story of the "liquidation" of the rainforest. The yellow of humanly modified landscapes, like a blight spreading across a field, has insinuated itself across almost the entire extent of the island. Clearly, the forest is *disappearing*. Where in 1954 only the eastern shores and southeast corner

VANCOUVER ISLAND ANCIENT TEMPERATE RAINFORESTS

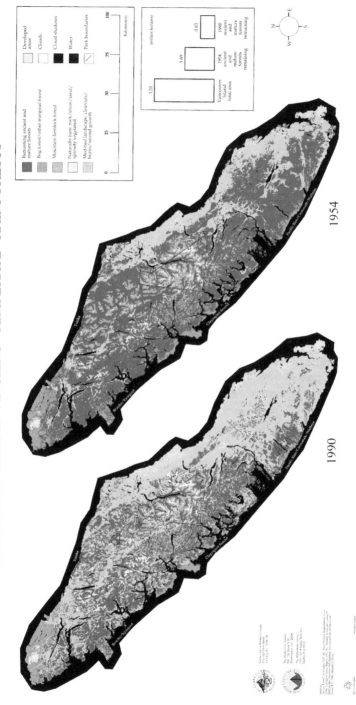

Figure 6.1. The disappearing forest. Reprinted with permission of Sierra Club of British Columbia.

of the island showed extensive modification, by 1990, what patches of green remained appeared as ever-shrinking islands in a sea of yellow— sad relics of a once extensive forest laid bare by the logger's ax.[3]

It is difficult to look at this image and not be moved. Indeed, it became as powerful an actor in BC forest conflicts in the 1990s as any politician, forest executive, or activist. When a slide of this image was shown at a 1993 rally in Vancouver, an audience of several hundred people erupted into spontaneous applause. Here was an image that established, irrevocably, the truth of what activists had been saying for years: the forest on Vancouver Island was vanishing. Without this image, environmentalists were comparatively weak. They could only point to *instances* of the forest's destruction, a strategy easily deflected by statements about the forest being an inexhaustible or renewable resource. Who could argue? One had only to look outside the window to see the green of the forest. What this image did was place environmentalists in a new position of *strength*. No longer a ragtag collection of local residents concerned about the incursion of industrial forestry into their backyards, the image consolidated the position of the BC environmental movement as guardians not of this forest or that forest, but of *the* forest. The image gathered all the scattered bits of the forest on Vancouver Island and turned them into a single whole. The truth of the forest's "destruction" was now visible: after years of struggle, nature (finally!) was on the side of the environmentalists. Until then, it had always lined up on the side of the foresters, who only had to point out their windows to where the forest stood.

There is a second, equally important reason why the image evoked applause: it is an *apocalyptic* image. If this was the forest speaking its truth, then it spoke in the language of crisis. Placed side by side, the images suggested a teleology. Indeed, viewers were implicitly invited to add two missing stages. The first, situated some time *prior* to the 1954 image, would have shown the "original" forest that supposedly existed before human (i.e., European) intervention. In this image, the island would appear an unbroken sea of green. The other, placed sometime in the (near) future, would complete the tragic narrative. Here the island would appear uniformly yellow. Lest there be any doubt about the moral of this tale, the guide published with the image makes the message clear:

> If logging continues to occur on Vancouver Island at historical levels, the Island's entire unprotected forest landscape will be completely denuded of its ancient temperate rainforest by the year 2022. Effectively,

the end of the ancient rainforest will occur much earlier, when the last of the remaining intact watersheds are opened to development. . . . Of the 174 primary watersheds in southern British Columbia larger than 5,000 hectares in size only six remain completely undeveloped today. (Sierra Club of Western Canada 1991, 4, 6)

In the face of an apocalypse of such magnitude, who could suggest anything else but to preserve what remains?

Apocalyptic environmentalisms are not in themselves wrong. In some cases, they risk presenting a sense of inevitability and thus promote quiescence, but they can also provide powerful reasons for political mobilization. It is important not to lose sight of the stakes involved in images like these, even as we explore their construction. Forestry is *not* benign, despite the claims of industry and the thousands of seedlings forest companies plant each year to replace the trees harvested. Indeed, in a province where information about forestry has often been difficult to obtain, images like these allow viewers to "see" spatial and temporal patterns, providing an effective—and alarming—counterpart to the comforting images of scientific forestry offered by industry and the state examined in chapter 2. Reproduced widely in BC and abroad, this particular image helped generate and consolidate international resistance to logging in BC's temperate rainforests. The success of BC's environmental movement, and the spread of antilogging protests to distant sites such as New York, London, and Bonn, stemmed in part from the ability to picture the forest industry's "assault" on the forest. Political constituencies, after all, are not ready at hand, merely waiting to be called upon; they must be constructed.

It is important to approach images like these in a critical way, for, although they appear as neutral mirrors of nature, they are noninnocent documents that construct the "truth" about the forest, position us as witnesses to a looming crisis, and call us into action. In short, these images establish an object of political calculation: the vanishing rainforest. Accordingly, this apocalyptic imagery has become as much part of how social natures on Canada's west coast are produced as the chainsaws and grapple yarders of industrial forestry. In the following sections, I explore the construction of this image. I will proceed in an agnostic manner, suspending for the moment belief in tragic tales of the forest's "liquidation." This decision to defer final closure on the truthfulness of the forest's disappearance should not be taken as a willful or irresponsible nihilism; instead, it is done in order to open space for imagining other possible stories foreclosed by this apocalyptic narrative. It is important to be clear about this, lest my argument be taken up as an apologetics for industrial forestry. The purpose of interrogating images as artifacts is not to reveal them as false, but rather to question the certainty and obviousness of their truths. This is not about error or misrepresentation. To evaluate images in terms of *mis*representation would be to tacitly accept what Timothy Mitchell (1988) has called a "modern enchantment" that assumes that truth lies outside or beyond representation. In such a metaphysics, representation becomes merely an issue of *accuracy,* along the lines of the statement "the truth is out there, if only we could get beyond the ideological lenses that exist like a screen between us and the world." By this view, ideology and power distort our understanding of the world, rather than being what makes such understandings possible to begin with. As Mitchell notes, to assume that power operates through *mis*representation (or mystification) is to leave *re*-presentation itself unquestioned; it is to forget that truth is not prior to, but always an effect of, its representation. Thus, rather than seek to replace one (false) statement with another (true) one, I instead ask how statements such as "the forest is disappearing" obtain their status as "truth." What representational practices—and what absences and silences—allow such a statement to appear certain?

These are not simply questions of epistemology; they are important political questions too. Images, and not just people, do a great deal of social work. To draw on Ian Hacking's (1983) felicitous phrase, representation begets intervention; the work going on inside the computer mapping lab matters because it shapes how we perceive and act on the world outside the laboratory (Latour 1988). The task of criticism, then, is to ask whether the stories the Sierra Club/Wilderness Society images tell are the only available and the ones we most need to hear.

Building the Mirror of Nature

Let me return again to the images of Vancouver Island. What makes them so powerful? I noted earlier that they resemble satellite photographs. Several aspects of this deserve attention. To begin, this has the effect of positioning the viewer in (outer) space, far above the island. With its irregular blue borders, Vancouver Island appears as if it had been cut out from a larger picture, taken from miles above the earth. Indeed, this "reality effect" is enhanced in the 1990 image, which includes clouds and their shadows. It is now commonplace to note the effect of this view from above, how it positions the viewer as a disembodied and disinterested observer, presents the scene below as a totality, and allows fantasies of intellectual and physical mastery (Haraway 1991; Rose 1993). Haraway (1991, 189) refers to this "seeing everything from nowhere" as a "god-trick"—an illusion of being able to transcend embodiment and occupy an unmarked position that itself escapes representation. Obviously, as a viewer, I am not literally suspended above the earth; this is merely an effect produced by the image itself. Yet, when looking at this image, it is hard to forget that what I am examining is simply a large piece of paper filled with a series of shapes, lines, and colors. It appears instead as an objective record of the world as it is.

Upon closer examination, it becomes evident that the images are actually computer-generated maps. Satellite photographs were important only in the initial stages of the project, when the Wilderness Society's mapping team used 1990 Landsat 5 satellite imagery to begin to determine the ground cover on the island for that year.[4] It is actually quite difficult to distinguish "ancient rainforests" from "modified landscapes" on the basis of satellite imagery alone. Both landscape types appear as a uniform color (green), and so-called old-growth forests do not stand out enough to be reliably identified by untrained viewers. Thus, in order to distinguish between so-called pristine and regenerated forests, cartographers had to turn to a vast array of other resources, including forest inventories produced by the BC Ministry of Forests. For my purposes, then, what is significant about the Landsat imagery is not that it presents Vancouver Island as it actually is, but that it translates the complex, three-dimensional spaces of the island into what Latour (1986) calls an inscription: a two-dimensional representation that can be read (in the comfort of one's office or home) and perhaps more important, related and combined with *other* inscriptions. Because the initial satellite images had been digitized, mapping forest inventory information onto the original image was a relatively simple—albeit time-consuming—procedure, allowing individual sites to be identified (and colored) according to type of forest cover (it now becomes possible to see that the decision to retain clouds was entirely strategic, giving the maps the appearance of "unmediated" satellite photos).

Of course, if one is going to map the remaining extent of the "ancient temperate rainforest," it is best to be clear about what properly belongs to these forests. Obviously, not modified forests, but what about mountain hemlock forests? Or bog forests? Both have somewhat different species compositions, and, perhaps more important, significantly *smaller* trees (despite their small stature, these trees can be quite old). How does one distinguish these from others? This leads to yet other sources, such as studies of "vegetation zones" by biogeographers and maps of biogeo-climatic units. Only *after* these data are compared and combined with all the other data is the final image possible. A simple, yet crucial, point can now be made: behind any 'inscription' lie many others (Latour 1987). What appears as a single image is in reality the result of thousands of acts of translation done by dozens, if not hundreds, of people (and more than a few machines). The "view from space" from which we began has suddenly become a lot more complicated.

I will return to this cascade of inscriptions momentarily. First, let me note that because the satellite images were digitized, their final appearance could be reworked. As I explained earlier, satellite photographs differentiate poorly between "ancient" and "modified" forests. Because retaining the original colors of the satellite photographs would reveal the island to be almost uniformly green—achieving the opposite effect from what was desired—the cartographers introduced a color scheme that would enable viewers to clearly distinguish so-called ancient forests from modified landscapes, mountain hemlock forests, developed areas, and so on. Ancient forests were transformed to appear a deep forest green; modified landscapes were made light yellow. In turn, bog forests were coded orange, mountain hemlock forests red, urban areas gray, and so on. The result was an alarming absence of green from the 1990 image.

What about the 1954 image? Although it appears the same, it was actually assembled in an entirely different manner. After all, no satellite images were available (a fact easily overlooked). Instead, cartographers had to rely on maps of forest cover produced that year from aerial photographs by the Ministry of Forests. Similar processes of translation occurred to make *these* forest cover maps. Individual aerial photographs (already a translation of three-dimensional space into a two-dimensional inscription) were first evaluated according to specific criteria for determining the height of trees. This information was then used to estimate the age of the forest, and these results were translated into the form of a map. It is these maps that the Wilderness Society cartographers retrieved from the Forest Ministry's archives and combined with their 1990 satellite photographs in order to produce the 1954 image. Again, because the satellite photographs were digitized, they could be used as a template, so as to provide a second image (1954) in exactly the same form as the first (1990).

Two aspects of this deserve additional comment. First, by translating the two images into the same form, it became possible to compare the two years. This is crucial; without these acts of translation it would have remained impossible to "take in" the forest crisis. In BC forest politics, nature would still be on the side of the forest industry, and the environmental movement would be much weaker than it is today. By rendering the two periods comparable, it became possible to view time and space "at a glance," and thus to construct a teleological narrative: *the forest is disappearing*. Somewhat paradoxically, this teleology was constructed retrospectively. The 1954 image was not simply found in an archive, dusted off, and compared with more recent data. It was produced in the present. Indeed, in my account I moved backward from the present to the past, because this is the same path that the cartographers took, using contemporary images and technologies to construct an image that could never have existed in 1954. Second, the fact that both images—one dating from before the space age, the other from after—were presented in the form of satellite photographs was no doubt a *strategic* decision, playing on the commonly held distinction between a map and a photograph. The former is increasingly viewed as a product of labor—and thus less able to appear entirely objective—whereas the latter, trading on the supposed innocence of the satellite photo, appears unmediated. Indeed, these images achieve a further level of authority simply by dint of their status as technoscientific objects. That these were strategic decisions, however, does not change the images' status, nor their truthfulness (*all* images are artifacts). It tells us only that the producers were savvy actors with a keen understanding of the role of imagery in late-twentieth-century technocultures.

Let me summarize my discussion so far. Far from mirroring nature, the Sierra Club/Wilderness Society image constructs a picture of the rainforests of Vancouver Island through a series of practices and technologies of inscription that translate what Bruno Latour (1987, 1993) calls the "testimony of non-humans" (trees, rocks, ice, water) into representations that can be read, compiled, compared, and translated. The result is to displace and simplify the complex landscapes of Vancouver Island into an image that can be taken in and understood at a glance; in essence, objects are turned into easily read signs. There is nothing magical about this; it is the product of countless mundane practices by many different actors: satellites, airplanes, cameras, cartographers, forest biologists, computer technicians, and printers. None of what I have said means that these images are untrue, false, or misleading. Nor does it mean that the cartographers who constructed the images were duplicitous. It simply helps us to recognize that, despite appearances, it is not nature itself that speaks in these images. I will explain why this matters shortly.

Immutable Mobiles: The Virtual Forest and Its Publics

Why concern ourselves with this image in the first place? There is a simple answer: it is important to examine it because *others* have. In the years after it was produced, the Sierra Club/Wilderness Society map quickly became an important actor in BC forest politics. Reproduced in pamphlets, hung on walls, shown at rallies, and reproduced in the pages of newspapers and magazines, it helped fuel a global campaign to save the "ancient rainforests" of Vancouver Island. This is the sort of work that is delegated to all inscriptions, although few are taken up so widely. These inscriptions matter because it is impossible to bring along the forests of Vancouver Island each time one wants to talk about them. One can appeal only to the virtual forest—to maps, figures, diagrams, and photographs. This is the case whether one is a forester, a government official, a company executive, or an environmental activist. Indeed, so accustomed are we to talking about the forest *through* these inscriptions that we forget that in their absence we would have a hard time saying anything meaningful at all about the forest, even if we were to bring part of the forest along![5] Inscriptions, then, are key to producing the objects about which further decisions and calculations can be made.

This merits further comment. Part of the social power of such images lies in the fact that they are both *mobile* and *immutable*. Even if I succeeded in taking part of the forest with me to support my position—a single tree, perhaps—I could only bring it to one place at a time. Images can be many places at once. Further, as arborists know, moving a tree can be a risky proposition. Even if it survives, it may be transformed in passage. Branches may break, leaves may turn color or fall off, and the soil surrounding the roots may shake loose. The final result can be disappointing. Inscriptions remain unchanged; they are *immutable* even as they travel across space and time.[6] Thus, inscriptions like the Sierra Club/Wilderness Society images are important not only because they make the "ancient rainforest" visible, but because they allow the truth of its disappearance to be disseminated across space. As Latour (1986, 11) explains, this has important ramifications:

> The links between different places in time and space are completely modified by this fantastic acceleration of immutable mobiles which circulate everywhere and in all directions. . . . For the first time, a lo-

cation can accumulate other places far away in space and time and present them synoptically to the eye; better yet, this synoptic presentation . . . can be spread with no modification to other places and made available at other times.

Crucially, immutable mobiles are neutral on the question of accuracy. They are about *consistency* alone: "no matter how inaccurate these traces might be at first, they will all become accurate *just as a consequence* of more mobilization and more immutability" (ibid., 12). This is not to suggest that accuracy or adequacy are unimportant, but rather that it is in part a function of repetition. Wherever the image travels, it becomes possible to "see" the forest's liquidation, and the more times and the more places this occurs, the more its narrative stabilizes as a fact.[7]

Drawing Lines in the Forest

What is gained by such an analysis? Most immediately, it allows us to see the statement "the forest is disappearing" to be contingent, and thus something that can be open to question. As we have seen, the Sierra Club/Wilderness Society images combined and translated material from multiple sources (satellite photographs, biogeoclimatic maps, forest inventories, air photographs), mixed these with the skills of technicians (photographers, computer programmers, cartographers), relied on the competencies of various instruments (computers, software, satellite technologies, cameras, printers), and drew on a set of guiding metaphors and concepts from such sciences as ecology. Each of these actors/actants and translations is a "black box" (Latour 1987). If one wished to make the effort (and had the resources), each could be opened to investigation. Forest inventories, for instance, are far from simple things. Translating trees into numbers and words is actually quite complicated, involving forest mensurationists, airplanes, cameras, stereoscopes, compasses, and more mundane items such as paper and string. The same can be said for biogeoclimatic maps and satellite photographs. Behind each inscription lie others, and between each type of information lie forms of *translation* that enable information to appear in different forms (cf. Latour 1999). The point is not to get lost in details, although it is important to recognize the stunning quantity of different practices that are embodied without comment in these images. Rather, it is to recognize that it is not nature that adjudicates between claims and facts. Facts are merely claims (statements) that have become stabilized. If at some point any of the techniques or procedures that were part of producing and stabilizing these statements are shown to be untrustworthy, these facts revert to claims.[8]

Here I wish to place in question only one aspect of these images—the distinctions drawn between different types of land cover, and the assumptions that lie behind them. As explained earlier, the map consists of a series of color-coded regions, in which each color represents a type of ground cover. There are only solid colors, no shades. Several aspects of this scheme deserve further comment. First, its effect is to emphasize differences rather than similarities. Thus, urban areas—although often highly vegetated and often remarkably complex ecosystems (Harvey 1996)—appear entirely unnatural. Likewise, "bog forests," despite containing many of the same species as surrounding forests, are presented as having no relation to contiguous tracts of ancient rainforest. Second, reinforced by the god-trick of objectivity, the viewer is left with the impression that these stark divisions can actually be found in nature itself. Yet, one does not simply step across a line from one biogeoclimatic type to another. These zones are not as distinct as they might first appear. They are differentiated by biogeographers on the basis of a small number of differences, even though many things remain the same. Further, ecological relations do not stop at these boundaries—different zones are tied together by myriad flows of energy and matter. This is true also of cities and their surroundings, blurring the boundaries between developed, modified, and natural landscapes.

This raises important questions about what principles inform *how* the map's lines were drawn, and what differences were thought to matter. If we look closely at the images, we see that the color scheme has been set up to emphasize one thing: the distinction between "ancient forests" and "modified landscapes," and the expansion of one at the expense of the other. On closer inspection, however, it becomes evident that "developed" areas have expanded too, yet the color scheme (light yellow/gray in the original) makes this difficult to take in. Why is this not emphasized? The answer, I suspect, is that the expansion of developed areas is everywhere at the expense of "already modified" forests, not the "ancient rainforest," and thus is considered relatively inconsequential.[9] As I will explain later, the ecological integrity of these modified landscapes is often viewed as already compromised; thus, there is less reason for alarm over a shifting urban/rural boundary (one rarely sees roadblocks at the edge of cities). Yet, the ecological transformations that come with urbanization are arguably as important, and have as far-reaching effects, as those that come with forestry. We can also ask why such a stark distinction is posited between ancient and modified forests. Are forests not just forests? As noted earlier, satellite photographs of Vancouver Island would reveal the island to be almost entirely forested, an even carpet of green. Only urbanized regions, the island's few farms, and sites of very recent forest industry would appear otherwise. To convey a story of disappearing rainforests, cartographers had to "recode" the visual image in order to draw the eye to the line that divided ancient from modified. This does not mean that the map is merely ideological. The distinctions that are made, the boundaries drawn, and the colors chosen are all defensible, in the sense that they are all based on differences in the landscape that, through a variety of techniques, can be located, measured, and mapped. What is important is not whether these distinctions *can* be made, but rather, why these differences, and not others, *were* made.

Why this boundary, and not another? Most preservationists would say that these maps draw the distinctions they do because *ecology* tells us that these are the ones that matter most. Indeed, at least since Rachel Carson's *Silent Spring* (1962), ecology has been the last court of appeal for environmentalists of all stripes (even as science and technology are viewed by many with skepticism; see Beck 1992). Yet, the appeal to ecology opens up as many questions as it answers. To borrow a phrase from Donna Haraway (1992), ecology is a discourse, not nature itself; its knowledges are at once cultural and political, even as they engage with, and are shaped by, encounters with humans, animals, and other organisms. Perhaps more important, ecology does not speak with one voice—it is internally heterogeneous and its concepts have changed over time. As I demonstrate in the next section, although these images assume an unambiguous ecological foundation, the science of ecology provides several different accounts of *which* differences in the forest matter and why.

Dreams of Unity: Equilibrium, Holism, and the Tragedy of Human Perturbance

To North Americans who came through public schools in the 1970s and 1980s, ecosystem ecology seems an almost unassailable body of theory. With its flows of energy and matter revealed in elegant diagrams of boxes and arrows, it appeared to be nothing less than nature's own language. Indeed, there remains something aesthetic about these diagrams, and the way that nature's machinery seems to work to preserve order and balance. The lessons that these diagrams relayed lasted long after details of individual case studies were forgotten: whether it was a forest, swamp, or pond mattered little; the boxes and arrows told us all that we really needed to know: nature's intricate balance was achieved through a symphony of all its parts; disrupt one and the ramifications would be felt throughout the whole. Although many ecologists now question the validity of the concepts and metaphors that dominated early ecosystem studies, its image of nature as a self-regulating, balanced system remains highly influential in our public cultures—indeed, so influential that people of my generation still find it unsettling to realize that the version of ecosystem ecology we learned in grade school held sway over the discipline of ecology for no more than a few short decades after World War II. For my purposes, historicizing ecosystem ecology can help open conceptual space for thinking nature's ontology differently from the familiar terms that ecosystem ecology provides. It also can help us understand why forest conflicts in British Columbia have taken the form they have, and perhaps allow us to imagine a re-fashioned environmental politics that draws different lines in the forest. In what follows, I focus somewhat selectively on three key aspects of post–World War II ecosystem ecology: its emphasis on understanding nature as a system; its assumption of balance and unity in nature; and the way that it understands human actions in the environment.

Nature External and Balanced

Let me begin by returning to Haraway's (1992) statement that biology (or ecology) is a discourse. What does this mean? At the very least, it suggests the existence of an ordered set of statements and concepts that govern what can be known and said about nature. It also suggests that these concepts are historical and partial, rather than timeless and universal. Indeed, as familiar as its concepts may seem, ecosystem ecology is actually quite recent, taking shape primarily after World War II. It has antecedents, of course, many dating to the nineteenth century and earlier. Joel Hagen (1992), for instance, suggests that the origins of ecosystem ecology can be traced back as far as Charles Darwin and Herbert Spencer. Darwin, he suggests, gave modern ecology its understanding of life as a web of complex relations, and provided a number of key ideas: that species do not exist independently, that they often form interacting groups, and that, even if indirectly, they work to regulate the population of other species. From Spencer, Hagen argues, modern ecology obtained other concepts, including the organismal analogies and metaphors that surfaced throughout its history: for example, as found in Frederic

Clements's notion of plant communities in the 1920s, or in Howard and Eugene Odum's influential notion of self-regulating natural systems in the 1960s. Hagen's history is no doubt too linear, giving the impression that ecology has existed as a unified, coherent body of ideas that gradually shifted through time.[10] A more genealogical approach—one that refuses to assume ecology as a single, or internally coherent, body of knowledge—would no doubt emphasize a more diffuse set of historical practices, from U.S. atomic weapons research to cybernetics, including any number of technical and intellectual innovations unrelated to the discipline (computers, radiometry), institutional structures that facilitated research (university and military laboratories), historical contexts (cold-war science), and cultural narratives that provided guiding metaphors (systems theory, notions of homeostatis, etc.).

For my purposes, what is most useful about Hagen's account is its focus on ecosystem ecology's guiding metaphors and concepts.[11] He argues, for instance, that organismic metaphors dominated ecological thought (and not *only* ecosystem ecology) throughout the twentieth century, from early ecologists to adherents of the Gaia hypothesis in the 1990s. Indeed, in the Gaia hypothesis, Hagen identifies two ideas that he claims are present to a greater or lesser extent throughout the modern history of ecology: that nature is self-regulating, and that this self-regulation is analogous to homeostatic mechanisms in organisms and cybernetic controls in automated machines. Hagen suggests that G. Evelyn Hutchinson made similar claims about the "biosphere" in the 1940s, and that notions of balance in nature were relatively widespread among ecologists even earlier. Frederic Clements (1928), for instance, is widely considered responsible for popularizing the notion of "communities" of organisms, and that these communities had "life cycles" (stages of succession) that, at maturity, culminated in a static form. For Clements, plant communities were complex organisms whose functioning could be explained in terms of the activities of their parts. Individual organisms acted on, and reacted to, each other (and the physical environment) in a self-sustaining manner. Thus, nature tended toward equilibrium as communities progressively stabilized their environments and formed barriers to invasion. "Such a climax is permanent because of its entire harmony with a stable habitat," Clements (1928, 99) wrote. "It will persist just as long as the climate remains unchanged, always providing that migration does not bring in a dominant from another region." In other words, this climax stage was capable of endlessly perpetuating itself, unless something interfered.

It would be misleading to assert that all ecologists at the time shared Clements's views (see Gleason 1926; Elton 1930). Yet, ideas of community and stability were widely influential, contributing a compelling intellectual approach to research for future ecosystem ecologists, and, perhaps more important, a set of mobile and easily understood concepts that could be translated to new sites and diverse research projects. Even though few ecologists after World War II continued to believe that a community or ecosystem *was* an organism, many believed that these "higher-level systems" behaved somewhat *like* organisms. Succession remained the paradigmatic example, but a host of other notions—the cycling of energy and materials, specialization and niches within food chains, energy flows, metabolism—permitted nature to be seen in terms of systems of interacting elements that formed a closely and intricately woven fabric.[12] These metaphors did not disappear with attempts in the 1960s to place ecology on the level of other hard sciences such as physics. Rather, they were reworked in new ways. During the 1950s and 1960s, for instance, ecosystem ecologists drew heavily on insights from cybernetics, the science for designing self-regulating machines such as guided missiles and thermostats. Cybernetic devices exhibited purposeful behavior, continually regulating their operation through a set of feedback loops that kept the machines they governed at equilibrium. Likewise, ecosystems, already understood through notions of balance and self-regulation, could now be pictured as a "feedback loop of energy" flows that kept communities in balance, a view of nature widely disseminated in the diagrams of energy flows produced by Howard Odum and others (see Figure 6.2).[13]

The point of this brief discussion is not to provide a complete history of ecosystem ecology but to highlight the holistic metaphors, and notions of order and balance, that have been tightly woven into its governing concepts. Ecosystem ecology in the years after World War II understood nature in terms of *functional relationships,* including at the level of ecosystems whereby the whole was seen to be greater than the sum of its parts; *equilibrium,* such that the introduction of change was seen as unnatural; and *self-regulation,* which allowed equilibrium to be seen as internal to ecosystems. It is worth noting that although these notions lent support to preservationists concerned about the disruption of nature's "natural balance" by humans, ecosystem ecology could also provide justification for forms of environmental engineering. Howard Odum's energy budgets, for instance, lent themselves to a managerial ethos just as easily as a preservationist ethic (cf. Worster 1988; Hagen 1992). Indeed,

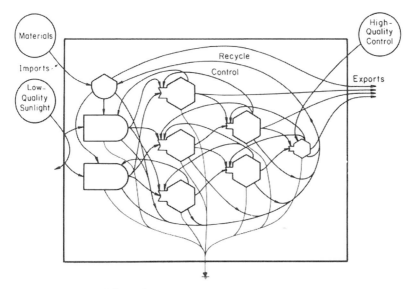

Figure 6.2. General form for an ecosystem. Source: Howard Odum (1983).

this is true of systems theory more generally, which set up nature (and society) as a series of quantifiable relations that could be regulated or engineered to achieve a desired result (see Figure 6.3). In one instance, Odum famously envisioned waste-management schemes that circulated municipal wastewater through existing local aquatic ecosystems. He argued that nutrients removed by microbes could serve as a base for vibrant food chains, including harvestable species such as crabs. Clean effluent could then be reused as drinking water. As Donald Worster writes (1988, 313), Odum's ecology was "almost perfectly tailored" for the needs of modern conservation: "A traffic controller or warehouse superintendent could not ask for a more well-programmed world . . . it was perhaps inevitable that ecology too would come to emphasize the flow of goods and services—or of energy—in a kind of automated, robotized, pacified nature."[14]

Worster's critique notwithstanding, ecosystem ecology has more often been a rich source of images and concepts for the environmental movement, especially for those interested in the preservation of wilderness. With its functional view of self-regulating ecosystems, it offered a new holism that was still deeply imbued with romantic and holistic tropes, recycling old notions of balance, divine providence, and natural order, and clothing them in the objective language of science (McIntosh

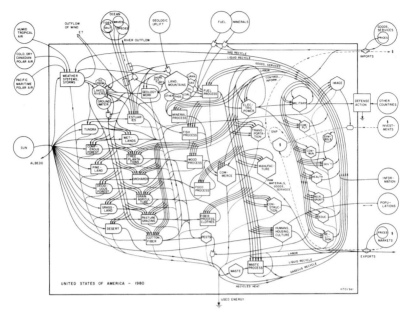

Figure 6.3. "The energy basis for the system of humanity and nature in the United States of America." Source: Howard Odum (1983).

1985, 1987; Botkin 1990; Demeritt 1994; Wu and Loucks 1995). Some, like Alston Chase (1995, 7), a persistent critic of radical environmentalists, suggest that this latent holism encouraged North America's environmental movements to appropriate ecology within a preservationist ethic:

> Instead of being reductionist, [ecology] was holistic; and rather than being morally neutral, it appeared to carry an important ethical message. Based on the notion that nature was organized into networks of interconnected parts called ecosystems, the new science seemed to say that conditions were *good* so long as ecosystems kept all their parts and remained in balance.
>
> Combining the ecosystem idea with the nature worship they had inherited from early preservationists, the more radical activists conceived a unique ideology, eventually called biocentrism. If everything is dependent on everything else, they reasoned, then all living things are of equal worth, and the health of the whole—the ecosystem—takes precedence over the needs and interests of individuals. . . . The metaphor of the ecosystem revived the notion of nature as purposive

and as the foundation of value. Since an ecosystem's "health"—that is, its stability—was the highest good, then any human activity that upset this balance was not merely mistaken but immoral.

Of course, Chase's critique is far from disinterested; it has at points provided support to corporations and wise-use groups seeking to deflate the influence of environmental critics. Yet, his acerbic comments helpfully focus attention on the convergence of ecosystem ecology and preservationist politics. This goes beyond a shared concern with the stability and balance of nature to include another notion: that nature's balance is *threatened* by human activities. In part, this view finds its way into ecosystem ecology through the field's tendency to study ecosystems separately from people. In other words, to the notion of nature as a self-regulating system, ecosystem ecologists have often, although perhaps unwittingly, added the notion that nature is a domain *external* to culture. It matters little whether this was deliberate (it is much more likely a result of institutionalized divisions between the natural and social sciences). Framed in these terms, people could be seen as potential agents of disruption rather than as part of the ecological relations that ecosystem ecologists studied. This was often further extended: if natural systems were intricate and fragile networks of relations where each part was dependent on, and maintained by, the whole, then it was only reasonable to view intervention by humans, however small, as a threat to nature's inherent balance.

Such views were popularized in the 1970s by influential writers such as Barry Commoner (1971), whose four "laws" of ecology—everything is connected to everything else; everything has to go somewhere; nature knows best; there is no such thing as a free lunch—drew explicitly on notions of nature's unity and externality. Commoner (1990, 14) later argued that "any distortion of an ecological cycle . . . leads *unavoidably* to harmful effects" (emphasis added), an argument consistent with his assertion that an ecosystem is always "consistent with itself" (Commoner 1990, 11), and that maintaining this steady state was a *moral* imperative. That this reiterates notions of an ordered and purposeful cosmos hardly needs mention. Nor is it hard to see the Romanticism that pervades such passages. But it is crucial to recognize that Commoner's barely veiled Romanticism—and the Romanticism of much of the environmental movement—is not something that has been mistakenly *appended* to ecology. As David Demeritt (1994) notes, the organicism of the Romantic movement was already intricately woven into ecological thought in the nineteenth century. Post–World War II ecosystem ecology, with its emphasis on self-regulating systems, and its externalizing of nature as something outside of, and threatened by, human culture, did little more than place further emphasis on ideas already latent within the discipline's central concepts (see also Wu and Loucks 1995).

We should not be surprised, then, that ecosystem ecology has provided grounds for the reiteration of the nineteenth-century Romantic notion that nature "cannot be disturbed in even the most inconspicuous way without changing, perhaps destroying, the equilibrium of the whole" (Worster 1988, 82), nor that post–World War II ecosystem ecology science has been enrolled, often passionately, in political projects to preserve "untouched" nature and minimize human disturbance in the environment. If we remember that the boundaries separating science, culture, and politics are porous (Martin 1996), or if we accept that science is a form of practice (Haraway 1997) that can never be completely disentangled from its enabling political, institutional, and cultural conditions, this makes a great deal of sense. To argue that science is culture does not invalidate its claims, however; it merely alerts us to the partial and power-laden nature of scientific knowledge, and to the possibility that nature may be compelled to speak its truths otherwise.

Nature Spoiled or Nature Saved: Drawing Lines in the Forest

We are now in a better position to understand the lines that the Sierra Club/Wilderness Society map draws in the forest. If nature is *external* and *balanced,* then by definition human disturbance destroys nature. Following human modification, the forest is no longer truly a "forest," it is a "modified landscape." In the stark lines drawn between ancient and modified forests, the Sierra Club and the Wilderness Society map a romantic ecology onto the landscape: "effectively, the end of the ancient rainforest will occur . . . when the last of the remaining intact watersheds are opened for development" (Sierra Club of Western Canada 1991, 4).

Holistic tropes like these are heard frequently in BC forest politics. Ancient temperate rainforests are seen as "interconnected webs" and "fragile networks" in which individual organisms and abiotic elements have specific functions. As forestry critic Herb Hammond (1991, 15) writes at the beginning of his widely read treatise on wholistic forest use:

> Forests are interconnected webs *which focus on sustaining the whole,* not on the production of any one part or commodity. . . . Each group

of organisms . . . is essential to the survival and healthy functioning of the forest. Plants and animals, the easily visible components of a forest, and microroganisms, the unseen components, are all manifestations of *a fragile network* of soil, water, sun and air. (Emphasis added)

My interest here does not lie with Hammond's focus on interconnectedness but with how this becomes subsumed into a vision of a totality in which all of its parts work to sustain the whole. Echoes of a planned and purposeful cosmos are more than faintly discernible. This rhetoric is repeated throughout Hammond's work, and the implications of "meddling" in nature are presented in remarkably apocalyptic terms:

Trees are biological pumps which pull water out of the soil through deep root systems, use the water for growth processes, and finally return it to the atmosphere as water vapor. Large trees can pump more than 100 gallons of moisture into the atmosphere every 24 hours. Without forests and older trees to pump and distribute water vapor, many areas once forested become considerably drier. *Parts of the Sahara Desert were once a forest.* Tree cover slows the melting of snow in the spring in forested watersheds. Snow melt water is thus released later from ridges and mountain slopes that have a mantle of trees than from slopes that are open. This hydrological effect results in a more even flow of water from a forest than from extensively cleared areas. Decaying wood in fallen trees and in the soil acts as a sponge. Water is slowly metered into surrounding areas from the forest "reservoir". Healthy tree cover avoids spring floods and fall droughts, and provides a water filtration system that far surpasses technological systems. (17; emphasis added)

Throughout Hammond's text, the forest is the site of finely balanced symbiotic relationships. Plants provide food and habitat, shade streams reduce moisture loss, protect young trees, and furnish nutrients for others: "from bacteria to bears, slugs to salmon, earthworms to eagles, a wide variety of microscopic and macroscopic animals depend upon a diversity of forest communities" (18). Animals, he explains, also provide functions needed to maintain the whole forest. For instance, the Columbia ground squirrel contributes to soil enrichment. Its burrows consist of a variety of chambers that are used for "food storage, sleeping quarters and bathrooms." The animal continuously brings vegetation to use for food and bedding. This all serves a greater purpose: "plant matter is thereby mixed with soil and mineral matter to provide a balance of soil nutrients for the development of life throughout a forest. Organic material found in the bathroom chamber is 'redi-mixed'" (19). What emerges is a picture of a complex, integrated, self-regulating whole. Every organism has a function, every action a telos (Figure 6.4). This extends, it appears, to Natives too.[15] Change occurs—squirrels burrow, plants and animals die, rivers erode their banks, Natives build longhouses—but the system maintains equilibrium: "diversity works to maintain balance and stability throughout the forest" (15). In short, nature—even with its regimes of disturbance and its "natural" human inhabitants—maintains a regulated balance: the sum of any ecosystem is always greater than its parts.

Figure 6.4. Nature as harmony. Source: Hammond (1991).

What is at stake here is not whether trees affect hydrology, or whether ground squirrels mix plant matter with soil and mineral matter. No doubt they do. What I want to draw attention to is the manner in which individual elements—plants, animals, and Natives—are viewed merely as instruments for the unfolding of a greater, purposive whole. It is precisely this, I argue, that licenses some environmental groups to draw stark categorical distinctions between natural ecosystems and modified landscapes, and to make this the key battleground for eco-politics.[16] For groups such as the Western Canada Wilderness Committee and the Sierra Club of Western Canada, once (modern) human activity *modifies* a landscape it can no longer properly *be* an ecosystem—its equilibrium is disturbed and the consequences reverberate throughout the system. Yellow and green cannot occupy the same place on the map.

With this in mind it should come as little surprise that BC forest politics has often centered on one practice above all others: clearcutting. Perhaps more than any other forestry practice, clearcuts appear to disrupt, if not completely destroy, nature's balance, rending the intricate fabric of the forest. Like a torn garment, it can never be stitched together in the same form. There can be little doubt that the horror of clearcuts is related in part to its visual impact. This was clearly the strategy in the book *Clearcut: The Tragedy of Industrial Forestry,* which documented with great effect clearcuts from across the Pacific Northwest (Devall 1993). It is evident too in the stark dualism drawn during a Greenpeace campaign in Britain (see Figure 6.5). Indeed, ecology and Edenic narratives are close cousins in these efforts, and these images evoke, in turn, a set of associated terms that have their roots in Western patriarchal cultures: virgin, unspoiled, untouched on one side, the "rape" of wild nature on the other. Each image and phrase speaks of an *irreversible* condition or a loss that can never be restored.

Pictures of clearcuts have been immensely important to antilogging efforts. Yet, they tell us nothing about the *ecology* of clearcuts. In a similar fashion, the Sierra Club/Wilderness Society map with which I began is not a map of ecological relations, or even of habitat. It does nothing more than distinguish those areas that still show no signs of human activity (green) from those where the so-called original nature has been disturbed (yellow). Its underlying assumption is that human disturbance equals the end of nature (clearcuts are only the most extreme case). Indeed, it is only when these are the differences that matter in the forest that critics such as Robert F. Kennedy Jr., writing for the U.S.-based Natural Resources Defense Council, are able to draw an equivalence

Figure 6.5. Nature lost/nature saved: the binary logic of industrial nostalgia. Photograph by Garth Lenz. Reproduced with permission of Greenpeace Canada.

between forestry (the use of forest resources) and deforestation (the destruction of forest ecosystems, usually through conversion to other uses). In British Columbia, Kennedy (1993, B4) writes, "*deforestation* rates rival those in tropical rainforests and threaten to *exterminate* North America's last accessible coastal rainforests and associated rivers and estuarine eco-systems" (emphasis added).[17] The Western Canada Wilderness Committee (1994) makes a similar claim about the disappearance of the forest: "Coastal temperate rainforests have always been a rare ecotype on Earth. After the last ice age they covered only 0.2 percent of Earth's land area. Today 90 percent of these wild forests are gone. The 10 percent that is left—one-quarter of it in B.C.—is disappearing at an accelerating rate."

What intrigues me is what is left unsaid, and, in turn, the problems this poses for a truly radical environmental politics. First, forests grow back. Despite the rhetoric of Canada's environmental movement, forestry is *not* the end of nature. The landscape is still an *ecological* landscape, even after roads are built and trees cut. Indeed, as the geographer David Harvey (1996) has pointed out in a stinging critique of the single-mindedness of North American environmentalists, *cities* are themselves ecosystems. The yellow areas on the Sierra Club map may be modified landscapes, but they are still predominantly forested. So also are large parts of the urbanized landscapes colored gray on the Sierra Club/Wilderness Society map, although when viewing the map this is almost impossible to imagine. In the rhetoric of the environmental movement, there are only two binary poles: nature spoiled or nature saved; yellow or green. Each is exclusive of the other, and all eyes anxiously follow this shifting boundary.

This does not mean that there are not good reasons to carefully differentiate between types of human disturbance. Quite the opposite: many intricate relations exist in the forest, and, depending on what we want from our forests (this is itself an ethical-political question, not something dictated by nature), these relations need to be carefully considered. Indeed, in some cases—for instance, if sustaining specific wildlife populations is our goal—logging may need to be halted entirely. Further, the forest that grows back after forestry cannot be the *same* forest, even if every attempt is made to mimic the old. This is true of any disturbance, of course, but it needs to be emphasized that the forest industry's comforting stories of *regeneration* are as flawed as the environmental movement's images of *destruction.* To assume our ability to "mirror" nature is as misleading as the view that human modification "destroys" nature. Edenic narratives and technophilic optimism are equally unhelpful.

A second, unstated assumption in these apocalyptic images is that the temperate rainforest is a *stable* ecosystem, rather than an environment in constant flux. I explore this in the next section, but the point I wish to stress in advance is that by holding tightly to a Romantic ecology in which nature is seen as static rather than dynamic, environmentalists may be inadvertently playing into an increasingly powerful antienvironmental backlash. While the rhetoric of nature's "fragile balance" has been wildly successful for gaining public support for preservationist causes, critics of the wilderness movement (including the forest industry) have been increasingly successful in their efforts to again enroll nature as *their* ally. How this has occurred is the topic I examine next.

Dynamic Ecology and BC's War in the Woods

Since the early 1980s, many ecologists have questioned whether equilibrium and homeostasis really do characterize how nature works. In place of confident assertions about the ordered, self-regulating, and teleological nature of ecosystems, advocates of "dynamic ecology" have placed an increasing emphasis on *contingency* and *change* (see Sprugel 1991; Wu and Loucks 1995; Scoones 1999). As Daniel Botkin (1990, 6) explains, "We have tended to view nature as a Kodachrome still-life, much like a tourist-guide illustration of La Salute; but nature is a moving picture show."

Several aspects of this shift deserve attention. First, as is common in the history of science, the announcement of a new paradigm in ecology has had the effect of showing the old ecology to be *ideological* rather than *scientific*. In Gaston Bachelard's words, "contemporary science is able to designate itself, through its revolutionary discoveries, as a liquidation of a past. Here discoveries are exhibited which send back all recent history to the level of prehistory" (Bachelard 1982, 137; translated and quoted by Young 1990, 49). Second, the matter of history is indeed central to the shift that I am tracing, although in a somewhat different manner. Critics of systems ecology argue that the old ecology tended to be *ahistorical* in two important respects. First, it rarely examined the history of ecology itself, and thus borrowed many of its models and metaphors from the development and application of systems theory in other disciplines without considering whether they were suitable for the study of ecological relations. If systems ecologists had read carefully the history of the discipline, the argument goes, they would have discovered vigorous debate around notions of equilibrium and disequilibrium many decades earlier, and might have been less likely to embrace the notion of self-regulating homeostatic systems provided by cybernetics (see Worster 1990). Ecologists such as Henry Gleason, for instance, severely criticized Clements's characterization of communities as "superorganisms" almost as soon as the work of the latter had appeared (Gleason 1926; Hagen 1992). Even earlier, W. S. Cooper (1913) had drawn attention to "disturbance" as a key aspect of vegetation regimes. Second, and more important, systems ecologists often ignored the *historical* nature of ecological relations themselves. This was a shortcoming of systems theory more generally, which tended to calibrate its models in abstract time rather than historical time (see Gregory 1985). In the case of ecology, Botkin (1990, 9) argues, historical analysis would have led ecologists to recognize that change rather than homeostasis was "intrinsic" at many scales of time and space.

This rethinking of ecology's central concepts has come about through the efforts of scholars working in a variety of different fields and temporal and spatial scales. It has drawn heavily, for instance, on the reconstruction of historical climate change by paleoclimatologists.[18] These records, constructed from pollen samples and often entered as data in powerful global climate models (GCMs), prompted Margaret Davis (1986, 269) to conclude that "for the last 50 years or 500 or 1,000—as long as anyone would claim for 'ecological time'—there has never been an interval when temperature was in a steady state with symmetrical fluctuations about a mean. . . . Only on the longest time scale, 100,000 years, is there a *tendency* toward cyclical variation, and the cycles are *asymmetrical,* with a mean much different from today." Such conclusions have had far-reaching effects for ecological thought. First, ecologists were forced to rethink whether the community or ecosystem was indeed the logical level of analysis. Add time to the equation, and many factors external to the specific present-time qualities of ecosystems appear more important for why a particular constellation of plants and animals is found in any given place and time. Second, ecologists shifted attention from equilibrium to change. Climate change now was seen as a driving force behind what was fast becoming a new fascination among ecologists: environmental *flux.*[19]

The emphasis on flux stemmed from work at smaller scales of analysis too. In the 1970s, research by ecologists increasingly turned to the study of disturbance. Fire, windstorms, invading populations, soil movements, floods, volcanic eruptions, climate change, even the daily activities of individual organisms (like squirrels!), were now seen to introduce continuous change rather than maintaining balance in environments that previously had been considered "steady-state." Ecosystems, stripped of their self-regulated homeostasis, became "mosaics of environmental conditions" characterized by heterogeneity rather than homogeneity and by change at every temporal and geographical scale. Some have described this as a "paradigm shift" in ecological thought, but it is important to note that notions of equilibrium have not disappeared entirely. Much turns on the scale and method of analysis. Sprugel (1991), for instance, notes that disequilibrium at one scale may result in stability at another scale, something ecologists refer to as a "shifting mosaic steady state" (see Watt 1947; Bormann and Likens 1979). Others have suggested a "multiple equilibrium" perspective based on nonlinear mathematics (Levin 1979). More recently, Jianguo Wu and Orie Loucks (1995, 451, 453) have argued for what they term "hierarchical patch dynamics" that may—or may not—lead to states of equilibrium:

> In hierarchically structured, patchy ecological systems, the phase change of individual patches at local scales and the pattern change in patch mosaics at broader scales together give rise to system dynamics. Thus, the dynamics of ecological systems are composed of the dynamics and interactions of constituent patches on different scales; this is an emergent property in that it is not simply the sum of the individual patch dynamics. . . . Nonequilibrium patch processes at one level often translate to a quasi-equilibrium state at a higher level. This homeorhetic, quasi-equilibrium state has been termed meta-stability . . . [and] illustrates the conceptualization of order out of apparently random fluctuation, where nonequilibrium dynamics at one scale can become the means of quasi-equilibrium at a higher level (either spatially larger or temporally longer).

There is little consensus on whether disturbances at one scale lead to stability at another. Complicating matters is the fact that dynamic ecologists and paleoecologists often work independently and have different research methods and materials: thus, while spatial scale matters a great deal to nonequilibrium ecologists, temporal scale is more important to paleoecologists. What can be asserted with some confidence, however, is that, at least among natural scientists, notions of an ordered cosmos have largely lost their hold.

What does this mean for forest politics in the Pacific Northwest? At the very least, it challenges the terms in which forest debates have been staged. Recent evidence from paleogeography, for instance, suggests that many previous assumptions about Pacific Northwest rainforests are untenable. Cathy Whitlock (1992) has challenged notions of the rainforest as a stable, self-regulating system. Based on the study of fossil records, Whitlock has reconstructed Pacific Northwest climates for the past twenty thousand years. This is complicated work that turns on a network of actors and techniques—pollen, radiocarbon-dating techniques, climate modeling, computer technologies—which allows core samples from various lakes in the region to be translated into numbers and then graphs. Through these graphs, past climatic conditions are inferred. Based on her reconstructions, Whitlock argues that the region has seen almost constant change. Around 20,000 BP, for example, the west side of the Olympic Peninsula in Washington was covered by a mixture of tundra and parkland vegetation. Between 10,000 and 5,000 BP, the region may have been considerably warmer but drier than today. Fire would have been a more important cause of forest disturbance, and vegetation patterns (and species) would have been considerably different. Whitlock suggests that it was only in the late holocene (5,000 BP to the present) that so-called modern vegetation patterns were established, a result of lower temperatures and greater precipitation. Only in the last three thousand years did a reduced fire threat allow mature forest stands to develop, prompting Whitlock (1992, 22) to conclude:

> [Old-growth forests] developed relatively recently on an evolutionary time scale and probably do not represent a coevolved complex of species bound together by tightly linked and balanced interactions. . . . It is clear that vegetation has responded continuously to a varying array of climatic conditions. . . . No millennium has been exactly like any other during the last 20,000 years.[20]

As if this is not enough to challenge notions of stability, Whitlock pushes her argument further, suggesting that "modern [forest] communities are loose associations composed of species *independently adjusting their ranges* to environmental changes on various time scales" (22; emphasis added). This is a marked departure from conventional representations of the "ancient" rainforest. When we consider that mature trees on Vancouver Island can live for more than a thousand years, the "primeval" forest appears somewhat less enduring and considerably more contingent: it is only a few generations of trees old (see also Sprugel 1991). Likewise, the notion of an intricately balanced web in which each part sustains the whole is seriously undermined when the forest is merely a snapshot in time.

The question of disturbance has seen vigorous debate. As noted, forests are now widely thought to be characterized by change as much as stability, although there is vast disagreement on the significance and scope of these changes. For some, disturbance regimes simply work toward stability at the scale of landscape, while others point to catastrophic events at the landscape scale as evidence that claims of "metastability" are equally untenable (massive forest blowdowns on Vancouver Island in the early twentieth century and the explosion of Mount Saint Helens in 1980 are frequently cited examples). These debates in ecology would matter little if, to borrow a metaphor from Emily Martin (1996), the citadel of science were not so porous. But, as Donald Worster and Daniel Botkin note from opposite sides of the debate, the shift from notions of "steady-state" ecosystems to one where change is the new catchphrase has opened a Pandora's box. Since the 1960s, ecosystem ecology has provided powerful normative statements. The notion of an ordered cosmos and its metaphor of a "web" of life provided a sense of what was natural or normal, and in turn provided a template for human behavior. Dynamic ecology introduces troubling questions. "What, after all," Worster (1990, 16) asks, "does the phrase 'environmental damage' mean in a world of so much natural chaos?" How does one measure the health of the planet as a whole if biological systems are inherently unstable and if evolutionary change continuously produces a constant stream of unique conditions? For others, the lesson was clear: "no single state of affairs," Alston Chase (1995, 114) writes, "can be either 'healthy' or 'unhealthy.' The earth can be 'healthy' for humans or 'healthy' for dinosaurs, but it is never just plain healthy. Habitat can be good for deer or good for owls, but never merely good for wildlife."

We may be uncomfortable with Chase's blithe dismissal of normative statements, but clearly the "ecology of chaos" makes it difficult to justify the stark boundaries on the Sierra Club/Wilderness Society map. This has not been lost on the forest industry in BC, which has quickly moved to appropriate dynamic ecology to place in question the claims of environmentalists. A key figure in this has been Hamish Kimmins, a forestry professor at the University of British Columbia and a self-styled demystifier of the "misrepresentations" of the environmental movement. Kimmins (1992, 20) writes:

> So often the argument is about the *present* condition of a particular forest ecosystem rather than how that ecosystem will *change over time.* . . . The concerns of environmentalists about forests have fre-

quently been presented in pictures of clearcuts, slashburned sites, or soil erosion taken in the initial weeks or months after the event, with little or no evaluation of how long such conditions will persist. This 'snapshot' evaluation of *what are in reality dynamic and ever-changing ecosystem conditions* can lead to a serious misrepresentation of the ecological impact of natural or management-induced disturbance.[21]

Likewise, Patrick Moore, a cofounder of Greenpeace in 1971 but, during the 1990s, chairperson of the industry lobby group COFI (Council of Forest Industries), argues that

> All forests, no matter how old they are, are growing on land that was once treeless. As climates have changed over the millennia, forests have come and gone, ever changing themselves as they respond to a changing environment. . . . The history of the ice-ages tells us some important things about the evolution and composition of forests. First, there is no ideal forest composition for any given climate or region. In the absence of human intervention a fairly wide range of conditions are capable of occurring and of sustaining themselves over long time periods. This may mean there is no need to be terribly rigid in defining what constitutes a native or "natural" forest for a given area. (Moore 1995, 12, 51).

Moore goes on to chide environmentalists for their dualistic vision of nature as excluding humans. This, he claims, inhibits people's ability to understand their place in the environment—a claim that hits its mark, but that in Moore's hands becomes an apologetics in support of the forest industry's continued authority in the woods.

Despite their mobilization of dynamic ecology in the interests of capital, critiques by Kimmins, Moore, and their supporters must be taken seriously. Kimmins, for instance, combines dynamic ecology with an understanding of human actions as *internal* to nature to develop a forceful argument that seriously undermines political claims based on Romanticism and classical ecosystem ecology. The outlines of this argument are predictable. Claims that forestry destroys nature are disputed on the basis that there is no "essence" to the forest: thus, the modified forest is no less an ecosystem than the previous forest it replaced. What is being replaced, after all, is not static, but rather an accident of historical conditions: "the single most fundamental characteristic of ecosystems," Kimmins argues, "is that they *change over time*" (1992, 88). The forest has no teleology, no final form to which it moves. Humans do not disrupt nature, they are merely an active agent within nature. Change is neither unnatural nor bad, nor is there an original nature that can be held out as an objective measure against which human environmental practices should be judged.

Kimmins readily admits that the modified forest will always be different from what it replaced, but suggests that this is by no means a travesty. What is important for Kimmins is not whether ecosystems are disrupted by humans, but precisely what kinds of disruptions take place. Here dynamic ecology provides Kimmins with an understanding of disturbance that refuses to make clear, unambiguous distinctions between "natural" and "human" changes. Kimmins notes, for instance, that forests—whether temperate rainforests, northern boreal forests, or the hardwood forests of the east coast—all have disturbance regimes. In the temperate rainforests of Vancouver Island, these disturbances occur in various forms and at a variety of scales, ranging from autogenic change caused by the plants themselves to fire and large-scale windthrow on the scale of several hundred hectares and, ultimately, to climate change measured in hundreds of kilometers and on the order of decades, centuries, and millennia.[22]

The important issue, Kimmins concludes, is not whether disturbance is caused by humans or nonhumans, but whether—if forest renewal is what is socially desirable—the ecological mechanisms for recovery have been damaged or preserved. Kimmins readily admits that the effects of clearcutting can be considerable, ranging from large changes in microclimates, the alteration of soil microbes and organic matter, reduced soil stability on steep slopes, relative advantages and disadvantages for various species of wildlife, changes in the quality, quantity, and timing of water flows in streams, even global warming. But, he explains, ecology cannot tell us whether these changes are good or bad. As he explains: "there is no basis in the science of ecology for saying a spotted owl in an old-growth forest is better or worse than a sparrow in a clearcut" (28).[23]

For Kimmins, this suggests a refocusing of environmental concern. Rather than dwell on the removal of trees from the forest, something he notes occurs every day in nature, Kimmins argues that attention must be paid to changes to soils and climate that ultimately determine the biological potential of the forest environment. These changes, he suggests, should be of more concern than shorter-term changes in plants, animals, and microbes caused by natural or human disturbance. In Kimmins's opinion, part of the problem is that what I have called Romantic ecology—which understands *any* interference as a threat to the fragile balance of nature's interconnected webs—fails to adequately consider questions of scale. Whether or not forest conditions are reestablished, and colonization by animals and plants occurs, he claims, depends on what is being done to the landscape as a whole. Here Kimmins both appropriates and complicates criticisms made by "new forestry" advocates such as Herb Hammond (1991). Hammond places considerable emphasis on the problem of the "fragmented" forest. The forest's fragmentation, he argues, makes the movement of species between habitats, or the possibility of recolonizing habitats after regeneration, more difficult. Fragmentation, by this view, threatens biodiversity. Kimmins argues that this is too simple, and suggests that foresters and ecologists need to distinguish between different types of diversity: *alpha* diversity (the number of species in a given ecosystem type); *beta* diversity (the difference in species composition between the different ecosystems found in a particular landscape); *geographical species* diversity (the variation in the species list across large regional geographical units); *temporal* diversity (the change in the species list and vegetation structure of a particular ecosystem over time); and *genetic* diversity (the amount of genetic variation within a particular species in a specific ecosystem and across its geographical range). By making these distinctions, Kimmins argues that alpha diversity at any given stage in the forest's development—what he implies most concerns environmentalists—may not be as important to the functioning of ecosystems as some think. This is owing in part to what he calls redundancies between species, and in part because, given proper forest management that takes into consideration the spatial and temporal patterns of the forest, modified forests can be recolonized by species from other nearby forested regions as they advance through different seral stages. "One is left to conclude," Kimmins writes, "that for most forests, the overall functioning of the ecosystem will generally continue regardless of a periodic loss of individual species because of disturbance" (164).

For my purposes, the validity of Kimmins's biodiversity typology, and the conclusions he draws from it, are less important than what it tells us about the politics of ecology in BC's war in the woods. Ultimately, what matters in Kimmins's account is less how one measures biodiversity than whose claims to the forest will be able to cloak themselves in the mantle of scientific rationality. In short, it is a question of which side can successfully enroll nature as its ally:

Environmentalists have frequently suggested that we should maintain "current" (or "pre-European" in the case of North America) levels of biodiversity and species ranges. This implies that the present condition of the world's ecosystems is the way nature intended it to be, an implication that is *scientifically insupportable*. Climates have always been changing and will always change. Many species have occupied their present geographical distributions for only the past few centuries or millennia, and are still changing their range in the wake of the last glacial period, or at least of the "little ice age" in the Middle Ages. Many ecosystems owe their present condition to wind, fire, and past timber harvesting or deforestation for agriculture. Their present species diversity and structure reflect human and natural history and not what the forest would look like after five hundred years of development in the absence of disturbance.

In most forests, biodiversity is not a fixed, God-given thing but a complex and ever-changing ecosystem and landscape characteristic. (Kimmins 1992, 165)

Nature, it seems, is back on the side of industry. If we are to believe Kimmins, the lines drawn on the Sierra Club/Wilderness Society map reflect *ideology*, not *reason*; there is no basis for them except in outmoded and discredited knowledges. The problem, according to Kimmins, is that environmentalists are poor ecologists, who, in their urban settings, "have lost touch with nature and have little understanding of the ebb and flow of natural change and its time scales" (22).

Drawing New Lines in the Forest

How are we to evaluate these shifting discursive and ideological terrains? Is dynamic ecology simply an apologetics for industry, a new, flexible nature for global capitalism? Donald Worster clearly thinks so, and worries that the new fascination with disturbance, disharmony, and chaos has resulted in a situation where conservation "is often not even a remote concern. . . . What is there to love and preserve in a universe of chaos?" (1990, 3, 16). This is no small concern. Today, dynamic ecology is enrolled to support statements that would have been highly contentious only a decade ago. Near the end of his book, for instance, Kimmins writes: "we know that nature periodically 'throws away' many ecosystem 'parts' as a result of disturbance. The resulting forests are often different to some degree from those they replace, but they generally work just fine" (1992, 166).

The impact of such views is hard to measure. The environmental movement continues to have great success portraying the rainforest through holistic metaphors and apocalyptic imagery: nature saved or nature lost. But there are, I think, signs that the concepts and images governing forest politics in British Columbia are becoming somewhat less fixed. Whether this signals a transition to a new forest politics is unclear, but it presents a moment both of danger and hope. With this in mind, I want to conclude by turning to a highly publicized study of forest practices in Clayoquot Sound. By examining its reframing of the forest, it may be possible to begin to imagine new ways of drawing lines in the forest.

Ecologizing Forestry, Socializing Nature

The Scientific Panel for Sustainable Forest Practices in Clayoquot Sound (hereafter Scientific Panel) was created in response to the controversy over the BC government's Clayoquot Sound Land Use Plan of 1993. Faced with massive opposition, and divergent claims over the ecological consequences of logging, the BC government brought together nineteen blue-ribbon experts to review current standards for forest practices in the region.[24] Although such inquiries are often little more than a form of state legitimation, the Scientific Panel attained a surprising measure of autonomy. Given the politically charged context—and the appeal to the authority of science by actors on all sides—the government had to justify appointments based on scientific credentials rather than political connections. The resulting panel consisted of internationally recognized scholars and included a number of individuals whose stubborn commitment to intellectual honesty and methodological rigor suffered no compromise. As a result, the panel was not afraid to challenge the very terms of reference provided by the government: charged with making recommendations for forest practices on lands designated "General Integrated Management Areas" (a move designed to legitimate the government's land-use plan), panel members chose from the start to study the region as a whole, and to make recommendations for Clayoquot Sound at the scale of landscape, basing their decision on the scale of ecological mechanisms in the region. While it is true that forestry was still assumed—and thus, as environmentalists pointed out, preservation was ruled out from the start—this too was subverted by the panel's call to replace sustainable yield with sustainable ecosystems as the goal of forest management. As we will see shortly, this meant that in some cases preservation could be the best "scientific" management strategy.

Much more could be said about the panel's makeup and the reception of its recommendations.[25] What interests me were the new principles that the panel argued should govern forestry practices in the region. As with the Sierra Club/Wilderness Society map, what was at stake was what kinds of lines would get drawn in the forest, and what sorts of nature would be produced. This is not merely an ecological matter, for in BC's forest economies, to ecologize forestry is at the same time to transform the very nature of work and thus the fabric of social and cultural life.[26] Let me turn to some of the most important aspects of the panel's reports. As already noted, central to its recommendations was a call to move forest management away from sustainable *yield* to sustainable *ecosystems.* In some respects, this merely built on a long-held view in conservation—that the sustainability of resources and resource values depends on maintaining ecosystem productivity and thus the landscape's biophysical connections. But the panel's recommendations significantly extended and deepened this. They consistently stressed that the "integrity" of ecological relations should take priority over forestry. The latter should proceed only if it did not threaten the viability of ecological relations across a variety of scales, and not at all if this could not be achieved. This challenged the very nature of forestry in the province. Whereas the calculation of an annual allowable cut (AAC) on any given tenure had previously been volume-based (an annual quota based on timber inventories and length of rotation), the panel argued that levels of cut should be spatially based and thus seen only as an output of a comprehensive planning program that took as its first principle the maintenance of ecosystem mechanisms, rather than an annual quota to be met.

Although this emphasis on ecosystem sustainability mirrors many of the concerns of the environmental movement, the panel elsewhere called into question the language of "primeval" forests. For instance, although set resolutely against the timber bias of current forest management, its recommendations invoked ecology as a science of relations but consistently avoided wrapping it in Romantic notions of a timeless, balanced nature, or using this as a basis for preserving the forest intact. Panel members placed considerable emphasis on interconnections *and* disturbance, depicting the forest as a shifting configuration of elements. While they drew attention to how individual species were sensitive to the character and composition of the forest, thereby raising important questions about how these relations would be transformed by forestry, they also argued that one of the most important distinguishing characteristic of any forest ecosystem is change. Thus, in a statement clearly directed at an ecopolitics focused on preservation, they argued that "there is a continual turnover of living organisms—even in what appear to be stable ecosystems. Forest trees are long-lived but not immortal" (Scientific Panel 1995b, 17).

The panel also refused to make a categorical distinction between human and nonhuman disturbance: green and yellow were not mutually exclusive. Although this was partly an effect of the panel's mandate to study forestry practices, the view was also consistent with the panel's *ecological* understanding of the forest:

> Wild forests renew themselves naturally in a manner that depends on the natural disturbance regime of their region. *Logging is a recent disturbance* that inevitably alters the pattern of renewal. Ecological knowledge can be used, however, to ensure that the changes caused by logging, and the forests that regenerate after logging, are not dramatically different from those created by the natural disturbance regime. Forest practices that *approximate* natural disturbance regimes help to retain ecosystem processes and maintain ecosystem productivity and connections. (Ibid., 17; emphasis added)

Several important implications derive from this statement. First, it shows a subtle shift away from disturbance-as-destruction to disturbance-as-renewal. Given this, the issue no longer is how to limit human disturbance (industrial nostalgia), but whether human disturbance can be managed to fit within a broad range and scale (temporal and spatial) of disturbance regimes, and patterns of forest renewal, that are found historically in the region. As the panel explained, retaining the integrity of ecosystems did not mean keeping all the pieces intact—an ethic of wilderness—but instead "[e]nsuring that ecosystem processes and states do not depart from the range of natural variability exhibited before logging; that is, maintaining functioning, self-sustaining ecosystems with characteristics similar to the original ones" (79). Although the language here veers back toward systems ecology and its notion of "self-sustaining" ecosystems, the more important implication is that the panel refused the forest industry's antinormative interpretation of dynamic ecology, even as it accepted the notion of continuous change. The panel defined change as natural—and did not rule out humans as agents of change—but sought to develop criteria by which to distinguish between good and bad change. In contrast to Kimmins and Moore, for whom the lesson of ecology is that there can be no basis for arguing in favor of one set of ecological relations over another (thus equating all change as the same), the panel suggested a simple baseline: observed variability. By this view, some environmental practices are acceptable, others not. To be sure, this carried a number of problems. It still assumed a nature/society dualism, in that natural variability is determined apart from human practices (although these practices may be made to mirror this variability). The question of history also troubles the recommendation: how far back does one go to measure "natural variability"? And, ultimately, it reaffirms the ecologist as the arbiter of good environmental practices, returning us to a question central to this book: who is authorized to speak for nature in struggles over the future of the forest?

Second, the panel sought to provide a more textured view of the forest by drawing attention to questions of *differentiation, scale,* and *process.* It divided the Clayoquot region into biogeoclimatic units, and subdivided each further into zones, subzones, and variants. In each, the panel analyzed disturbance regimes and forest renewal patterns. These were found to vary geographically. Thus, rather than generalizing all human disturbances in the forest as the same (whether seen as good or bad), the panel suggested that forest planning must account for *spatial variations* in disturbance regimes. Panel members also focused on the relative importance of different ecosystems and different aspects of specific ecosystems. Some sites and processes were considered more important than others. Hydroriparian ecosystems, for instance, were singled out as of particular importance, because they are essentially "the skeleton and circulation system of the ecological landscape," where events upstream can influence downstream characteristics and organisms (xiii).[27] The scale of these biophysical processes depends fundamentally on the process examined. It follows, then, that planning must take into account not only how areas of the forest differ, but also how different forest spaces are related.

To return again to Romantic ecology and the lines on the Sierra Club/Wilderness Society map, this differentiated forest is no longer one where all anthropogenic change is equally bad. Different practices have different effects, and attention now comes to settle on the consequences of specific forest practices across different temporal and spatial scales. This is also different from claiming that all change is the same simply because the forest is in flux. Differences matter. The forest that grows back after logging is not the same as what regenerates after windthrow. The latter differs in significant ways from what emerges after fire or slope failure. Logging riparian ecosystems has different effects than logging has at other sites. Large breaks in the forest canopy have consequences much different than do small breaks. Whereas industry apologists such as Moore simply added clearcuts to a long list of similar forest disturbances, the Scientific Panel asked pointed questions about when and how such disturbances occur, and the extent and rate at which disturbance is allowed to happen.

Ultimately, this has ramifications for forest planning. If, as the panel argued, the starting point for forestry decisions is sustaining the productivity and connections of the ecological landscape, then planning must shift away from an approach that is organized around the maximization of timber values. It must also shift from administrative to ecological boundaries, adopting physiographic or ecological land units (such as watersheds) as the basis for planning. In turn, rather than plan for harvesting in terms of volume *removed,* the panel suggested that foresters focus on the *rate* (percentage of area cut per unit time) and geographical *distribution* of timber harvesting. Planning, in other words, should focus not on the trees harvested, but on the forest left behind. Moreover, because ecological processes occur over time and space, planning for any one unit must always look beyond the site to take into account its context.

To achieve this, the panel recommended a new planning regime that focused on three nested tiers: subregional, watershed, and site. The first—*subregional*—would focus on formulating and monitoring overall land-use objectives and address issues that cross watershed boundaries (networks of reserves; habitat; linkages between watersheds, and so on). The second—*watershed*—would focus on determining which areas could be harvested and which not, identifying a riparian ecosystem, locating unstable terrain, noting significant habitat, identifying vulnerable species, protecting First Nations sites, managing for scenic values, and so on. The third—*site*—would focus on the type of management activity at specific locales. Here small-scale features would be taken into consideration (e.g., ephemeral streams) and silviculture planning would occur, including harvest methods and strategies for regeneration. This is no longer an undifferentiated resource landscape. But neither is it one where green becomes yellow the moment the chainsaws arrive.

Landscape Ecology: Nature after the "End" of Nature?

The point of the preceding discussion has not been to determine whether the Scientific Panel's recommendations were the right ones. Rather, I have sought to call attention to the shifting constellation of ideas that are reshaping how the forest is being understood and people's place in it imagined. In part, this is a result of changing concepts in ecology, in part, the result of new valuations of the forest. It is, most certainly, a historical shift, in the sense that it is not something dictated by nature, but instead an emergent set of ideas and practices forged in the crucible of culture and politics.

Notably, the panel's recommendations paralleled developments in landscape ecology—a growing field of research and professional management. Founded to a large extent on the legacies of ecosystem ecology and its insistence on interconnections, landscape ecology seeks to reintegrate landscape within the horizon of planning and thus to move beyond the specific qualities of individual sites in order to take a wider view that understands the intertwining of social and ecological processes across time and space. What this permits, as Alexander Wilson (1991) notes in his rediscovery of Ian McHarg's *Design with Nature,* are approaches that bring questions of ecology back into how we think about and plan the spaces in which we live and work. Wilson points to "restoration ecology" as a promising example of how "landscape work" can bring together communities, activists, and professionals around the questions of place, landscape, and nature. Restoration ecology, he writes, is "dedicated to restoring the Earth to health. Restoration is the literal *reconstruction* of natural and historic landscapes. . . . [it] seeks out places to repair the biosphere, to recreate habitat, to breach the ruptures and disconnections that agriculture and urbanization have brought to the landscape" (113, 115).

Admittedly, this is a somewhat ambivalent discourse that can lend support to widely different social and ecological projects. When landscape ecology becomes about *restoration,* thorny questions come into play: Restore what? In whose interest? What is native, original, or natural, and who defines this? In rural areas, practitioners have often seen their role as that of restoring pre-Columbian natures, mirroring the same distinctions between pristine and spoiled natures that I have argued are so disabling. Indeed, the vocabulary often employed—*exotic* species, *feral* animals—reveals a preoccupation with policing the pristine (and a view that environmental agency lay entirely with European settlers). This has old roots. As Alston Chase notes, when Aldo Leopold stated in 1934 that he wanted "to reconstruct . . . a sample of original Wisconsin," he meant "what Dane County looked like when our ancestors arrived here during the 1840s" (quoted in Chase 1995, 112).

Wilson proposes a different reading. He understands our ideas about nature to be always "culturally mediated," and thus the question "Which nature?" must be foregrounded rather than assumed. Further, in Wilson's view, landscape ecology necessarily assumes the interwining of social and ecological processes: it begins with an earth irrevocably marked by human presence, and asks questions about how to live in such a condition. This is not about finding in nature a blueprint for social life. Rather, restoration recognizes that once lands have been "disturbed"—worked, lived on, meddled with, developed—they require human intervention and care:

> We must build landscapes that heal, connect and empower, that make intelligible our relations with each other and with the natural world: places that welcome and enclose, whose breaks and edges are never without meaning. Nature parks cannot do this work. We urgently need people living on the land, caring for it, working out an idea of nature that includes human culture and human livelihood. . . . Unlike preservationism, [restoration ecology] is not an elegiac exercise. Rather than eulogize what industrial civilization has destroyed, restoration proposes a new environmental ethic. Its projects demonstrate that humans must intervene in nature, must garden it, participate in it. Restoration thus nurtures a new appreciation of working landscapes, those places that actively figure a harmonious dwelling-in-the world . . . the boundaries of the garden have become less distinct. (Wilson 1991, 17, 115)

There are still echoes here of a Romantic view of nature (see Smith 1996).[28] Read generously, though, we can find in Wilson's sense of restoration ecology an attempt to channel the emerging discipline in particular directions. It cannot be an elegiac exercise; landscapes are invariably working landscapes, and the boundaries between the pristine and the modified are no longer so clear. We can perhaps extend his vision to the temperate rainforest, and to the findings of the Scientific Panel. The panel's report was a scientific document, but it was also an important *social* document. Approached through Wilson's lens, we can see that it was not just about trees, but about working out an idea of nature that included humans. By this view, humans do not interfere in nature, nor are they seen as invariably destructive. Questions come into view that are not about anxiously policing lines in the forest, but instead imagining the forms of social and ecological organization that can produce social natures that are at once just and sustainable.

Mapping Futures

This chapter, then, has been about mapping. Not just about mapping the forest and its extent, but about mapping futures—the future of the forest and the future of society. In British Columbia, these cannot be pulled apart. Ultimately, this mapping turns on, and contributes to, a shifting constellation of ideas about nature and society and about their relation. Is nature an ordered, self-regulating entity or continuously in flux? Are humans part of this nature or aliens intruding from the outside? Does human modification of the forest destroy nature, or does it merely reconfigure its elements and its relations? What lines in the forest matter? What changes are good? Which are bad? And how do we evaluate this? I have sought to show that our maps of the forest are profoundly social and political, even those that appear most dispassionate. This does not mean that they are false, only that they are artifacts of discourse and practice that can be analyzed for their enabling conditions and founding presuppositions. Imagining the forest's future means not simply mapping nature, but attending closely to how we come to draw lines in the forest.

In the background lies a larger question. What constitutes a radical environmentalism at the beginning of the twenty-first century? Is pristine nature what environmentalism should be about? Is a radical ecology focused on preserving the forest radical enough? An increasing number of critics—many of whom consider themselves environmentalists—are saying no. Indeed, the 1990s witnessed significant attempts by scholars and activists (especially those in the environmental justice movement) to shift North American environmentalism away from its preoccupation with untouched wilderness (or "big forests," in the language of David Foreman). This has taken several forms. Some have worried that preservation—precisely because it is built on notions of homeostasis—is actually "antiecological" and leads to the practice of fighting change, even though change is precisely what many ecologists now claim distinguishes how nature works. As Cathy Whitlock (1992, 22) has noted in relation to Pacific Northwest forests, "conservation efforts that emphasize the preservation of communities or vegetation types will probably be unsuccessful because future climate changes quite likely will dismantle the community or vegetation type of concern."

Perhaps more disconcerting, by drawing lines between 'pristine' and 'modified' environments, and then imagining ecopolitics in terms of policing traffic between the two, an environmentalism predicated on nature as external leaves us without ways of imagining how to live *in* nature. William Cronon (1995) has been among the most eloquent critics of this tendency:

> If we allow ourselves to believe that nature, to be true, must also be wild, then our very presence in nature represents its fall. . . . If this is so—if by definition wilderness leaves no place for human beings, save perhaps as contemplative sojourners enjoying their leisurely reverie in God's natural cathedral—then also by definition it can offer no *solution* to the environmental and other problems that confront us. To the extent that we celebrate wilderness as the measure with which we judge civilization, we reproduce the dualism that sets humanity and nature at opposite poles. We therefore leave ourselves little hope of discovering what an ethical, sustainable, honorable human place in nature might actually look like. . . . In its flight from history, in its siren song of escape, in its reproduction of the dangerous dualism that sets human beings outside of nature—in all of these ways, wilderness poses a serious threat to responsible environmentalism at the end of the twentieth century. (Cronon 1995, 80–81)

Perhaps it is time to dispense with "wilderness," with notions of a static, unchanging, prehuman nature, and time to think carefully about connections between landscapes, to see the forest as a shifting mosaic of differences rather than a static essence, to see nature as including human practices rather than excluding them. Perhaps it is time to be more realist(ic) than the Sierra Club/Wilderness Society map, replacing its yellow regions with the subtle shades of green found on satellite photographs, shades that suggest differences and similarities, ruptures and connections, and, more important, the need to take responsibility for, and engage in dialogue over, the kind of natures we wish to inhabit, and how we might achieve them. As Cronon (1995, 85–88) explains, "We need an environmental ethic that will tell us as much about using nature as about not using it. . . . If living in history means that we cannot help leaving marks on a fallen world, then the dilemma we face is to decide what kinds of marks we wish to leave."

Conclusion

Reimagining the Rainforest

> We do not therefore speak of a dualism between two kinds of "things," but of a multiplicity of dimensions, of lines and directions.
>
> —Gilles Deleuze and Claire Parnet, *Dialogues*

What kind of marks do we wish to leave? This is the question that impresses itself upon us most urgently today. If nature is neither external nor purposeful, then perhaps environmentalism must be reconceived in terms of nature's production, not its preservation. This notion, developed most fully within Western Marxism, takes humanity's "species-being" as evidence that we *cannot not* leave marks on the world (Schmidt 1971; Smith 1990; Cronon 1995). How this occurs is not given in advance; it is the outcome of debates in ethics, and perhaps more important, an outcome of the social organization of production (O'Connor 1998). By this view, capitalism gives to nature its specific form. Emphasizing nature as an artifact rather than an essence does not mean that we are freed from attending to ecological relations, or from responsibility more generally; it merely refuses to posit ecology or nature as a realm separate from the human relations with which it is always articulated. To imagine otherwise is to miss a great deal of what is going on. Thus, far from starting from the assumption that nature and culture need to be held apart, the pressing task of criticism and politics falls on determining what form, and with what social and ecological relations, we wish to continue remaking the natures of which we are inevitably a part. Ultimately, if we take writers such as Neil Smith and William Cronon seriously, it comes down to recognizing our responsibility for the hybrid natures that we make and understanding the ways in which our everyday practices are implicated in social and ecological relations both near and far. There is no room for nostalgia here.

In these last pages I wish to retain this emphasis on social nature—and on thinking about individual and collective responsibility—yet push the notion of nature's "production" somewhat further. Some worry that views expressed by writers such as Cronon, Smith, Haraway, and others simply reinforce an *instrumental* relation to the nonhuman world, thereby further contributing to nature's destruction (cf. Soule and Lease 1995). Social nature, by this view, hides an anthropocentrism that merely recenters humanity as the privileged agent of history. The problem from this perspective is that there is simply far too much *culture* in these authors' natures. I too have concerns that the language of "production" can become an apologetics for exploitation, yet even though sympathetic to arguments for an ecocentric ethics, I find it difficult to imagine a relation to the world that is not in some sense instrumental. This does not preclude granting to nonhuman nature value beyond its utility for humans—something that finds political and cultural expression in the animal rights movement, and in our decisions about what we use and consume—but I would argue that any such revaluing must be understood as an outcome of political struggles, and not because of the discovery of a transcendental truth written in the order of things. All valuing is, finally, a human valuing (Harvey 1996).

My challenge to such writers as Cronon or Smith is a different one. Far from placing too *much* attention on culture at the expense of nature, I argue that they have not yet taken the culture of nature seriously enough. What might it mean to focus more attention, rather than less, on the culture of nature? And why might this lead to a better ecopolitics? We can answer this by returning to Cronon's important essay on wilderness. Cronon is correct, I think, to see wilderness as an ideological concept. Nature, he argues, is not external, it includes humans; thus, in turning to wilderness we are blinded to the reality of what nature really is. This is an important critique, especially in light of the preoccupation with "wild" spaces within North American environmentalism. But to some extent it remains caught within the same framework that it seeks to displace. If we accept Cronon's argument, the problem with wilderness is that its advocates have mistaken wilderness for nature, sending environmentalism down the wrong path in a quixotic search for an imaginary object. Thus, by focusing on wilderness (the absence of culture), we lose sight of the "truth" of nature (its social production), and therefore are at a loss on how to proceed in a world where nature is always already marked by human actors. The result, Cronon argues, is a confused environmentalism that expends immense resources preserving those sites that can most signify wildness, while remaining unable to imagine an environmentalism for all those other areas—the vast majority of the world—that cannot appear pristine. Crucially, for Cronon, *environmental history*—grounded in political economy and ecology—arrives on the scene to correct the mistake, revealing the truth about the physical world, its ecological relations, and its social transformation at the hand of humans. Reoriented thus, environmentalism can be put back on track, engaging with the central question of our time: how are we to live?

Of course, Cronon is not so naive as to think that history is a simple matter of recording the past. To write environmental history is already to narrativize, to produce an order out of the myriad events and actors that shape the natures that we inhabit today (Cronon 1992). Still, his critique of radical environmentalism—like so many others—turns on *demystification*; Cronon understands wilderness as a fetish that stands in or substitutes for something else (origins), and takes on the task of the critic to reveal the truth that is hidden behind its ideological veil. For Cronon, wilderness is too "cultural," it fails to see the world for what it really is. I wish to suggest something different. One of the objectives of this book has been to suggest that there is no secure place of knowledge from which the truth about how the world is, and what our responsibility should be, can be known for all time. Culture is not something displaced or unmasked in the name of a higher, more foundational, even more *original* truth about "social" nature. Rather, what projects of critique reveal is that at every turn what counts as nature is, unavoidably, an *effect* of culture (and power), and that this trace can never be completely excised. There is no site outside culture and language from which to fix once and for all nature's truth or to adjudicate competing epistemological and political claims. As I have sought to show, the "forest" is something that takes shape as a thing, something that enters history through, rather than despite, our words and concepts. It is decidedly *not* something discovered (by the disembodied scientist, the disinterested ecologist, the adventure traveler, or the environmental historian). This does not deny the materiality of the physical world, nor does it ignore the fact that not all the actors in the drama of nature's remaking are human. The insistent materiality of the physical environment, and its more than passive presence in our lives, is continuously recorded in the physical, institutional, and cultural organization of social life, from the way a road curves around a mountainside, to how a river continually carves its course through city and country, to the unique forms of technical and managerial rationality that develop as a consequence of working with trees rather than, say, coal, oil, or corn. My point lies elsewhere: there are many forests, not one; there are myriad ways in which the physical worlds of the west coast are imbued with meaning and intelligibility, not a single unassailable truth that once found will show us the way forward. Criticism is thus not about unveiling the ideological, it is about attending to how statements get made, taken up, and turned into facts. It asks what is gained by this, and what is perhaps left out. Ultimately, it reminds us that our work is never finished.

Refashioning our lenses in this way allows other questions to come into view. These questions do not demand that we dismantle our barricades on the front lines of ecological politics so much as prod us to consider what we have built our roadblocks from, what concepts and ideas inform our actions, and who we find as our allies. They draw attention to the cultural politics of nature, and the political and economic interests that different constructions of the temperate rainforest work to secure. Once we realize that culture is necessary for us to talk about nature, rather than an ideological mask that must be stripped away, and once we see that politics does not begin with the decision to defend the rainforest, but is already part of how something called the rainforest comes to be seen as a thing worthy of defense, ecopolitics can never be the same.

Toward a Politics of Nature: Tracing Cognitive Failures

It should be clear by this point that my claim that the rainforest is a cultural construction does not mean that we should stop talking about rainforests. This is the first mistake made by critics of constructivism, who think that it pulls the rug out from beneath ecopolitics. We can see why this concern is misplaced through an analogy. It is now a commonplace that both gender and sex are constructs, yet this does not mean that we abandon use of the terms generated *(male, female)*. Rather, it compels us to consider *how* gender and sex difference is constructed, what regulates these identities (discursive practices, law), and with what effects. Ultimately, Judith Butler (1993) argues, this opens space for the articulation of other, subjugated positions. The same is true of the forest. Consider the forest found in the pages of MacMillan Bloedel's documents (discussed in chapter 2). Industrial forestry, I argued, obtained its authority on the basis of a set of discursive displacements that constructed and gave meaning to one entity—the "normal forest"—within a historically specific system of signification (fiber, value, productivity) and that normalized *this* forest rather than another. Indeed, so hegemonic was this forest that it was—and still is—seen by many as essential to the health and welfare of British Columbia and its residents. By this view, the forest is little more than a machine for making fiber—and in the years after World War II it was a common assumption that by harnessing and rationalizing these productive forces, a second set of objects and relations—society—could be placed on a rational, orderly basis too. In framing the forest in this way today, forestry corporations (and the Ministry of Forests) displace trees from their ecological and cultural relations, and deny *other* systems of signification that imagine the forest very differently. To say this in other words: industrial forestry in BC occupies a conceptual and cartographic space opened up through a series of highly successful cognitive failures (the erasure of First Nations, ecological relations, etc.).

These cognitive failures are successful only as long as they remain transparent. Further, precisely because industrial (colonial) forestry can continue to operate only by disavowing these cognitive failures, it is always precarious (see also Braun and Wainwright 2001). At any moment, that which is disavowed (i.e., the territorialities of First Nations) may return to haunt the colonial order, and reveal its constitutive displacements as a type of violence enacted on people and things. Viewed in this light, we might suggest that the alternative accounts of the forest provided by ecologists and First Nations have been effective to the degree that they have rendered these founding displacements visible. In the case of ecologists, to speak of the temperate rainforest, rather than timber, is to disrupt a system of signification that underlabors for extractive capital. The "rainforest" is corrosive to industrial forestry precisely because it insists on understanding the forest as a dense web of relations rather than a stand of trees. The rainforest, however, is also a deeply cultural concept—the outcome of an array of scientific, cultural, and political practices. To speak of the rainforest is not to arrive at the final truth of nature. This is key to the arguments of this book, and to a refashioned environmentalism. Precisely because nature is something that must be *represented* (it cannot simply speak for itself), the *act* of representation becomes that much more important, for it necessarily constructs that which it speaks for. What "deconstruction" teaches is that there is no way around this. Thus, rather than take the rainforest as an ontological given, it is perhaps useful to consider the rainforest in terms of what Gayatri Spivak (1988b) has called a "necessary theoretical fiction," something that must be posited in order to stage a critique, even as we recognize its contingent nature.

By invoking the rainforest (as an intricate web of relations), it becomes possible to shed light on the cognitive failures that underwrite and authorize a regime of resource abstraction, and to rename its practices as a violence to animals, organisms, and peoples alike. BC environmentalists have done this with great success. Yet, if criticism stops here, it has failed to push thought—and politics—far enough, for the rainforest is it-self a product of its own successful cognitive failures. We need look no further than the Sierra Club/Wilderness Society maps discussed in chap-ter 6. These maps have been immensely effective, enrolling large con-stituencies in projects to save the forest from an industry jealous for fiber. What I left unremarked in my comments in chapter 6 was that these im-ages obtain their rhetorical force through their own displacements. What enabled people living in Vancouver, Los Angeles, or London to be inter-pellated as defenders of the forest was not simply the fact that "im-mutable mobiles" now allowed them to see the forest's destruction from afar, but also the erasure of *locality* in these maps. In order for the image to tell its singular story of the rainforest's destruction at the hands of a ra-pacious industry (and, by extension, humanity), the forest had to be ab-stracted from its specific local and historical contexts and reconstituted as a thing entirely separate from them. One has no sense from these maps that what is happening at different places on the island may be the result of very different kinds of practices, involve very different actors and rela-tions, and have varying ecological and social effects. There is only one story: nature's immanent destruction at the hand of humans. Perhaps worse, at no point are the local communities whose residents are closely tied to the forest resource allowed to enter the frame. Most of the com-munities in areas coded green—and labeled the "ancient rainforest"—are Native, and there was no room on these maps for their needs and aspira-tions. Indeed, as we saw in chapter 3, these communities are often simply conflated *with* the primeval forest in a logic that goes further to equate logging with cultural genocide, because primeval nature is, after all, the "natural" home for aboriginal peoples.

The image—and the interpretive guide that accompanied it—speaks of the forest alone. This cognitive failure permitted a global environ-mentalism to speak authoritatively in nature's name. It accounts in part for the strong reaction to environmentalists found in many forest-dependent communities in BC, where workers and their families are understandably anxious when people far away start speaking "for" the natures that are central to their health and livelihood (although it must be stressed that a much *greater* threat in the 1980s and 1990s came from restructuring in the forest industry). It also accounts for the distrust of environmentalists often expressed by First Nations. This should not pre-clude the involvement of distant actors, but it should caution us to consider the implications of our actions, and the constructions of the

forest that these actions are based on. It is with these sorts of displacements in mind that the maps of culturally modified trees (chapter 3)
and the images of social nature painted by Yuxweluptun (chapter 5) can
be seen as yet another set of important efforts to resignify the forest.
Crucially, these are not innocent representations, either. They too are effective only to the extent that they remain transparent. As has become
evident in struggles within and between First Nations (a book that waits
to be written), these images also leave out a great deal (see Sterritt 1999).

My point is not merely the obvious one—that no final "truth" of the
forest can ever be reached—nor is it that the forest is a space of multiplicity, as if that by itself were a progressive statement. Rather, it is to
recognize that the stabilization of nature's truth at any given moment is
both an effect of power and something that serves particular interests. It
is to recognize that cognitive failures matter, whether they be those that
enable the ecological and social projects of the state and extractive capital, ecologists, or First Nations. Efforts to *disrupt* hegemonic relations in
the forest must be attentive to their own closures. To say this differently,
although it may be necessary to construct the rainforest as a prior ontological category in order to displace other systems of signification, this
act of construction must be made "scrupulously visible" (Spivak 1988b)
in order to avoid repeating the game of knowledge as power.

It is with this in mind that the Scientific Panel discussed in chapter 6
takes on added importance. It did much more than challenge the displacements of industrial forestry and the Romantic ecology of the Sierra
Club; it also refused to abstract the forest from its cultural surrounds.
Among the experts chosen to sit on the panel were four members drawn
from three Nuu-chah-nulth bands in the Sound, with the result that a
substantially different forest was brought into view. This is not the place
to take up the complex and difficult question of how different knowledge systems can be related and compared (see Turnbull 1997).[1] My interests lie instead in how the panel helped push forestry onto a different
footing. Owing at least in part to the presence of Nuu-chah-nulth representatives, the panel responded to the premier's call to "make the forest
practices in Clayoquot not only the best in the province, but the best in
the world" in a way that the premier had likely not anticipated: as something that was possible *only* if the interests and concerns of the people
living in the forest were addressed. What began as a technical and managerial problem was transformed into a moral and political one. Working
with a very different sense of the forest from industrial forestry, the
panel essentially argued that forest practices were *poor* if they abstracted

resources from their cultural surrounds. Ecologizing forestry was crucial, of course, but this was an ecology that included rather than excluded people.

I do not wish to make more of the panel's modest accomplishments than is warranted. Its terms of reference meant that it was obliged to presume the presence of logging, and thus was constrained to determine only the most ecologically viable *methods.* Further, it remained within a discourse of state territorial sovereignty, even as the presence of the Nuu-chah-nulth members called it into question. Still, there was something insurgent about its findings. One of the important projects initiated by the panel was an inventory of the plants and animals important to the cultures and economies of the Nuu-chah-nulth. These were correlated with specific sites, giving culture a spatial expression within what was previously an "empty" forest. As in the case of the map of culturally modified trees discussed in chapter 3, the panel reinscribed Nuu-chah-nulth territorialities onto a landscape that had been discursively emptied more than a century earlier. No longer wilderness, it now became impossible to make recommendations for forest practices—or even to imagine forestry at all—without the input of the Nuu-chah-nulth.[2]

The panel provides a sense of what it might mean to unthink the colonial logic of forestry and environmentalism on Canada's west coast. Despite its limitations, the panel's final reports can be read as a (positive) response to what Homi Bhabha (1994) called the "displacing gaze" of the colonized, a gaze that disturbs a hegemonic science, culture, and politics. This suggests a very different sense of responsibility to what Cronon sketched in his critique of wilderness. For Cronon, responsibility meant taking our connections in the world seriously, and recognizing links between the everyday and the eternal, the local and the global. Such a cognitive mapping is essential. But there are many other responsibilities that must be considered, and not the least among these is a vigilant tracking of the operation of *power* in all its guises and strategies. Avoiding this, ecopolitics becomes irresponsible and its acts of ethical responsibility risk new, more subtle, forms of domination.

Assemblages: Toward a Philosophy of Immanence

My focus in the preceding section was on the (im)possibility of arriving at any final definition of the forest (and, in turn, how a good conscience never rests). This moves environmentalism some distance away from a theory of representation and its assumption of a world of things more or less adequately mirrored by language, and toward a reflexivity about its governing concepts and metaphors. But throughout this book I have also hinted at something else: that perhaps nature and culture are simply the wrong categories by which to think about the landscapes of Canada's west coast. One of my objectives has been to track the *career* of these terms; that is, how something called nature and something called culture coalesce from within the shifting material-semiotic fields that compose life in the region. Perhaps nature and culture have no prior, independent existence, but are merely rhetorical containers into which different wordly elements are somewhat arbitrarily assigned: the result may be conceptually tidy, but it is a temporary measure at best, and, as we have seen, soon everything is a mess again.

Perhaps there is no easy way of dividing the world into determinate sets of things: nature there, culture here, and in between a zone of illegitimate, hybrid couplings. I owe this insight in part to Bruno Latour (1993), who has argued persuasively that it is a peculiarly modern conceit to believe that the world can be divided in two. For Latour, and also for Donna Haraway (1994), the world exists only as so many imbroglios in which culture, politics, machines, institutions, and organisms are mixed together. Our mistake, both argue, has been to try to untangle these knots with the hope of recovering some lost essence. In so doing, we miss much more than we see. Both authors believe that there is not much use in trying to recuperate these terms; at best, nature names a desire, not a place (Haraway 1991).

It is perhaps Gilles Deleuze and Félix Guattari (1977) who have made the most sustained and rigorous attempt to displace these terms and to replace them with a new vocabulary altogether. For them it would not be enough to speak of tangles, knots, or imbroglios; these still give a sense of *combination* and *mixing* whereby previously distinct realms are increasingly brought into relation, until they bleed into each other so completely that they become indistinguishable. There is a sense in this that perhaps at one time it *did* make sense to talk about the two separately. Deleuze and Guattari argue that this would be to remain within "identity thinking," modified now only by the sense that the mixings are so profound and so pervasive that singular, static identities are no longer possible. There is still a kernel of nostalgia in the language of mixing. Instead, Deleuze and Guattari argue for a philosophy of immanence, a position that has antecedents in Spinoza and Nietzsche and that has much in common with many poststructuralisms. Although initially counterintuitive, their central insight is actually relatively simple (although its implications are not). Far from the world being composed of discrete things that are subsequently brought into relation, such things exist in the world only by virtue of the relations that constitute and sustain them. What exists are only "flows" and "intensities" that give rise to provisional and temporary effects (stabilities) that we mistake as static, ahistorical, ontological forms: bodies, the state, machines, nature. In their view, things in this world are only transitory moments within the relations that define them, but which, ultimately, remain external to them. Such things may become relatively fixed, but there is no transcendental force, or extrahistorical order, that determines their existence in advance. Thus, it makes no sense to speak of nature and culture, or a "nature-culture dialectic," or even of the popular notion today of the "coproduction" of nature and culture (Norgaard 1994). For Deleuze and Guattari, there are only heterogeneous "assemblages" within which things as such exist, but that as a whole and in their parts have no necessary or final form. Thus,

> there is no such thing as either man or nature now, only a process that produced the one within the other and couples the machines together. Producing-machines, desiring-machines everywhere, schizophrenic machines, all of species life: the self and the non-self, outside and inside, no longer have any meaning whatsoever. . . . [M]an and nature are not like two opposite terms confronting each other—not even in the sense of bipolar opposites within a relationship of causation, ideation, or expression (cause and effect, subject and object, etc.); rather, they are one and the same essential reality, the producer-product. Production as process overtakes all idealistic categories and constitutes a cycle whose relationship to desire is that of an immanent principle. (Deleuze and Guattari 1977, 2–3)

The predominance of *production* metaphors (especially the machinic) may seem a bit odd, given efforts to counter the productivist tendencies of Western thought (and global capitalism), in which nature is merely a bundle of resource for human development. These metaphors are meant do other work here—to point to the "always becoming" nature of the world. Thus, assemblages are machinic because they everywhere give rise to new forms and combinations. They are not speaking here (only) of production in the industrial sense, machines in the technological sense, or desire in any simple Oedipal story. Assemblages are machinic because they are productive of identities; assemblages are always in motion, giving rise to new and novel forms. Although this appears impossibly abstract and metaphorical, the authors mean this to be understood in the most insistently material ways. The problem, in their view, is that until now we have not been materialist enough, because by speaking of things and structures we fail to recognize their historicity and their provisional fixity as effects.[3] Categories such as nature, culture, animal, man, woman, machine, bodies, desire, economy are merely abstractions that do not pre-exist their construction. They have no existence apart from the relations that define them. There is nothing transcendental that drives the whole.

This may not be the foreign language that it first seems. Ecologists have argued for some time that entities in nature are nothing apart from the relations—flows and intensities—that sustain them. In this sense, we might say that nature is machinic (but in no way merely mechanical). Indeed, a number of commentators have noted the similarities between Deleuze and Guattari's work and ecological thought (see Kuelhs 1996). The major difference from ecology is that Deleuze and Guattari refuse a priori distinctions between nature and culture: theirs is a monism that sees *all* ecological, social, cultural, and political forms as historical and relational effects. The forest is a shifting set of relations that has no given form. Likewise, the logger who works in the woods is defined both in body and mind by the web of relations (which we assign as political, ecological, cultural, economic) in which he or she is a part. In turn, as I have shown at various points, *being Nuu-chah-nulth* also has no essence; it is defined anew in the present, within the heterogeneous forces re-shaping west coast landscapes (including their own practices of mapping examined in chapter 3).

There is now a large and increasingly familiar set of metaphors that strive to capture this sort of relationality: networks, webs, assemblages, rhizomes, cyborgs, topologies, cartographies. Each in its own way seeks to avoid falling back into a nature–culture dualism; each emphasizes the contingency of the present, and asks how it is that some things come to be fixed. For my purposes, the notion of assemblages is attractive be-cause it captures both the materiality of the world and its contingency. It retains a sense of the verb—to assemble—within its ontology. Assemblages may be material, but they are hardly fixed once and for all. Each has many rhythms, not one; each moves at several speeds rather than a single; each gives rise to new possibilities of becoming. This does not mean that there are no permanences—landscapes change slowly, identi-ties sediment and harden, and institutions become seemingly intractable. As Deleuze and Guattari (1977, 20) famously put it: "there are knots of arborescence in rhizomes." Movements get blocked, affects fixed, forms and systems organized. This is not always bad; we need organization, institutions, and regularities of all sorts in order to act in the world, individually and collectively. What the metaphor of assemblage refuses is an ahistoricism that takes the contingent as the eternal; it reminds us that ecosystems, forest services, technologies, economies, and traditions all dissolve and reform, become deterritorialized and reterritorialized.

What makes these metaphors so powerful is that they pry open the present in order to interrogate its conditions and to imagine other possible futures. Assemblages, rhizomes, networks—these terms provide critical leverage. "What must be compared," Deleuze and Parnet (1987, 134, 135) argue, "are the movements of deterritorialization and the processes of reterritorialization which appear in an assemblage. . . . it is in concrete social fields, at specific moments, that the comparative movements of deterritorialization, the continuums of intensity and the combinations of flux that they form must be studied." Indeed, if the world was not characterized by flux, if that which drives change was transcendental, otherwordly, perhaps magically written in the stars (or, for that matter, in nature), there would be no need for politics. Indeed, there would be no *possibility* for politics. The world would collapse into the eternity of the same. Politics, Deleuze and Parnet argue, is "active experimentation." On the one hand, it is about asking of assemblages: What type is it? How is it fabricated? By what procedures and means? And, on the other, it is about opening space for thinking, doing, and being otherwise. It is a politics with a purpose, but without any certain or final outcome, and perhaps it is this paradox that ecopolitics on Canada's west coast must courageously embrace.

New Trajectories of Social/Nature

This sort of "active experimentation" is precisely what is occurring today on Canada's west coast. Indeed, it always has been occurring, for historical efforts by indigenous peoples, modern states, industrial capital, and environmentalists to fix nature have merely been elements in this process, even if understood otherwise. George Dawson's maps of primitive and national space, Justice Sloan's fantasy of the forest-as-machine, ecologists' dream of unity and balance, even Nuu-chah-nulth concepts of *hishuk ish ts'awalk* ("everything is one") and its system of hereditary ownership and control of traditional territories *(ḥaḥuulhi),* have contributed to producing the social natures present today on Canada's west coast. Today, new concepts and actors are remaking the temperate rainforests of Vancouver Island and its social and ecological relations in new, unpredictable ways. In the years since most of the research for this book was done, treaty negotiations between First Nations and the federal and provincial governments—although painstaking and slow—have progressed. In northern British Columbia, the Nisga'a concluded treaty negotiations in 1998, leading to new levels of Nisga'a control over politics and resources in the area. These are controversial agreements, assailed both by conservative factions in white society and by First Nations groups who argue that too much compromise has been made and who refuse the extinguishment of Native title. Although such treaties "fix" responsibilities and relationships, they are perhaps better seen as part of the ongoing reworking of nature, culture, and politics in the region.

Similar changes have been occurring on a smaller scale on the west coast of Vancouver Island. Here, decades of activism by the Nuu-chah-nulth and allies, including groups such as the Western Canada Wilderness Committee, have led to new modes of governance, and new ventures in the woods. Today, decisions concerning the forests of Clayoquot Sound are not made only in distant boardrooms in Victoria and Vancouver (or, more recently, Seattle). The Nuu-chah-nulth are now full participants in governing bodies such as the Central Region Board, which plays an important role in how, when, and where forestry operations take place. Recently, environmentalists have nervously watched from the sidelines as the Ahousaht, Hesquiaht, Tla-o-qui-aht, Toquaht, and Ucluelet First Nations entered into a joint venture with MacMillan Bloedel, in a bid to reassert their place as both occupants of, and agents in, the forest. At present the Ma-Mook Development Corporation has plans to log in the northern part of Clayoquot Sound. These developments do not overnight resolve more than a century of colonial displacement. Reimagining the forest is not the only politics needed for a complete decolonization. Indeed, a case can be made that recent developments may merely incorporate the Nuu-chah-nulth *into* the existing state-industry forestry system, thereby allowing the externalization of nature achieved through capitalist enterprise to continue apace, and benefits to continue to flow to the state and transnational forest companies (Kuehls 1996). Yet, there is a clear sense that the hold of discourses of "natural resources" and "primeval nature" on the West Coast of Vancouver Island is lessening. Environmentalists have responded in different ways, some defending their notion of nature as separate and eternal, and thus shifting their critical attention from forestry companies to the activities of First Nations, others seeking to imagine alternative futures no longer governed by the colonialist tropes of the past. This uneven response points perhaps to a moment of conceptual crisis—and thus a potential moment of transition—in BC ecopolitics. The discourses that have governed forest politics since the Sloan commission are shifting. New forest imaginaries are coming into play, new political alliances being forged. The challenge for all involved—First Nations, environmentalists, the state, academics—will be to engage in the sort of active experimentation needed to build social natures on Canada's west coast that take seriously the cultural politics of the forest and its colonial pasts. In this moment of danger is perhaps a moment of great hope.

Notes

1. The Intemperate Rainforest

1. Protesters were charged with criminal contempt of court for failure to obey a court order prohibiting them from blocking the road. The court order was the result of earlier procedures in which MacMillan Bloedel had filed civil suits against environmentalists blocking logging roads. After the suits were filed, the company asked the court for an injunction barring further road blockades while it prepared for the court trial. The deferment of the civil suit therefore became a way of extending the length of the injunction. Many protesters noted the irony that the injunction had been issued on behalf of a corporation in a conflict over lands the corporation did not own. See Hatch 1994.

2. Protesters were sentenced to fines of up to three thousand dollars and imprisonment of up to sixty days.

3. Valerie Kuletz (1998) documents the social construction of "sacrificial landscapes" in the deserts of the American Southwest, as part of the United States' military-industrial complex. I suggest that this sacrificial logic is a *structural component* of the nation-state within global capitalism.

4. Pulp and Paper Workers of Canada (PPWC) and International Woodworkers of America (IWA). These unions held quite different positions. The IWA consistently resisted calls to reduce levels of harvest, which would have had serious impacts on its constituency working in the woods. The PPWC, on the other hand, represented workers in the forest industry who were further removed from the woods, and not as directly affected by reductions in the logging of old-growth forests. Its position at the time was more aligned with the concerns of environmental groups in the province.

5. The Nuu-chah-nulth (literally, "all along the mountains") live along the west coast of Vancouver Island and consist of fourteen groups organized under the umbrella of the Nuu-chah-nulth Tribal Council. The council is divided into northern, central, and southern districts. Clayoquot Sound lies

in the central district and includes the traditional territories of five bands (Tla-o-qui-aht, Ahousaht, Hesquiaht, Toquaht, and Ucluelet). In anthropological literature, the Nuu-chah-nulth have long been referred to as the "Nootka," an appellation commonly attributed to Captain Cook.

6. British Columbia has numerous environmental groups, and their views vary considerably. Key actors include the Western Canada Wilderness Society, Sierra Club of Western Canada, Wilderness Society, Friends of Clayoquot Sound, and Valhalla Society. The Rainforest Action Network has been most active since the second half of the 1990s and represents a coalition of green movements in the province. The Western Canada Wilderness Society and Rainforest Action Network have been the most consistent in their attempts to articulate environmental issues with First Nations land rights.

7. The New Democratic Party took power in the provincial elections of 1991. The land-use plan of May 1993, developed and approved by the NDP government, was the catalyst for the massive protests of that summer and reinforced divisions between labor and ecology activists in the province.

8. For an effort to articulate the basis for an environmental democracy in BC, see Mason (1999).

9. The one exception to this was Meares Island, which was excepted from the plan on account of an unresolved court case involving the Tla-o-qui-aht, whose village, Opitsat, is located on the island, and MacMillan Bloedel, the company with state-sanctioned logging rights to the island's forests.

10. Nuu-chah-nulth leaders simply rejected the whole decision, claiming that the province had no jurisdiction over these lands. Since the Clayoquot land-use decision was announced in 1993, a series of further agreements have been appended that seek to incorporate Nuu-chah-nulth members into decision-making processes in the region. This has included Nuu-chah-nulth elders being included as "experts" in a scientific panel investigating forestry practices in the region (see chapter 6), and equal Nuu-chah-nulth and provincial membership on a panel reviewing land-use plans for the region. Both developments occurred only after significant resistance on the part of the Nuu-chah-nulth. The region as a whole remains subject to land-claims negotiations.

11. Although increasingly commonplace in critical studies of the environment, a few writers have stood out as key figures in the turn to understanding nature in terms of the interweaving of nature, culture, and politics. These include Michel Callon (1986), Donna Haraway (1989, 1991,

1997), Bruno Latour (1993, 1999), and Alexander Wilson (1991). For a survey, see Castree and Braun (1998).

12. Specifically *capitalist* productions of nature have been most thoroughly examined through numerous articles in two important journals: *Antipode* and *Capitalism, Nature, Socialism.*

13. A number of critics, both inside and outside Western Marxism, have indicted historical-materialist accounts for failing to adequately recognize the "materiality" of nature, or, in other words, the ways that nonhuman nature is an agent and presence in history (see Worster 1988; Benton 1989; Castree 1995; Fitzsimmons and Goodman 1998). Too much attention, they feel, is placed on human actors alone, such that it appears that there are no constraints placed on how nature can be produced. Clearly, there are grounds for such a charge: Marx viewed human labor as *the* "motive force" in nature's "metabolism." But there is no reason why social nature must be anthropocentric in this sense. Although not addressing Marx directly, writers such as Michel Callon (1986), Bruno Latour (1993), and Donna Haraway (1991, 1997) have insisted that not all the actors in nature's production are human. Yet the language of nature's "agency" is misleading. Consistent with his ontology of "networks," Latour uses the term *actant* to draw attention to the ways that agency is distributed across—and through—networks of relations, rather than arising from some site outside these networks, or as something that is an inherent quality of things (including people). Further, there is no way to describe this agency, nor to know the impact of human activities on a nonhuman world, apart from specific situated knowledges. This does not mean that agency does not exceed the human; rather, it points to the risk that in talking of nature's agency we may reintroduce an unacknowledged realism to the study of society–nature relations (see Lenoir 1994). Haraway's (1991) metaphor of the trickster points to this distribution of agency, without giving it a fixed address.

14. Robyn Eckersley (1992) argues that Marxist conceptions of nature reinforce an *instrumental* relation to nonhuman nature, and are thus irrevocably complicit with the *domination* of nature. This is an important argument, one that injects a needed measure of humility into what risks becoming an exuberant—and violent—humanism of the worst sort. Yet, as important as this critique is, it is difficult to imagine a relation to the nonhuman world that can be neither anthropocentric nor instrumental, simply because our relation to things is always already mediated through language and practice (see Bennett and Chaloupka 1993). In this sense, although biocentrism can be a viable ethical position (i.e., the decision to take into

account a series of relations described to us by the science of ecology, and that transcend the interests of the "merely" human), it cannot claim a transcendental foundation (i.e., we cannot claim to know what it means to "think like a mountain"!).

15. It is worth noting that an ecopolitics based on social nature is not easily mapped onto the "shallow" versus "deep" ecology debate because it rejects the central notion that underlies the debate: that nature is "external."

16. This by no means implies that culture is *prior to* nature in an ontological sense. This would merely invert, rather than displace, the original opposition (Derrida 1982). The argument here is an epistemological and political one about the possibilities of knowledge and associated plays of power.

17. Curiously, although Heidegger uses spatial *metaphors,* he paid very little attention to the *spatiality* of "being." As will become clear in this book, what counts as the rainforest—and our interests and investments in these forests—emerges from multiple spatiotemporalities.

18. Heidegger's argument has important—and often troubling—implications for ethics. For Heidegger, ethics is problematic to the extent that it fails to examine its own conditions of possibility, or, in other words, so long as it remains forgetful of the world's *worlding* (see Heidegger 1977a). Heidegger called for a more "originary ethics" that involved thinking the clearing—*Dasein*—in which ethics occurred. John Caputo (1993) argues that Heidegger's "stepping back" from ethics in the name of something "more primordial and originary" is itself problematic, because it still rests on an account of origins and beginnings.

19. Heidegger (1977b) felt that this "forgetting" was particularly acute in modernity. The danger was that this gave to modern representation the character of *gestell* (stamping), whereby representation functioned to fix identities as immutable. Gayatri Chakravorty Spivak (1988a) captures this conjoining of knowledge and power particularly well in her notion of "epistemic violence."

20. In more recent work Smith (1996) has modified this position, noting that the "conceptual" construction of nature is also important, and that the production of nature is as much a cultural as it is an economic process.

21. Donna Haraway's desire to "queer" what counts as nature fits well within this theme. Like Foucault's "analytics of truth," she explores what functions of power our truths about nature serve, what they conceal, and what actions or practices they enable or authorize. Such critique does not prescribe political positions or specify desirable futures, but neither is it disinterested. Quite the opposite, it takes what is closest and most pressing and

converts it into an expanding series of questions in order to provide new sites for political agitation or the pursuit of alternatives.

22. Perhaps symptomatic of critical theory and cultural studies more generally, very little has been written within postcolonial studies on the relation between colonialism and nature. However, see Grove 1995, Arnold 1996, Beinert and Coates 1995, Neumann 1998, Prakash 1999, and Sivaramakrishna 1999.

23. Crucially, different diasporas must also be understood in terms of forms of class mobility within a globalizing capitalism, whereby certain postcolonial subjects are more free to move than others. See Dirlik 1994, Valie and Swedenburg 1996, and Mitchell 1997.

24. Antitreaty positions have been a mainstay of the Liberal and Reform parties in BC, as well as the Reform Party nationally (renamed Canadian Alliance in 1999).

2. Producing Marginality

1. Pearse Commission public hearings, Victoria, 30 October 1975. Until the 1980s, anthropologists referred to the indigenous peoples of the west side of Vancouver Island as the "Nootka," an appellation commonly attributed to Captain Cook. The West Coast District Council of Indian Chiefs is now known as the Nuu-chah-nulth Tribal Council and consists of representatives of the council's fourteen member tribes from the west coast of Vancouver Island.

2. Two other First Nation groups also made submissions to Pearse. These were the Skidegate Band Council (Haida) and the Nicola Valley Indian Administration (Nlha7kápmx). Both made oral and written submissions.

3. In a report released six months after the decision, the provincial ombudsman claimed that the Nuu-chah-nulth had not been adequately consulted during the events leading to the development and release of the controversial land-use plan (Government of British Columbia 1993a). Following the release of the report, an interim agreement was signed between the central-region tribes of the Nuu-chah-nulth Tribal Council and the provincial government that incorporated the Nuu-chah-nulth as comanagers of certain regions. Interim agreements are increasingly prevalent as First Nations and the provincial government seek ways to manage resources on traditional Native territories while treaty negotiations are ongoing.

4. The 1945 Sloan Commission Report provided recommendations to the provincial government for setting in place a system of sustained-yield forestry in British Columbia. Most of these were incorporated into the Forest Act in 1947.

5. About the same time as the Clayoquot Sound decision, the Canadian government mounted an international tourism campaign that drew explicitly on the mythic place of its indigenous peoples as progenitors and spiritual resources of the "nation."

6. _Hahuulhi_ is the name given by the Nuu-chah-nulth to the system of resource ownership, control, and use practiced by its various constituent groups. Resource-procurement sites were owned by individual chiefs and this was recounted and reinforced in oral traditions during feasts and other cultural gatherings. Along with ownership came certain responsibilities (see Scientific Panel for Sustainable Forest Practices in Clayoquot Sound 1995a).

7. The abstraction and displacement of the local into the global has become a well-rehearsed theme and it is not only _aboriginal_ communities that are marginalized by such processes (see Hecht and Cockburn 1989). However, in the case of British Columbia, for this abstraction and displacement to proceed, a _Native_ presence must be at once erased or marked in ways that de-link Natives from their surrounds. This occurs in different ways from the marginalization of _other_ social groups.

8. Nor do I subscribe to the position that there is a prior Native identity and politics that lie outside the subject-constituting processes of a capitalist modernity. See Spivak (1988a).

9. For a discussion of the politics of the "tribal slot," see Li (2000).

10. This rhetoric is prevalent among members and supporters of the BC Liberal and Reform parties and the Canadian Alliance at the federal level.

11. In his _The Mirror of Production,_ Jean Baudrillard (1975) shows how Marxist theory cannot escape the orbit of a bourgeois capitalist imaginary. In a similar way, green movements are irrevocably (and necessarily) tied to systems of nature's production in industrial capitalism. Thus, the preservation of nature comes to mirror the relentless commodification of nature, as capital stalks the earth in search of (surplus) value. Indeed, one of the ironies of the environmental movement's attempts to preserve wilderness is that it can do so only through the language of "value" (see Demeritt 1997; Harvey 1996).

12. It can be argued that all cultures, and not only those of advanced capitalism, displace nature into abstract systems of commodities and exchange (see Appadurai 1986).

13. In 1999, MacMillan Bloedel was purchased by Seattle-based Weyerhaeuser, consistent with recent trends toward consolidation in the global forest industry.

14. The focus on MB was the result of a number of factors: as one of the few large corporations working in the forest that began as a local firm, it

had heightened visibility; its forest tenures included some of the most spectacular stands of old growth on the coast; these tenures were located in regions accessible to urban dwellers in Vancouver and Victoria; and the company had at different times aggressively challenged environmentalists, resulting in considerable media exposure. During the writing of this book, MB softened its antagonistic stance to the environmental movement, especially in Clayoquot Sound, and began to explore corporate ventures with the Nuu-chah-nulth. The fate of these new directions was uncertain after the company merged with Weyerhaeuser in 1999.

15. MB found itself in the position of defending its position as "steward" of a public resource owing to the nature of BC's forest tenure system. Much of the coastal rainforest remains public lands, but these are divided into "tree farms" and leased to forest companies for lengthy periods (the tenure is perpetually renewed unless the companies do not meet certain forest management objectives).

16. The BC Forest Alliance, an umbrella group representing forest companies active in the province, took a lead role in promoting the industry's positions. Complicating the scene of forest politics in BC, the head of the BC Forest Alliance for much of the 1990s was Patrick Moore, one of the original founders of Greenpeace.

17. The level to which these itineraries were consciously deployed by the company, or the extent to which they simply existed as forestry's unconscious, is open for debate. Certainly, with the increased involvement of public-relations firms during the period, companies such as MB became acutely conscious of the many different levels at which politics infused forestry discourse.

18. This was one of a series of publications distributed by MB during the 1990s.

19. Greenhouses purported to be growing genetically modified (GM) seedlings have been attacked by Canadian activists, mirroring campaigns in Britain and the United States that have destroyed GM food crops.

20. MB closed its visitor centers in the late 1990s.

21. MB made learning about the forest and forest practices fun. Further, this public face allowed the company to appear open and honest, responding to the widely held belief that major decisions about BC's forests occurred behind closed doors and that forestry companies were not forthright about their intentions.

22. Habermas's argument was first articulated in *Toward a Rational Society* (1971) and *Knowledge and Human Interests* (1972). Here Habermas develops an argument that knowledge is grounded in two quasi-transcendental cognitive interests: social labor and social interaction. The first is organized

around instrumental action (and thus involves the realization of a technical interest that relies on "empirical-analytical" sciences). The second is organized around communicative action (and thus involves a practical intent that relies on historical-hermeneutical sciences). Habermas argued that problems arose when technical interests came to replace practical interests.

23. Haraway argues that gender figures centrally here too, because women are often seen to be too invested in the world to be objective.

24. The question of waste in timber harvesting has long been a contentious issue. Initially, it became an issue for conservationists worried over the long-term sustainability of harvest levels (Hays 1959). More recently, the "gospel of efficiency" has taken a new turn. Environmentalists argue for increased efficiency and wood recovery (both in the forest and in mills) as a way of reducing the spatial area needed for industrial forestry.

25. In 1993, at the height of BC's war in the woods, a forest industry lobby group estimated that forest products accounted for more than half of the province's manufacturing shipments, that the industry accounted for 17 percent of the province's GDP, and that 16 percent of the province's workforce was directly or indirectly employed in the forest sector (Forest Alliance of British Columbia 1994).

26. As Bruce Robbins (1993) notes, appeals to a "public interest" are equivocal. On the one hand, in Western democracies the "public" has served as a rallying cry against private greed, propertied interests, and bureaucratic secrecy. But, as in the case of BC forestry, it has equally served to silence so-called minority concerns (see also Fraser 1991; Polan 1991).

27. The normalization of nature in forms of modern power remains undertheorized. Michel Foucault (1979, 1980), for instance, rarely looked beyond human subjects, bodies, and institutions, but clearly the normalization of life that he documented with such brilliance—its ordering and disciplining through modalities of power, knowledge, and spatiality—extends to and incorporates not only human subjects but "nature" itself. The regulation of populations and economy in BC required not simply the exploitation of the forest but its *construction* in discursive practices that at once constituted, rendered available, and rationalized the forest within an administrative apparatus, making it adequate for models of social and ecological productivity. For a discussion of Foucault and nature, see Darier 1999, Rutherford 1999.

28. These reserves were established in the absence of formal treaties between First Nations and the government of Canada (see Tenant 1990).

29. These commissions receive yet another treatment in Harris (forthcoming).

30. In his *Islands of Truth* Daniel Clayton (1999) traces the construction of the west coast as an "imperial space" through the transatlantic circulation of knowledge/power by which encounters on the west coast dating back to as early as the late eighteenth century were remembered and reassembled as systems of knowledge in Britain, eventually contributing to imperial projects in the region in the nineteenth century.

31. Euro-Canadian opinion on the size, or reasonability, of Indian reserves was not monolithic. One of the first IRC commissioners, Gilbert Malcolm Sproat, became a sharp critic of the process, and was eventually replaced by commissioners more in line with the opinions held by government leaders in Victoria.

32. I borrow the phrase "modest witness" from Shapin and Schaffer (1985). See also Haraway (1997).

33. The erasure of Native presence—textually and physically—occurred in many ways, and was uneven across the Americas. Historians of the American West, for instance, have emphasized how the frontier mythology was central to the removal of Native people from their lands (Drinnon 1980; Slotkin 1985; Limerick 1987—Limerick argues that this continues to underwrite American imperialism). In Canada, frontier mythologies did not take hold in any comparable way. Regardless, what I trace here is not the evacuation of the real into mythology (and thus into the realm of the untruthful), but how locating the real or the truthful through representational practices became aligned with colonialism. In a sense, the subtitle to Drinnon's book—*The Metaphysics of Indian-Hating and Empire Building*—aptly captures this conjoining of knowledge and power in the marginalization of Natives, even if his account does not work directly with this constellation of ideas.

34. For a good discussion of the role of Native actors in the production of imperial knowledge, see Raffles (forthcoming).

35. It is no accident that Dawson first traveled to the west coast as part of a joint British and American survey of the international boundary between Canada and the United States.

36. Although these details appear mundane, they are far from unimportant, for they give a sense of the complex cultural and political relations that Europeans and Natives alike were compelled to negotiate in their travels and economic activities.

37. Not unexpectedly, what emerges in Dawson's journals is a much more developed sense of the social and cultural networks, and technical practices that facilitated his journeys, including the labor and knowledge of Native actors.

38. In his journals, these follow each other.

39. Dawson's photographs are collected in the National Archives of Canada, Ottawa, Ontario.

40. Dawson's study of the Haida was originally published as an appendix to the GSC's *Report of Progress for 1878–79,* but was later reprinted separately in *Harper's Magazine.*

41. It is worth noting in passing that Dawson was part of a committee that brought Franz Boas to the west coast.

42. My brief sketch of the history of geology draws largely on Rudwick (1976, 1996), Porter (1977), Guntau (1978), Secord (1986), Landau (1987), Stafford (1990), and Stoddart (1995).

43. Writing in the *Canadian Pacific Railway Report,* Dawson (1877, 227, 234) notes of western anthracitic coals: "Valuable deposits may, however, yet be found in the carboniferous formation proper of the far west; and where, as on some parts of the west coast, the calcareous rocks of this age are largely replaced by argillaceous and arenaceous beds, the probability of the discovery of coal is greatest. I believe, indeed, that in a few localities in Nevada, coal shales, used to some extent as fuel in the absence of better, are found in rocks supposed to be of this age. The discovery of certain fossils in 1876 in the limestones of the lower Cache Creek group now allow these, and probably also the associated quartzites and other rocks to be correlated with this period. . . . Rocks of the same age with the coal-bearing series of the Queen Charlotte Islands are probably present also on the Mainland, where fossils indicating a horizon both somewhat higher and a little lower in the geological scale have already been found, and apparently occur in different parts of a great conformable rock series."

44. A *British Colonist* (Victoria) editorial from 27 July 1863 makes this explicit: "Every school in the colonies where boys are taught should make these branches [geology and mineralogy] part and parcel of its curriculum. Small cabinets of rocks and ores could be easily made or imported for the purpose of giving the pupils a practical acquaintance with the subject matter of those sciences. . . . The mountains, the hills, and the rocks of the island and the mainland would be no longer trodden over in ignorance without attention. . . . Combining this acquaintance with theory they may learn from books, they would in their prospecting tours be alive to metaliferous indications, and would no longer walk blindfolded, passing unconsciously material for untold wealth, as must now be often the case" (De Cosmos 1863, 2).

45. For a detailed discussion of this "modern theology," see Timothy Mitchell (1988).

46. This is evident in an analogy that appears early in Sloan's (1945, 19) text that is worth quoting in full: "If there were a mountain near Vancouver with a gently ascending slope, the climber would find as he progressed upwards that beyond the 2,000-foot line a gradual change in the forest species was encountered. He would notice the Douglas fir was thinning out and the stand was now made up of cedar, hemlock, and balsam, in that order of importance. Still climbing, he would find himself in a forest of hemlock, cedar, spruce, and balsam. Higher up his forest would now be hemlock, balsam, spruce, and cedar. Soon the cedar is left below and the hemlock, spruce, and balsam remain in that order. Should he persist in his climb, he would get into scrub and non-commercial mountain species.

"Now, let us conceive of our gradually ascending slope, not as a mountain near Vancouver, but as the coastal plane of the province, stretched out from south to north. Let us assume our climber is traveling north up the latitudes instead of up the mountain. He would come upon the same general classification of forest-cover in the same order of species as he encountered on our imaginary Vancouver mountain." Although it reads as a departure from the more scientific tone of the rest of his report, the passage deserves more comment than historians have given it. Sloan begins with a metaphorical mountain, and then switches to the reality of the coastal rainforest. Crucially, both the metaphorical and the real are abstractions, in which the forest is presented as full of timber but devoid of people. By staging the forests in this way, Sloan was able to generalize a model of sustainable forestry—and an accompanying tenure system—across the *entire extent* of known exploitable forest reserves.

47. Arguably, this displacement became more acute in the 1980s and 1990s as easily accessible areas within specific "working circles" were exhausted and forestry moved to ever more remote sites.

48. Even if taken up, Pearse's recommendations would have made little impact on how deeply Native communities were infused by colonial relations. "Band management" was, at best, euphemistic, because the financial affairs of bands were administered by a paternal federal government.

49. It should also be noted that by employing such an approach I depart in significant ways from the current fascination in British Columbia with roundtables on the environment, which assume that such arenas provide possibilities for "ideal speech situations" (see Mason 1999). Although these arenas do often increase possibilities for participation, they do not by themselves mitigate the relations of power that are inscribed into public debate through the categories and identities by which conflicts are organized and

understood. By establishing their resolutions as products of "open" public processes, existing relations are often legitimated.

50. The case, known as *Delgamuukw v. Her Majesty the Queen*, sought to establish questions of sovereignty over the traditional territories of the Gitksan and Wet'suwet'en peoples. Originally rejected by Judge McEachern of the BC Supreme Court, many portions of his ruling were overturned in an important Canadian Supreme Court ruling in 1998. See also Persky (1998).

3. "Saving Clayoquot"

1. See Drushka, Nixon, and Travers (1993), Berman (1994), and Rajala (1998).

2. Here I leave aside the problem of overcutting in the 1970s and 1980s, which exacerbated tensions in the 1990s and placed even more attention on remote forests.

3. The designation "wild side" used by environmentalists refers both to the grandeur of the region's natural forces and to the limited scale of development in the region, in comparison to the east side of the island facing Georgia Strait.

4. The forests of Clayoquot Sound are situated in two Tree Farm Licenses (TFLs), each controlled by one of the companies.

5. Most prominent in these efforts was the Clayoquot Sound Sustainable Development Task Force. Representatives withdrew from this group when it became clear that logging would continue as the task force met. For discussion, see the essays in Berman (1994).

6. Forestry companies and the BC government, in contrast, only slowly came to recognize that globalization was much more than an *economic* phenomenon.

7. One of the most effective transnational strategies used by British Columbia's environmentalist movement was to make visible the ecological relations of consumption for various forest products. Greenpeace, for instance, was highly successful in its campaign directed at the readers (and publishers) of German news magazines, resulting in many publications switching to alternate paper sources. Arguably, Greenpeace succeeded whenever it could raise the specter of a boycott.

8. The internationalization of British Columbia's environmental conflicts raised a number of important issues, such as how conventional social and political identities such as nation, community, citizenship, and sovereignty were being reconfigured by new spaces of publicity, questions about who is included and who is excluded from these spaces, and concerns over

the effect that the displacement of local politics into global circuits of information, images, commodities, and capital has on local political autonomy. In terms of the latter, residents of forestry-dependent communities expressed resentment at the intervention of "outsiders."

9. These questions are consistent with much ecophilosophy that finds in Western industrialized societies a narrow instrumental relation to the nonhuman world (see Naess 1989). Against this instrumental reason is posited a holistic or spiritual relation that is assumed to be more primary or original (Devall and Sessions 1985). Others have questioned the possibility of transcending instrumental relations, arguing that our understanding and experience of nature is always bounded in *language* and *practice,* suggesting instead that all that can be fostered within these power-charged fields is an *ethical* relation to the nonhuman. These questions are addressed in the collection of essays edited by Bennett and Chaloupka (1993).

10. I use the dual term *sight/site* to intentionally foreground representation as involving both "seeing" and "making objects to be seen" (Shapiro 1993, 129). In this first sense, *sight* refers also to the corporeality of sight, positioning representation resolutely in the historical, embodied viewer, rather than prior or external to the particular historical, social, political, and technical dynamics within which observers are constituted (Crary 1990). The second sense—*site*—refers to the historical practices by which things come into presence as visual objects. By writing of vision as the product of many practices, the still widely held belief that seeing is both passive and transcendent can be shown to be ideological in that it refuses to acknowledge how vision is located and enabled in specific material and discursive practices.

11. To start from this position (Native land rights) does not mean that the position is unassailable. Rather, it exists as a necessary political fiction that enables critique.

12. In the United States, the Sierra Club pioneered the method during the 1930s, when it published Ansel Adams photographs in a bid to preserve Kings Canyon, California, from resource development. Although such volumes are effective, the costs of producing them are enormous. The publishers of *On the Wild Side*—Western Canada Wilderness Committee—claimed that they lost money on the venture, although this is hard to measure: it is impossible to determine to what extent money the organization raised through donations could be attributed to the ways in which the volume raised awareness.

13. Cameron Young makes exactly this point in the book's introductory pages. "This is Adrian's book, a testament to his commitment to the wild side of Vancouver Island. Over the years he has assembled a spectacular

portfolio of wilderness photographs of the coast, and believed that a selection of this work might reach a wide audience with a fundamental message: the remaining wilderness areas on the west coast of Vancouver Island are in urgent need of protection. . . . I was honoured when asked to write the supporting text" (6).

14. For further critical discussion of photography, see Barthes (1981) and Sontag (1977).

15. Cultural geographers have led attempts to analyze how landscapes, fields of vision, or seeing more generally are constituted within cultural and social dynamics. See Cosgrove (1984), Cosgrove and Daniels (1988). Gillian Rose (1993) argues that this literature has all but ignored important questions concerning the gendered construction of both visual fields and viewing subjects.

16. In this chapter I explore how the environmental movement enacts closure around "wilderness." This is not meant to suggest that books such as *On the Wild Side* are closed to contesting interpretations. No doubt Native readers would read the book against the grain, but not necessarily in the same way that I do (and, given the contested character of nature, wilderness, and tradition in Native communities, these readings would invariably be multiple). The book, however, is clearly directed at an urban, middle-class audience for whom wilderness preservation is a compelling issue, and which has the political clout to push for state intervention. When, in the preface to the book, Robert Bateman speaks of "we Canadians," and "our generation," and calls readers to "draw the line" for the defense of wilderness on Canada's west coast, it is no doubt this audience that he has in mind. First Nations people or forestry workers would find it somewhat difficult identifying with this "we."

17. At two places in the book, images of nude women are included (one paddles a canoe, the other bathes beneath a waterfall). No such images of men are included.

18. Francis Bacon's sexualized language is discussed in Merchant (1980). See McClintock (1995) for a discussion of the role of sexualized landscape imagery in European imperialism.

19. There is no mention of the economic relations that enable Dorst's public function as nature's representative beyond a note that Dorst is dedicated to a simple life and that he makes ends meet by selling photographs, wood carving, and contract work for the Canadian Wildlife Service.

20. What counts as Clayoquot Sound varies considerably. For census purposes, the village of Ucluelet is included, as are also the First Nations communities nearby. However, the Clayoquot Sound that is contested by

ecology groups is usually taken to be the region that extends north from Tofino. The population estimates I provide cover this area only.

21. Nuu-chah-nulth communities were previously more numerous (and also seasonal in character). Centralization has been a result of several factors—population decline in the nineteenth century, changing economic practices, and efforts by the government to locate Natives in permanent communities.

22. The equation of wilderness with roadless territory is common in American environmentalism. The implication is that roads open a region to mass use, and thus once a road enters a region, its "nature" is destined to be brought within a cultural rather than natural economy.

23. The Clayoquot Sound environmental campaigns had a shifting cast of characters, but three groups in particular were active throughout the period: the Friends of Clayoquot Sound (FoCS), based in Tofino; the Western Canada Wilderness Committee (WCWC), based in Vancouver; and Greenpeace (operating mostly out of its regional office in Vancouver). It would be a mistake to assume that all three groups—and the people working for them—shared the same views, but there was considerable cooperation between them. Greenpeace, for example, often took its lead from FoCS.

24. Chief Justice Allan McEachern's comments were more extensive: "it would be inaccurate to assume that even pre-contact existence was in the least bit idyllic. The plaintiff's ancestors had no written language, no horses or wheeled vehicles, slavery and starvation were not uncommon, wars with neighboring peoples were common, and there is no doubt to quote Hobbs [sic], that aboriginal life . . . was, at best, 'nasty, brutish, and short.'" (*Delgamuukw v. Her Majesty the Queen* 1991, 129).

25. These are highly contested figures. More problematic, the interest in pre-contact population levels risks reinforcing the assumption that with contact—and rapid depopulation—Native cultures were destroyed. The interest in population quickly becomes an obsession with counting dead Indians, and it is questionable whether reconstructing the former Native life is of much use in anticolonial struggles in the present.

26. Arguably, this has led to the gendering of North American environmentalism, whereby the environments of "home" are the concerns of women (and thus devalued), while wilderness is the domain of "male" environmentalism.

27. This does not mean that there is no such thing as indigenous peoples. Rather, it is to say that the articulation of indigeneity occurs within the field of ideology and politics. There are no guarantees that any given

people will have this identity ascribed to them, or claim it themselves (see Li 2000).

28. The book's dedication also collapses indigeneity and nature, and thereby ascribes to the Nuu-chah-nulth the *same* wilderness ethic as held by the Wilderness Committee: "This book is dedicated to the original inhabitants of Vancouver Island's west coast, the Nuu-chah-nulth, and to all those who *share with them* a love for places *yet untrammeled, wild and free*" (emphasis added).

29. Arguably, the same would be true for photographs in a court case, although as Solnick (1992) shows in the case of *Delgamuukw v. Her Majesty the Queen* (1991), certain forms of vision are often privileged over others.

30. Locke (1967 [1680]) argued that personal property existed because all people had a property in their own person. Therefore, the labor of their body and the work of their hands were such that whatever one "removed from out of a state of nature" and "mixed labor with" was deemed their property. Thus, if land showed no signs of labor, it could not reasonably be claimed as property. Tenant (1990) claims that these Lockean assumptions were widely held in early BC history, and in support of his position quotes from an 1849 letter by Archibald Barcley of the Hudson's Bay Comapany: "With respect to the rights of the natives, you will have to confer with the chiefs of the tribes on that subject, and in your negotiations with them you are to consider the natives as the rightful possessors of such lands only as they are occupied by cultivation, or had houses built on, at the time when the Island came under the undivided sovereignty of Great Britain in 1846. All other land is to be regarded as waste, and applicable to the purposes of colonization."

31. These rhetorics were reproduced in a land-claims trial in northern BC (see Monet and Skan'nu 1992; Solnick 1992).

32. Drucker's work was based primarily on research done in the 1930s.

33. Those most active in the field include John Dewhirst, Arnaud Stryd, Richard Inglis, James Haggarty, David Huelsbeck, Yvonne Marshall, Alexander Mackie, and Gary Wesson.

34. This is consistent with efforts to recognize alternative modernities that are not simply delayed, deficient, or substitute strains of a European norm (Gaonchkar 1999).

35. Premier Harcourt's tour of European cities was intended to forestall a boycott of BC forest products that Greenpeace had threatened to organize. Watts's statements became one of the most provocative aspects of the trip, and the most widely reported in the BC press.

36. Watts's decision was no doubt strategic. Perhaps more important, his decision revealed significant differences within the Tribal Council, and

drew attention to the fact that the various groups that composed the Nuu-chah-nulth were faced with very different economic, political, and ecological conditions.

4. Landscapes of Loss and Mourning

1. Adventure is not unique to the present, although arguably it occurs today in new forms. For discussions of adventure, masculinity, and race at the end of the nineteenth century, see Haraway (1989), Seltzer (1992), Jasen (1995), and Phillips (1997).

2. The group consisted of eight men and five women. Three men and one woman came alone (all were in their thirties). Two other men—a stockbroker and a doctor—came together from Nevada. The remaining clients were couples, with a teenage daughter of one of the them being the thirteenth client. The oldest couple was in their fifties. All clients on the trip were white. Because of the nature of the expedition—clients paid handsomely for the experience—I had agreed with the company that my presence would be as an observer only. As a condition of my access to the expedition, I also agreed that the company would not be publicly identified. These conditions limited my observations to the group's movements, practices, and conversations that occurred as a matter of course during the journey. No formal interviews were conducted.

3. The woman who owned the inn had, like many others, moved to the region in the 1970s shortly after a new highway made the region accessible and the region south of the town was declared a national park. She opened the inn in 1988, and eventually added a bookstore, kayak rental shop, and coffee bar.

4. This was evident at two points: on the second-to-last day of the trip, the group remained land-bound because of fog. Earlier in the week, an effort to navigate the reefs protecting Cow Bay on Flores's west side was abandoned owing to the combination of moderate swell and worries over remaining daylight. The fact that the author had navigated this route the week before in much more severe conditions only underlined the fact that risk on this expedition was carefully managed. Although virtually all adventure travel companies carry insurance against accidents, serious incidents can undermine the company's reputation, and all efforts are made to avoid these.

5. This first leg of the journey is carefully planned to evaluate the group. Although paddlers are asked to rank their abilities prior to departure, such rankings are not considered reliable by the guides.

6. The Christie School is now a substance-abuse treatment center. These old residential schools are important local and historical landmarks and sites of some of the most violent chapters of European colonialism in the

region. Not only were students at these schools removed from their families and forbidden to speak their Native languages, but many were also subject to physical and sexual abuse. To date they have proved difficult to incorporate into itineraries of adventure travel.

7. It is worth noting that unguided *independent* kayakers not part of commercial ventures were more readily accommodated because they signified "hard" travel and thus bestowed authenticity on the commercial expeditions.

8. This is a gendered discourse, to be sure, although, as was evident in this trip, men and women could equally recognize themselves in its terms.

9. Among the most common complaints made in follow-up surveys was that the company's journeys rarely delivered on the promise of seeing whales.

10. In a sense, his shortwave radio and preoccupation with the market helped underline the argument that I make later in this chapter: adventure travel is an *expression* of a capitalist modernity, not its transcendence.

11. This fantasy has been accommodated and commodified by First Nations entrepreneurs in the region, who take clients into the waters of Clayoquot Sound in rustic longboats.

12. As Ian Munt (1994) notes, many alternative travel companies present their journeys as "educational."

13. The guides had also completed an intensive weeklong course on marine biology held at the nearby Bamfield Research Station. Significantly, the guides had no such training in decoding the region as a *cultural* landscape.

14. One of the ironies of adventure travel is that it often is even less integrated with local economies than so-called mass tourism. Many independent (i.e., noncommercial) adventure travelers on Canada's west coast subscribe to notions of self-reliance. What little they take is usually well planned in advance, and taken from home. With the exception perhaps of a postjourney meal before returning to the city, this type of travel remains almost entirely within metropolitan economies.

15. The question of sovereignty is treated in different ways by adventure travel companies. The company that ran the tour described here sought to instill among its clients a respect for Native sites, and asked the clients not to disturb them or to stray into areas that belonged to the Ahousaht as "reserve lands" (often fishing sites at streams entering the Sound's waters). Off-reserve cultural sites—such as "ruins"—were considered public. Unspoken in this discourse of respect is the reaffirmation of state sovereignty over all "nonreserve" areas of the Sound. A map at the back of the inn and kayak shop where clients spent their first night accomplished much the same, color-coding camping beaches in terms of whether they were on reserve

lands or federal lands. Paddlers were asked to avoid the former, but were free to land anywhere on the latter.

16. Most clients spent more money en route to the Sound than they did on the expedition itself. Although it is possible that some of the clients purchased Native arts and crafts before leaving Clayoquot Sound, thus contributing to the Native economy, most clients left the region immediately after the end of the trip.

17. These posters were interesting in themselves, in that AIM is primarily viewed as a movement limited to the United States. The presence of these posters on the west side of Vancouver Island thus troubled discourses of Canadian sovereignty in two ways simultaneously.

18. The nature of the journey—six days with strangers—and the presence of one client openly hostile to "liberals" tended to mute discussion of politics. The guides were also careful to avoid situations that might lead to factionalism.

19. It is important to recognize that many metropolitan subjects do *not* recognize themselves in adventure travel's often masculinist and racialized discourses. I thank Caren Kaplan for consistently stressing this point.

20. During summer 2000, NBC aired a documentary that proposed exactly these explanations.

21. Munt's argument echoes the views of a number of prominent social and cultural theorists, including Pierre Bourdieu, John Urry, and Anthony Giddens. Each in his own way focuses on the question of identity in capitalist modernity. In the most general terms, they argue that with the shift from traditional to posttraditional (i.e., consumer) societies, identity has been dis-embedded from place and tradition and appears as a problem that must be addressed by the individual. Thus, the question, "Who am I?" in capitalist modernity is no longer one given in advance, but instead continuously destabilized and reformulated amid shifting cultural and social fields. Identity, then, becomes a practice rather than a location.

22. The enduring distinction between travel and tourism, for instance, has carried with it assumed differences between authentic and inauthentic, and between individual and mass, with all the class codings that these differences entailed (see Fussell 1980; Culler 1988; Urry 1990). This explains in part why travel (in contrast to tourism) must continuously discover new unspoiled frontiers. The popularity of travel to Egypt among the bourgeoisie in the nineteenth century, for instance, occurred not solely because tropes of Orientalism allowed Europeans to weave imaginative geographies around these sites (Gregory 1995). Rather, this occurred also because a previous marker of class difference—the Grand European Tour—had become

too popular with the advent of rail travel, and thus devalued by its association with "mass" society. Today, differentiations between tourism and travel are played out at new sites and commodified in new ways—for instance, through the specialist tour operator who provides individuated trips to discriminating consumers (see Barrett 1989).

23. Bourdieu's notion of distinction through consumption is often interpreted in an overly voluntarist manner, where subjects strategically choose between forms of consumption. In his concept of habitus, Bourdieu makes it clear that taste is not an open field, but rather something that works through custom or learning to the point where these choices are essentially unconscious. One does not jump between cultural practices any more than one jumps between economic class positions.

24. REI now operates its own adventure travel company (REI Adventures) that offers trips ranging from Antarctic cruises to trekking the Inca Trail in Peru.

25. The journal *Ecotraveler,* for instance, had a regular column titled "Balancing Act" that addressed the question of "whether travel in remote corners of the world does more harm than good."

26. Likewise the photo of the guide—together with a description of his qualifications—underlines the "authenticity" of these supposedly unmediated encounters. The guide, through his identification and intimacy with the area, becomes nature's representative, the decoder of a nature that the client—only recently arrived from the metropolis—is not yet prepared to interpret. Hence, the text underlines the guides' experience, usually measured in years spent in the area or number of trips taken.

27. This search for the premodern lies side by side with the anxiety over its destruction. "Our routes," Ecosummer's advertisement explains, "are constantly changing as one of the aims of this program is to map and document the terrain and the inhabitants of this unknown land which may soon fall prey to a rapidly expanding world." Where Ecosummer sees itself and its clients in relation to this predatory modernity is unclear.

28. The first image, it should be noted, has its own, more subtle erotics—the canoe penetrating the pristine water, the pursuit of elusive wildlife—even if at first glance it has only to do with nature.

29. This returns us to the question of who can recognize himself or herself in adventure travel's narratives. See note 19.

30. Many catalogs of adventure travel companies rate their trips according to degrees of difficulty.

31. This does not contradict the well-worn argument that the tourist consumes signs—in this case, primitiveness, the pristine, and the authentic.

Instead, it situates these acts of consumption within wider discursive and libidinal economies. For discussions of the semiotics of tourism, see Culler 1988, MacCannell 1976.

32. I distinguish my use of the notion of spatial consciousness both from the sense developed by some Marxist critics in which spatial consciousness stands in for *class* consciousness (Fredric Jameson's "cognitive mapping") and from the notion of "imaginative geographies" used by others to speak of the discursive, or ideological, construction of geographies through specific ethnocentric lenses. My point is that a spatial consciousness—much like a temporal consciousness—is constitutive of both the subject and social reality.

33. The fascination with ruins that I noted among travelers in Clayoquot Sound might tempt us to understand adventure travel through Walter Benjamin's discussion of melancholia, but this would be misleading. For Benjamin (1977, 56), melancholia "embraces dead objects in its contemplation," but it does so because it is the transitoriness of *time* that is cherished by the melancholic spirit. For Benjamin, ruins were allegorical; they revealed the temporal, momentary quality of things, showed history to be dialectical, and demonstrated transience (decay) in nature. Ruins belonged to the same category as corpses. In adventure travel and ecotourism, the ruin signifies differently—not as a sign of transience in (human) nature but as a sign of the destructive temporality of a colonizing modernity.

34. The same notion of temporality informs the fetishization of heritage and the rustic and picturesque. See Nochlin (1989).

35. My argument is not that discourses of modernity-as-decline are correct. Indeed, there are good reasons to argue the opposite. Cornelius Castoriadis suggests that these narratives mark not the decline of the West, but, as summarized by Elliott (1996, 10), "the failure of social and political thought to address the imaginative opportunities ushered into existence by modern social institutions and their worldwide spread." My interest lies in the way that these cultural narratives frame the modern and instill a desire for return. In an important sense, adventure travel finds its conditions of possibility precisely in the reduction of discourses on modernity to an either/or, for/against debate.

36. As Elliot (1992) explains, both Adorno and Marcuse drew on the early "biological" Freud.

37. Although rarely discussed together, distinct parallels can be located between Adorno's critique of the self in "totally administered societies" and Lefebvre's (1991) argument that the passage from the premodern to the modern was accompanied by a shift from absolute to abstract space, and thus to the *decorporealization* of the body.

38. Perhaps this is why adventure travel concerns itself not only with its destination, but also with its mode. The point is not only to view the premodern but to be immersed in it, to have a kinesthetic relation to it. Modern methods of travel—planes, trains, and automobiles—are seen as alienating individuals from the physical world; conversely, premodern modes of travel return the traveler to the body and instinct. Thus, activities such as trekking, kayaking, even sailing become privileged vehicles for fantasies of return whereby the body is brought in tune with natural forces. Indeed, similar to the nineteenth century, the effort of travel (travail) becomes again part of the commodity sold. An article in *Ecotraveler* informed readers that one of the advantages of kayaking was that it introduced uncertainty—it provided for a "confrontation" with the unknown, and required "self-reliance" rather than the "crutches" of civilization, which simply insulated the traveler from his or her environs. Self-propelled travel brings self and nature into a unified harmony: "even as a beginner," the author writes, "I felt an intimacy with the water" (Glickman 1995, 8).

39. Anne Beezer (1993) writes that adventure travel is foremost concerned with "stripping away" the "supports" of Western culture. This is consistent with my argument. However, I am less willing to follow Beezer's subsequent claim that adventure travel thus represents an instance of postmodern cultural relativism whereby it becomes a journey toward a new selfless self that is "responsive" to cultural otherness. It seems far more likely, given the continual policing of the great divide between the modern and the premodern, the civilized and the primitive that characterizes adventure travel that the goal behind stripping down the self is precisely to *find* the self rather than lose it, and thus, rather than responsiveness to otherness, adventure travel is deeply narcissistic.

40. In *The Savage Mind*, Lévi-Strauss explicitly questions the opposition between nature and culture. Yet, as Derrida (1976, 105) notes, Lévi-Strauss at once conserved and annulled inherited conceptual oppositions. His thought "stands on a borderline: sometimes within an uncriticized conceptuality, sometimes putting a strain on the boundaries, and working toward deconstruction."

41. Significantly—for it makes an equivalence between the natural and the primitive explicit—Lévi-Strauss describes himself as "plunging" into the primal rainforest in order to meet up with one of the tribes he studied, the Tupi-Kawahib.

42. John Urry (1990) suggests that a growing "posttourist" sensibility is pervading postmodern forms of tourism. The posttourist, he writes, playfully accepts inauthenticity, is aware of the multiple options for the direc-

tion of the tourist gaze, dismisses "high" and "low" cultural distinctions, and knows that tourism is a game (for a summary, see Buzard 1993, 336). In brief, the posttourist engages the tourist experience in an ironic mode. Although it is true that some adventure travelers play ironically with tropes of imperial nostalgia, it appears, by and large, that most attempts to return to a past fullness remain under the thrall of nostalgia.

43. Here I am drawing explicitly on Freud's notion of the pleasure principle as the avoidance of "unpleasure" where unpleasure is understood as that which disrupts constancy.

44. As discussed in chapter 3, what counts as an "economically viable" forest in capitalist forestry is determined within a matrix of commodity prices and costs of production. Historically, what is included in the "viable" forest has shifted along with changes in technology, prices, and costs of production.

45. This is not to say that these communities have been left entirely outside the new tourism economy, although almost all Native development catering to tourists has occurred in or near Tofino.

46. Indeed, a number of companies have started to avoid Clayoquot Sound because the quantity of travelers was undermining the semiotics of adventure.

47. The protection of "viewscapes" is not necessarily a wise decision ecologically. Indeed, by reorganizing production around aesthetics, decisions on such things as tree harvesting may be displaced from ecological rationalities, to aesthetic rationalities, with shoreline "beauty strips" retained while more sensitive or critical landscapes are radically remade under the logics of extractive capital.

48. The trail had been used for generations by the Ahousaht, but had fallen into disrepair as changing transportation technologies and shifting economic and political relations reconfigured Ahousaht spatial practices.

49. The effects of adventure travel vary widely in the Sound, both between Native communities and within them. Residents of the Tla-o-quiat village of Opitsat, for instance, are more integrated into Tofino's tourism economy, and within Native communities clan and status have some bearing on who most benefits from this sector of the economy.

50. These figures were provided by community members.

51. The Western Canada Wilderness Committee benefited from this arrangement too. Its posters and literature were on display in the "Arts of Paawac" center, and its involvement with the project helped deflect criticism that its objectives in the Sound bypassed Native concerns.

52. Organizers estimate that revenue was distributed in the following

manner: 15 percent to the booking agency; 40 percent to the *Spirit of Marktosis*; 35 percent to guides; 10 percent to the Wild Side Heritage Trail. Additional revenues from sale of arts and crafts went directly to local artists.

53. This small trail is a remarkably dense material-semiotic path. Funding for the trail was obtained from a variety of sources: the Canadian government (Youth Service Canada); the Clayoquot Interim Measures Agreement (an administrative unit set up in the wake of Nuu-chah-nulth threats to sue the provincial government over its failure to consult the Nuu-chah-nulth in the events leading to its ill-fated land-use plan); Forest Renewal B.C. (a Crown corporation established in the early 1990s to reinvest state forestry revenues in forest communities); Long Beach Model Forest Society; MacMillan Bloedel Ltd.; and the Western Canada Wilderness Committee.

54. Kennedy has visited the Sound on several locations, and, with his family, was among the first to walk the completed trail.

55. Interview with Susan Jones, June 1998.

56. The Ahousaht are no strangers to visual technologies, and over the past few years have developed a GIS (geographic information system) project in order to map their traditional territories. For several years, the "Wild Side" homepage was maintained by the Western Canada Wilderness Society.

57. Interview with Susan Jones, June 1998.

58. The village of Marktosis remains one of the few Native villages that lie on islands off the west coast of Vancouver Island, or that are inaccessible by road. Other villages and tribal groups, either owing to economic conditions or under coercion by the state, have moved to mainland locations. In Nootka Sound immediately to the north, the Department of Indian Affairs moved the Muchalaht from their traditional location at Yuquot (the site of first contact between Captain Cook and Maquinna) to a new site near the white Canadian town of Gold River. This was accomplished by withdrawing services, most importantly, the local school. The villages of Hesquiaht, Opitsat, and Kyoquot are essentially the only other Native villages on the west coast of Vancouver Island that remain viable without road access.

59. Many Ahousaht band members fit Smith's (1994) description of "bifocal" or "transnational" subjects who continuously think and live two sites simultaneously.

60. One woman explained that some men in the village were now concerned that the women were not staying home enough.

61. One of the creative aspects of the trail project was that it brought Native and non-Native youth together during the building of the trail. As stated in the press release announcing the 1996 trail-building project issued by the Ahousaht Band Council (1995), "our youth are in desperate need of

the healing and hope that this project's cooperative, hands-on work and career planning will bring."

5. BC Seeing/Seeing BC

1. As I explain later, Yuxweluptun's work is not simply referential, and in this exhibit the curator, Andrew Hunter, went to some length to establish Yuxweluptun as an artist who does much more than "reflect upon" Carr's legacy.

2. Between 1927 and 1986, the National Gallery of Canada made no purchases of Indian art, preferring to look to Europe for its acquisitions and cultural roots. See Nemiroff (1992).

3. Hunter was not the first to draw these artists into relation. See Linsley (1991).

4. Members of the Group of Seven included Lawren Harris, Tom Thomson, J. E. H. MacDonald, Arthur Lismer, Frederick Varley, Frank Johnston, and Franklin Carmichael.

5. This view was sanctioned by none other than the governor-general of Canada, Vincent Massey (1948): "[This] group of gifted artists . . . turned their backs on Europe—quite deliberately—and surrendered themselves to their own environment, striving to uncover its secret. The inevitable happened. The Canadian landscape took possession of them. They abandoned the methods and techniques which were alien to Canada, and recorded its beauty faithfully in the clear lights, bold lines and strong colours which belong to it." It was F. B. Houser (1926), however, who first articulated this view, writing that the modern Canadian School "was inspired as the result of a direct contract with Nature herself" (24). This had required "a new type of artist; one who divests himself of the velvet coat and flowing tie of his caste, puts on the outfit of the bushwacker and prospector; closes with his environment; paddles, portages and makes camp; sleeps in the out-of-doors under the stars; climbs mountains with his sketch box on his back" (15).

6. Moray (1993) provides an excellent summary of early appraisals of Carr's art.

7. Similar rhetorics are found in Shadbolt's (1975, 4) earlier account of Carr's life, where, after tracing Carr's many contacts, Shadbolt concludes that, "despite all this, the popular view which sees her as simply the west coast painter of Indian subjects and rainforests is, though incomplete, justified. . . . She remained a non-intellectual who responded primarily to the spirit of things and people and had to learn intuitively."

8. This story of a "retreat" to Victoria is somewhat misleading. Carr remained in contact with a vibrant artistic community in Seattle, including

the Cubist painter Mark Tobey, continued to exhibit, and had contact with eastern curators and ethnologists.

9. In a similar fashion, Maureen Ryan (1992) argues that the work of nineteenth-century artists claimed "aesthetic proprietorship" over Native lands, and "could naturalize in aesthetic forms a racial theory that had served to marginalize a population of the country whose presence and title to the land had been at odds with Anglo-European expansion and settlement within the colony" (145). See also Dawkins (1986). Rosemary Coombe (1998) has explored the trope of appropriation, a key issue in Crosby's critique of non-Native artists and writers. For Coombe, cultural appropriation is something that must be evaluated *contextually* and in terms of *relations of power*. For instance, Coombe explores the ways that subaltern groups use mass-media texts, celebrity images, trademarks, and so on, to forge identities and communities. Understood in this context, cultural appropriation is a subversive act within "media-saturated consumer societies" (209). As Coombe notes, however, this celebratory notion of appropriation is cast in doubt in the case of the appropriation of Native artifacts. Although concerned that First Nations critiques of appropriation often rely on tropes of authenticity, and notions of property borrowed from the West, she argues that First Nations claims of cultural authenticity should be seen to foreground "the[ir] inability to name themselves and a continuous history of having their identities defined by others" (227). Coombe concludes that by contextualizing debates over appropriation, it becomes possible to "enact and practice an ethics of appropriation that attends to the specificity of the historical circumstances in which certain claims are made" (230).

10. Carr's departures from what was considered proper objects for women to paint is one basis on which she has been celebrated as an early feminist. For a discussion of limits to women's artistic endeavors, see Pollock (1988) and Wolff (1985).

11. It is important to emphasize that the gaze of the *viewer* of this archive of images is not singular for the same reasons. In the words of Deborah Poole (1997, 18), "once unleashed in society, an image can acquire myriad interpretations or meaning according to the different codes and referents brought to it by its diverse viewers." This further complicates any effort to achieve final closure on the meaning of Carr's artistic productions.

12. Newcombe was an amateur ethnologist and collector who was often employed by anthropological expeditions. Carr had some contact with Newcombe, and later became a close friend of his son William.

13. Such images have regained a measure of popularity today, and are found on postcards and popular books on Native history. How they are

read is not always clear. Although easily incorporated into an imperial nostalgia that mourns that which has been destroyed, such images are also often approached with a knowing irony, as if part of the pleasure of viewing them comes from knowing better.

14. The photograph was likely taken by Edmund A. Schwinke, Curtis's photographic assistant. See Holm and Quimby (1980).

15. It is likely that the number of totems and other figures was greatly reduced by the time of Curtis's travels, owing to the scramble for artifacts that occurred on the coast in the decades prior to his trip (see Cole 1985).

16. It is entirely possible that these milled-lumber structures were constructed after Newcombe took his photograph. But the larger point—that Newcombe tended to emphasize the totems at the expense of the broader village scene—remains intact. It should also be noted that Curtis's project was somewhat less invested in this primitive/modern divide toward its end (Jackson 1992). That the village was empty was likely a result of its use as a set in one of Curtis's films.

17. It is worthy of note that the practice of erecting totem poles had expanded markedly after contact with Europeans and that what was for a time the object that most signified "authentic" Northwest Coast Indian culture in Western imaginations was actually a product of complex transcultural exchanges.

18. In her "Lecture on Totems" (Carr 1913), delivered at the time of the exhibition of her totem-pole paintings in 1913, Carr described her intentions in similar terms: "The object in making this collection of these totem pole pictures has been to deposit these wonderful relics of a passing people in their own original setting. The identical spots where they were carved and placed by the Indians in honour of their chiefs."

19. One of the objectives of the 1927 exhibition, planned by Maurius Barbeau and Eric Brown and titled *Exhibition of West Coast Art—Indian and Modern,* was to display "exotic artifacts" of the west coast Native people as part of the art history of Canada, and thereby to offer them as inspiration to modern Canadian painters and as a resource for designers. Moray notes that Carr's Indian paintings were well suited for this purpose, and could be displayed and discussed as a bridge between a primitive Native past and a modern Canadian present (see also Morrison 1991). For an account of similar developments in the United States, see Rushing (1995).

20. The "emptiness" of Carr's villages has been explained away as a result of disease and depopulation, and because Carr often visited the villages during summer when residents had moved to sites of seasonal labor (including the industrial operations that she so studiously avoided).

21. This is not to dismiss the fact that gender was a key issue that shaped Carr's experience and the role it likely played in her "eccentric" view of west coast Natives and nature. However, as Mills (1991) and Kollin (1997) both explain, a focus on "female exceptionalism" can just as easily obscure the way that women were producers of signification for imperialist projects. See also Blunt (1994a, 1994b).

22. Tippett's (1979) biography is perhaps the most thorough in its discussion of the social networks and material practices of travel.

23. The growth of steamship service occurred for several reasons. First, by the close of the nineteenth century, advances in technology had made such travel relatively inexpensive, and far quicker than before. Second, in conjunction with improving transportation links, lumber camps and fish plants grew in number along the coast, resulting in a growing network of towns and villages stitched together by new social actors—machines. There is a wealth of material on these early steamship routes, much in the form of memoirs, others in the form of histories of individual companies (see O'Neill 1960; Rushton 1969, 1974, 1980; Hacking 1974; Turner 1977; Munro 1988). With the exception of Harris (1997), who focuses primarily on interior routes, few attempts have been made to organize this material into a comprehensive study.

24. Discussions of the west coast are often remarkable for their reification of region, or its reduction to a space delineated by physical geography. That the west coast is equally a product of political economy, shaped in relation to, and often directly as an extension of, wider cultural, economic, and technological forces gets lost in this essentializing move.

25. This does not mean that these communities were now excluded from processes of modernization, only that it placed new demands on Native mobility.

26. White communities were more likely to be described in terms of their level of development, pace of growth, and services available; in other words, by the distance they had traveled toward an imagined destination—the metropolitan center. Native villages were appraised through an entirely different lens. Ballou's (1890, 194) description was in many ways typical: "From time to time small native villages are seen on the islands and the mainland, all typical of the people, and quite picturesque in their dirtiness and peculiar construction. Some of the cabins are built of boards, but mostly they are rude, bark-covered logs. In front of these dwellings stand totem poles, presenting hideous faces carved upon them in bold relief, together with uncouth figures of birds, beasts, and fishes. A portion of these tall posts are weather-beaten and neglected, significantly tottering on their founda-

tions, green with mold, unconsciously foreshadowing the fate of the aboriginal race." Linda Nochlin (1989) notes that the picturesque in nineteenth-century art found its themes in the marginal and the peripheral, the rough and ready and the unrefined. Potentially subversive of dominant culture (and classical painting), the picturesque could also be deeply nostalgic, focusing on that which was vanishing or becoming obsolete.

27. For a good summary of the trade, see Hinckley (1965). These trips were popular with educated classes and artists, a point remarked upon in the memoirs of many travelers. Nor were women excluded. This is borne out in the description provided by Septima M. Collis, who traveled the route in 1890: "The pleasure is much enhanced by the fact that those who are your fellow-passengers are apt to be ladies and gentlemen, by which I mean persons whose good breeding naturally tends to a regard for the comfort of their companions; and among them you will find men and women, young and old, of bright intelligence, who, devoting their time to travel, are full of fact and anecdote—scientists, savants, authors, and artists of renown from all parts of the earth" (Collis 1890, 193). It was the practice on some of these inland cruises to segregate social groups. Native men and women were provided only with deck fare passage (available to non-Native men too), whereas non-Native women were permitted only to purchase cabin fares.

28. This does not mean that all travel experience is determined in advance. Indeed, travel can be, and often is, an uncanny experience that disturbs conventional views. See Gregory (1995).

29. "Kodaking" became popular along the Inside Passage during this period, not only because the technology had democratized photography, but also because the region now was set up for the viewer as a visual resource, framed to signify the primitive and pristine. See Collis (1890) for a description of "kodaking."

30. Carr's anxiety to define herself as a traveler rather than a tourist belies the fact that during part of the time that she worked on her ethnographic project, she was paid by the Grand Trunk Pacific Railway to travel on its trains and steamships. See Tippett (1979, 289 n. 13). As important, the argument that Carr's travels and work challenged conventional roles for women needs to be approached with caution. Many of the paths Carr traveled were readily available to women, especially women from the upper classes to which Carr belonged, even if she disowned the connection. When she departed from these paths, she was usually under the (paternal) "care" of white traders, missionaries, and relatives.

31. Established first as a museum of natural history, it very shortly began

to add Native artifacts. It was common at the time for ethnological collections to be treated as components of natural history collections.

32. Moray (1993) provides the most extensive discussion of Carr's efforts to sell her canvases to the museum.

33. The similarity with Robert Bateman's preface to *Clayoquot: On the Wild Side* should not go unmentioned (see chapter 3).

34. It is instructive that Carr identifies this site incorrectly, a sign of her lack of attention to differences among and between Native groups, something that she shared with other tourists but not ethnologists, for whom the differences were crucial.

35. The preface to Curtis's (1907) collection was written by none other than Theodore Roosevelt, the American president most associated with the frontier and its heroic masculinity: "In Mr. Curtis we have both an artist and trained observer, whose pictures are pictures, not merely photographs; whose work has far more than mere accuracy, because it is truthful. All serious students are to be congratulated because he is putting his work in permanent form; for our generation offers the last chance for doing what Mr. Curtis has done. The Indian as he has hitherto been is on the point of passing away. His life has been lived under conditions through which our own race passed so many ages ago that no vestige of their memory remains. It would be a veritable calamity if a vivid and truthful record of these conditions were not kept. . . . [Mr. Curtis] is a close observer, whose qualities of mind and body fit him to make his observations out in the field, surrounded by the wild life he commemorates. He has lived on intimate terms with many different tribes of the mountains and the plains."

36. While in France, Carr studied with a number of artists, including one—William Gibb—who became her mentor, and who was in turn a friend of Picasso and Matisse. During her stay, Carr abandoned the documentary, naturalist, and conservative realism that characterized her early work in favor of a newly acquired knowledge of Postimpressionism, particularly Fauvism, and the early stages of Modern Primitivism. Carr's years in France have been documented by Tippett (1979), Shadbolt (1979), Thom (1991), and Moray (1993).

37. The German critic Kasimir Edschmid (1918) summarized this well in his discussion of Expressionism: "We no longer have the concatenation of facts: factories, houses, illness, whores, outcry and hunger. Now we have the vision of them. *Facts have significance only so far as the artist's hand, reaching through them, grasps at what lies behind.* Art, which seeks only the essentials, excludes the incidental" (quoted in Sotriffer 1972, 6).

38. Carr often portrayed herself—and has been portrayed by critics and

historians—as nonintellectual, and therefore unconcerned with the changing currents of aesthetic theory. According to Shadbolt, Carr "responded primarily to the spirit of things and people and had to learn intuitively" (1975, 4). This "nonintellectual" posture was in line with other Post-impressionist painters (see Sotriffer 1972).

39. Deploying tropes of antiquity and mirroring the language of grave robbers, the *Toronto Star* (9 January 1928) described the show as "a Canadian tomb of Tutankhamen." This exhibit has been analyzed at length by Moray (1993), Morrison (1991), and Nemiroff (1992).

40. Shadbolt notes that Carr owned a well-thumbed copy of Ralph Peason's *How to Look at Modern Paintings,* which gave particular attention to Cubism, which Carr used to great effect in her depiction of both artifact and forest. The American painter Mark Tobey is widely considered to have been instrumental in Carr's Cubist turn.

41. Carr's advancing age and deteriorating health must also be seen as important factors that made it difficult for her to continue her study of artifacts in the Native villages of the coast. Yet this is not explanation enough for why her Native themes ended. The region surrounding Victoria contained ample Native material in the 1930s, only, just as when Carr began her career decades earlier, it was difficult to depict these Native groups in the same romantic and nostalgic registers that Carr used on her northern material. Tellingly, in the early 1940s, confined to her studio, Carr returned to Native themes, drawing again on her already large archive of images from earlier trips north.

42. In Carr's voluminous writing, there is at least one passage in which she registered an emotive response to the violence of logging: "There's a torn and splintered ridge across the stumps I call the 'screamers.' These are the unsawn last bits, the cry of the tree's heart, wrenching and tearing apart just before she gives that sway and the dreadful groan of falling, that dreadful pause while her executioners step back with their saws and axes resting and watch. It's a horrible sight to see a tree felled, even now, though the stumps are grey and rotting. As you pass among them you see their screamers sticking up out of their own tombstones, as it were. They are their own tombstones and their own mourners" (1966, 132–33)

43. For a summary and critique, see Agarwal (1992).

44. Carr's sexuality has been the object of intense speculation, ranging from arguments that Carr was sexually abused as a child; that she was simply a particularly prudish product of Victorian culture; and that she was a closeted lesbian.

45. In the past years, BC's landscape traditions have been increasingly

challenged in a number of exhibitions. These included, among others, two shows at the Morris and Helen Belkin Art Gallery on the University of British Columbia campus: *Born to Live and Die on Your Colonialist Reservations,* which featured the work of Coast Salish artist Yuxweluptun, and *Capitalizing the Scenery: Landscape, Leisure and Tourism in British Columbia 1880s–1950s,* an exhibition organized by art historian John O'Brian and his students. Another show, *Out of the Garden: The Contemporary British Columbia Landscape,* shifted the gaze from romantic and lyric landscapes to modern, humanized, and technologized ones.

46. For a critique of the privileging of the hybrid, exile, and migrant in contemporary cultural theory, see Kaplan (1996) and Clifford (1997). To criticize postcolonial theory for its emphasis on mobility, however, is not the same as dismissing postcolonial criticism—as do Ahmad (1992) and Dirlik (1994)—because it is written by those whose mobility has enabled them to occupy privileged positions in the U.S academy.

47. The VAG's *Topographies: Aspects of Recent B.C. Art,* staged in 1995, goes some way toward exploring this fractured visual field.

48. Yuxweluptun's work has received increased critical attention. Especially useful are the essays written for his solo show at the Morris and Helen Belkin Art Gallery, Vancouver, in 1995 (Yuxweluptun 1995).

49. Importantly, Yuxweluptun has also maintained links to rural Native communities and currently lives in Fort St. James.

50. Both parents were involved with the North American Indian Brotherhood. For a time his father headed the Union of BC Chiefs and his mother worked with the Indian Home-makers Association of BC (see Townsend-Gault 1995).

51. The resurgence of traditional Northwest Coast Indian art has been highly controversial. Although individuals like Bill Reid and Robert Davidson have been instrumental in reinvigorating traditional artistic styles and techniques, they have also been criticized for placing Northwest Coast Indian art in a straightjacket of tradition. Whether these criticisms are fair is not entirely clear, but certainly one of the great ironies of this resurgence is that it has leaned heavily on the work of non-Native art historians such as Bill Holm, whose highly influential studies of Northwest Coast Indian art have served to codify its forms and styles (see Holm 1965).

52. Robert Linsley (1995) makes the point that Yuxweluptun's landscape paintings recall multicolored resource and land-use maps. What makes his paintings remarkable, then, is precisely that these resource landscapes are no longer seen to exist outside social relations as "natural" resources, but are infused with politics and power.

53. Although it has not received much attention, Yuxweluptun's intervention in the politics of Clayoquot Sound—traditional Nuu-chah-nulth territories—is not without controversy.

54. This mirrors Frida Kahlo's refusal to accept the surrealist label: "I never painted dreams. I painted my own reality" (quoted in Riding 1998).

55. Townsend-Gault (1995) refers to this as "reciprocal appropriation."

56. This statement should be read as in part a reaction to the way in which Native traditions were revived in the 1960s as a catalog of forms that appeared timeless and unchanging (see Holm 1965).

6. Picturing the Forest Crisis

1. The study of these tangled knots of science, culture, and politics has given rise to the burgeoning field of "science studies." See Latour (1987, 1993, 1999); Rouse (1993); Haraway (1994, 1997); Martin (1996); Hess (1997).

2. The image was jointly produced by the Sierra Club of Western Canada (Victoria, B.C.) and the Wilderness Society (Seattle).

3. It is worth noting that much the same progression interests industry. In other words, both the environmental movement and the forest industry are interested in drawing the same kinds of distinctions (old-growth versus second-growth; bog and alpine versus temperate rainforest, etc.), but for very different reasons. "Old-growth" forests show little or no annual increase in wood fiber, whereas the mean annual incremental increase of wood fiber in "second-growth" forests increases rapidly until a certain age. Thus, because the forest industry seeks to capitalize on the forest's productivity, the forest industry is most interested in transforming the same "decadent" old-growth forests that environmentalists seek to preserve into highly productive second-growth forests. A rational progression from dark green to pale yellow is therefore seen as a corporate (and public) good.

4. No single Landsat image contained the entire island, so five separate photographs had to be processed and classified individually and then combined in order to create the single image.

5. Famously, the Western Canada Wilderness Committee took a massive tree stump on a tour of several European cities. This was highly effective, but only as an *illustration* of the disappearing forest, for which images—photographs, maps, and so on—were critical.

6. Because Latour finds his examples in print culture, he misses one of the important ways that immutability is itself a construction. With the global spread of electronically mediated communication, databases, and computer programs, images can be changed with the touch of a keyboard.

7. In *Beyond Sovereign Territory*, Thom Kuehls (1996) suggests that environmental issues undermine notions of national sovereignty: "descriptions of political space that isolate politics to particular institutions that contain sovereign authorities and are themselves contained within particular territories barely begin to capture the diversity of the space of (eco)politics" (x). This is an important challenge to traditional conceptions of sovereignty, which fail to recognize that ecological relations do not stop at international boundaries. Although Kuehls does not explore this, it is equally true that in ecopolitics sovereignty is undermined by immutable mobiles whereby local ecological issues are displaced into global arenas. The internationalization of Clayoquot Sound, for instance, was facilitated in large measure by the circulation of images such as those discussed in this chapter.

8. Latour (1987) argues that the stability of facts is not given by nature, but occurs when it has become too costly to contest their truth claims. This should not be taken to mean that the material world has no part to play. Rather, the costs of producing *other* knowledges hinge in part on enrolling the support or testimony of things. This may involve expensive instruments and labor, but it also means that things may simply betray the scientist, refusing to be enrolled in the manner desired. In this sense, nature does not speak its truth, it merely refuses the truth it has been asked to speak.

9. That developed areas are colored a dull gray reflects perhaps the common understanding of the city as a degraded, compromised, even degenerate place.

10. The French historian of science Gaston Bachelard is perhaps most reponsible for the view that science cannot be understood through a progressive, linear history. Robert Young (1990, 50) summarizes Bachelard as arguing that "The temporality of science cannot be accommodated to the rhythms of traditional historiography . . . [which understands history] solely in terms of precursors and anachronistic anticipations of modern ideas in early thinkers, as if science unrolled smoothly and inevitably from year to year. The problem with this approach is that it overestimates the extent of narrative continuity in the history of science, which . . . works rather by sudden disruptions, discontinuities and entire reorganizations of its principles. . . . Major transformations cannot be mapped onto the model of a continuous history, for its stress on putative anticipations fails to account for the way in which the whole form of knowledge can be transformed and a new understanding created."

11. Among other intellectual historians who have written histories of ecology, see McIntosh (1985, 1987); Worster (1988, 1990); Bramwell (1989); Botkin (1990).

12. As Hagen notes, individual ecologists could often hold contradicto-

ry views. The animal ecologist Charles Elton, to whom we owe the term *food chain,* could write of nature as consisting of "automatic balanced systems" while also holding that "'the balance of nature' does not exist and perhaps never has existed" (quoted in Hagen 1992, 57, 58).

13. See Alston Chase (1995) for a good discussion of cybernetics and ecology.

14. In an eloquent passage, Worster (1988, 313) captures not only the conservation ethic that systems theory could underwrite, but also its aesthetic: "[In] Eugene Odum's flow-charts of an ecosystem, all the energy lines move smartly along, converging here and shooting off there, looping back to where they began and following the thermodynamic arrows in a mannerly march toward the exit points."

15. This image lends itself to various readings. Although it subsumes First Nations as "natural" cultures who are part of nature's cosmic order, the presence of First Nations—and the marking of this presence through the *use* of the forest—destabilizes the notion of "external" nature that is so central to ecosystem ecology. Humans are present in the landscape, and this introduces the question of what *sorts* of practices are "properly" ecological and which are not, a complex question that always risks placing in question the status of the "ecological" itself.

16. To be fair to Hammond, he follows his discussion of the finely balanced, fragile forest ecosystem with a discussion of "alternative forestry" rather than strict preservation, but throughout it is the notion of nature as a stable, purposive whole that informs his comments.

17. This same discourse of disturbance as destruction underlies the comparison between Canadian forestry practices and Brazil's forest clearing for *other* land uses. In this case, the distinction "modified landscape" is duplicitous. Not all landscapes are modified in the same way.

18. As important as paleontological records have been for the recent emphasis on disequilibrium, it has not been the only source for this change. Population and evolutionary ecologists, for instance, with their emphasis on individuals, have questioned whether ecosystems exist in reality, or whether it is more accurate to see nature as a continually shifting mosaic of individual organisms, each acting in terms of its own survival. Seen in this light, it is difficult to see the ecosystem through "organismic" metaphors, or to see ecosystems as having "goals" (i.e., stability). Finally, although Howard Odum was himself one of the strongest advocates of ecosystem ecology, his reduction of ecology to a network of energy flows opened the doors to the recognition that ecosystems were not closed—that is, that there were continual inputs and outputs, and thus, that homeostasis was not an essential component of ecosystems.

19. Worster (1990, 9), complains that in the 1980s "it was as though scientists were out looking strenuously for signs of disturbance in nature."

20. It should be noted also that evolutionary biology—with its emphasis on populations—has put in question the notion that an ecosystem is the site of "coevolution." Rather, an ecosystem is no more than a snapshot of multiple evolutions operating on various temporal and spatial scales. Because the concept of ecosystem treats historical entities through ahistorical models, it is possible to argue that ecosystems exist only as freeze-frames, and that any map of interrelations and energy flows is outdated the moment it is drawn.

21. Kimmins's account is somewhat inconsistent—elsewhere, he reiterates notions of orderly forest succession in order to comfort readers disturbed by images of logging. The invocation of dynamic ecology in his case is clearly designed to challenge the scientific authority of the environmental movement.

22. Most windthrows on the west coast of Vancouver Island are small in scale (individual trees or clusters). Fire, the main cause of disturbance in other parts of the temperate rainforests, is rare on the west side of the island, and when it occurs is rarely extensive.

23. Alston Chase (1995) presents a similar argument that questions the environmental movement's quick adoption of the notion of biodiversity. The preservation of such species as the spotted owl is necessary, preservationists argue, because the loss of any one species threatens the whole: "biodiversity resembles a hammock," writes prominent BC environmentalist Tzeporah Berman. "As destructive industrial practices like clearcutting dramatically alter existing ecosystems, species go extinct, the hammock unravels. Eventually the hammock can no longer hold anything" (Berman 1994, 6). Chase, on the other hand, points to numerous examples where ecosystems are relatively simple yet remarkably stable.

24. The panel was announced by Premier Mike Harcourt on 22 October 1993 and charged with "scientifically reviewing current forest practices standards in Clayoquot Sound and recommending changes to existing standards to ensure that these practices are sustainable" (Scientific Panel 1995b).

25. Because science was established as the highest court of appeal by all actors, the government was in a position where it would have found it difficult to dismiss the panel's recommendations. Had it rejected the recommendations, the government would have faced the daunting task of also rejecting the science of internationally renowned scientists. Thus, it came as little surprise that when the Scientific Panel released its final report in April 1995, the government accepted all ninety-seven of its recommendations.

26. Strangely, the panel was formed with no *social* scientists, reaffirming the idea that forestry is an ecological and technical issue, not a social or political one.

27. The organismic metaphors—"skeleton," "circulation system"—are clear echoes of ecology's legacy of holism.

28. Although he is in broad agreement with Wilson's premise that "North American nature is a socially constructed environment," Neil Smith (1996, 44–46) is sharply critical of Wilson's appeal to "restoration." Smith finds in Wilson's work "a yearning for an unnecessarily nostalgic and one-dimensional re-immersion in nature," betrayed by notions of repairing "ruptures," "mimicking" nature, and restoring a "harmonious dwelling-in-nature." "If not quite elegiac," Smith writes, "this vision nonetheless embodies a romantic universalist view of human society and nature," one that, further, "utterly forgets the vicious externalization of nature as object of capitalist labour. And forgets too that universal nature is every bit as much a capitalist as a pre- and post-capitalist project."

Conclusion

1. As Latour (1993) explains, Western science is predicated on a great divide between Western and non-Western ways of knowing, in which the former is understood as *separate* from culture and politics, and the latter always entangled with belief (hence, merely "ethnoscience"). This opens a yawning gap between "our" science and "their" belief. One of the effects has been that while it has been possible to write ethnographies of the cultural practices of the Other, including their "ethnoscience," the ethnography of Western scientific practices has only recently become possible. It is precisely this view of a great divide that has allowed science to proceed as somehow external to society, and has also authorized "our" scientists to look over the shoulders of other societies in order to correct their beliefs with our truth. The inclusion of four Nuu-chah-nulth members made the operation of this ideological divide far more problematic. Still, the difficult translation work necessary for displacing these divides was, for the most part, avoided. The traditional ecological knowledges (TEK) of the Nuu-chah-nulth were generally treated as "inventory" science, much like botany in the eighteenth century, rather than as aligned with the law-building sciences of today. When the final report was issued, Nuu-chah-nulth knowledge was generally dealt with in a separate volume.

2. It is important to keep in mind that the incorporation of the Nuu-chah-nulth *within* forestry does not necessarily imply the challenging of state sovereignty. Indeed, although he too readily discounts the very important progress made through interim agreements that give the Nuu-chah-nulth

much more control, Thom Kuehls (1996) is undoubtedly correct to identify these—and the training programs initiated in the late 1990s—as part of a regime of "governmentality" that seeks to constitute Natives as productive forest workers.

3. Bruno Latour (1986) makes a similar charge, suggesting that Marxist thought has not been materialist enough because it too quickly turns to abstractions such as the economy to explain causation, rather than attending to specific practices.

Bibliography

Agarwal, Bina. 1992. The gender and environment debate: Lessons from India. *Feminist Studies* 18: 119–58.

Ahmad, Aijaz. 1992. *In Theory: Classes, Nations, Literature*. London: Verso.

Ahousaht Band Council. 1995. Press release. October 16.

Althusser, Louis. 1969. *For Marx*. Trans. Ben Brewster. London: Allen Lane.

Althusser, Louis, and Étienne Balibar. 1970. Trans. Ben Brewster. *Reading Capital*. London: New Left Books.

Appadurai, Arjun. 1996. *Modernity at Large: Cultural Dimensions of Globalization*. Minneapolis: University of Minnesota Press.

———— (Ed.). 1986. *The Social Life of Things: Commodities in Cultural Perspective*. Cambridge: Cambridge University Press.

Appelhof, Ruth. 1988. *The Expressionist Landscape: North American Landscape Painting 1920–1947*. Seattle: University of Washington Press.

Arcas Associates. 1989. *Patterns of Settlement of the Ahousaht (Kelsemaht) and Clayoquot Bands*. Port Moody: Arcas Associates.

————. 1986. *Native Tree Use on Meares Island, BC*. Report in four volumes for the Ahousaht and Clayoquot Indian Bands. Port Moody: Arcas Associates.

Arnold, David. 1996. *The Problem of Nature: Environment, Culture and European Expansion*. Oxford: Blackwell.

Arshi, Sunpreet, Carmen Kirstein, Riaz Naqui, and Falm Pankow. 1994. Why travel? Tropics, en-tropics, and apo-tropics. In George Robertson, Melinda Mash, Lisa Tichner, Jon Bird, and Barry Curtis (Eds.), *Traveller's Tales: Narratives of Home and Displacement*. London: Routledge. 225–44.

Bachelard, Gaston. 1982. *L'Engagement rationaliste*. Paris: Presses Universitaires de France.

Badlam, Alexander. 1891. *The Wonders of Alaska*. 3d ed. San Francisco: Bancroft.

Bal, Mieke. 1991. The politics of citation. *Diacritics* 21: 25–45.

Ballou, Maturin. 1890. *The New Eldorado: A Summer Trip to Alaska.* Boston: Houghton Mifflin.

Barnes, Trevor, and Roger Hayter. 1997. *Trouble in the Rainforest: British Columbia's Forest Economy in Transition.* Victoria: Western Geographical Press.

Barrett, R. 1989. *The Independent Guide to Real Holidays Abroad.* London: The Independent.

Barthes, Roland. 1981. *Camera Lucida: Reflections on Photography.* Trans. Richard Howard. New York: Hill and Wang.

Baudrillard, Jean. 1975. *The Mirror of Production.* Trans. Mark Poster. St. Louis, Mo.: Telos Press.

Beck, Ulrich. 1992. *Risk Society: Toward a New Modernity.* London: Sage.

Beezer, Anne. 1993. Women and adventure travel tourism. *New Formations* 21: 119–30.

Beinert, William, and Peter Coates. 1995. *Environment and History: The Taming of Nature in the USA and South Africa.* London: Routledge.

Bell, Daniel. 1976. *The Cultural Contradictions of Capitalism.* New York: Basic Books.

Beltgens, Paula. 1995. Resource information training program. *Cultural Survival* 18: 21–22.

Belyea, Barbara. 1992. Images of power: Derrida/Foucault/Harley. *Cartographica* 29: 1–9.

Benjamin, Walter. 1977. *The Origin of German Tragic Drama.* Trans. John Osborne. London: New Left Books.

Bennett, Jane, and William Chaloupka (Eds.). 1993. *In the Nature of Things: Language, Politics, and the Environment.* Minneapolis: University of Minnesota Press.

Benton, Ted. 1989. Marxism and natural limits: An ecological critique and reconstruction. *New Left Review* 178: 51–81.

Berkhofer, Robert F. 1978. *The White Man's Indian: The History of an Idea from Columbus to the Present.* New York: Knopf.

Berman, Marshal. 1982. *All That Is Solid Melts into Air.* London: Verso.

Berman, Tzeporah. 1994. Takin' it back. In Tzeporah Berman (Ed.), *Clayoquot and Dissent.* Vancouver: Ronsdale Press. 1–8.

Bernstein, Richard. 1992. *The New Constellation: The Ethical-Political Horizons of Modernity/Postmodernity.* Cambridge: Polity Press.

Bhabha, Homi. 1994. *The Location of Culture.* London: Routledge.

Bill, Andrew. 1994. Exotic destinations. *Natural History,* March.

Blackman, Margaret. 1992. Of "peculiar carvings and architectural devices": Photographic ethnohistory and the Haida Indians. In Elizabeth

Edward (Ed.), *Anthropology and Photography: 1860–1920*. New Haven: Yale University Press. 137–42.

Blanchard, Paula. 1987. *The Life of Emily Carr*. Seattle: University of Washington Press.

Blunt, Alison. 1994a. Mapping authorship and authority: Reading Mary Kingley's landscape descriptions. In Alison Blunt and Gillian Rose (Eds.), *Writing Women and Space: Colonial and Postcolonial Geographies*. New York: Guilford Press.

———. 1994b. *Travel, Gender and Imperialism: Mary Kingsley and West Africa*. New York: Guilford Press.

Bormann, F. Herbert, and Gene E. Likens. 1979. *Pattern and Process in a Forested Ecosystem: Disturbance, Development and the Steady State Based on the Hubbard Brook Ecosystem Study*. New York: Springer-Verlag.

Botkin, Daniel. 1990. *Discordant Harmonies: A New Ecology for the Twenty-first Century*. New York: Oxford University Press.

Bourdieu, Pierre. 1984. *Distinction: A Social Critique of the Judgement of Taste*. Trans. Richard Nice. Cambridge: Harvard University Press.

Braidotti, Rosi. 1993. Embodiment, sexual difference, and the nomadic subject. *Hypatia* 8: 1–13.

Bramwell, Anna. 1989. *Ecology in the Twentieth Century: A History*. New Haven: Yale University Press.

Braun, Bruce. 2000a. Producing vertical territory: Geology and governmentality in late-Victorian Canada. *Ecumene* 7: 7–46.

———. 2000b. Rereading Sloan: Metaphor, the "forest" and the long shadow of colonialism. Paper delivered at Green College, Vancouver.

Braun, Bruce, and Noel Castree (Eds.). 1998. *Remaking Reality: Nature at the Millennium*. London: Routledge.

Braun, Bruce, and Joel Wainwright. 2001. Nature, postructuralism and politics. In Noel Castree and Bruce Braun (Eds.), *Social Nature: Theory, Practice, Politics*. Oxford: Blackwell.

Breen-Needham, Howard (Ed.). 1994. *Witness to Wilderness: The Clayoquot Sound Anthology*. Vancouver: Arsenal.

Brown, Robert. 1864. *Vancouver Island Exploration Expedition*. Victoria: Harris and Company.

———. 1869. *Memoir on the Geography of the Interior of British Columbia*. Holograph, Brown Collection, I, 10. Public Archives of British Columbia.

Brown, Wendy. 1998. Genealogical politics. In Jeremy Moss (Ed.), *The Later Foucault*. London: Sage. 33–49.

Butler, Judith. 1990. *Gender Trouble: Feminism and the Subversion of Identity.* New York: Routledge.

———. 1993. *Bodies That Matter: On the Discursive Limits of "Sex."* New York: Routledge.

Butler, Judith, Ernesto Laclau, and Slavoj Žižek. 2000. *Contingency, Hegemony, Universality.* London: Verso.

Buzard, James. 1993. *The Beaten Track: European Tourism, Literature, and the Ways to "Culture," 1800–1918.* Oxford: Oxford University Press.

Callon, Michel. 1986. Some elements of a sociology of translation: Domestication of the scallops and the fishermen of St. Brieux Bay. In John Law (Ed.), *Power, Action and Belief: A New Sociology of Knowledge?* London: Routledge. 196–229.

Caputo, John. 1993. *Against Ethics: Contribution to a Poetics of Obligation with Constant Reference to Deconstruction.* Bloomington: Indiana University Press.

Carr, Emily. 1913. "Lecture on totems." Unpublished manuscript. Public Archives of British Columbia.

———. 1946. *Growing Pains: The Autobiography of Emily Carr.* Toronto: Oxford University Press.

———. 1953. *The Heart of a Peacock.* Toronto: Oxford University Press.

———. 1966. *Hundreds and Thousands: The Journals of Emily Carr.* Toronto: Clarke, Irwin.

———. 1971. *Klee Wyck.* Toronto: Clarke, Irwin and Company.

Carson, Rachel. 1962. *Silent Spring.* Boston: Houghton Mifflin.

Carter, Paul. 1987. *The Road to Botany Bay: An Essay in Spatial History.* London: Faber.

Castree, Noel. 1995. The nature of produced nature: Materiality and knowledge construction in Marxism. *Antipode* 27: 12–28.

Castree, Noel, and Bruce Braun. 1998. The construction of nature and the nature of construction: Analytical and political tools for building survivable futures. In Bruce Braun and Noel Castree (Eds.), *Remaking Reality: Nature at the Millennium.* London: Routledge. 3–42.

Chambers, Iain. 1994. *Migrancy, Culture, Identity.* London: Routledge.

Charleson, Karen. 1992. Parks: Another insult to Natives. *Vancouver Sun,* January 15. A19.

Chase, Alston. 1995. *In a Dark Wood: The Fight over Forests and the Rising Tyranny of Ecology.* Boston: Houghton Mifflin.

Chatterjee, Partha. 1993. *The Nation and Its Fragments.* Princeton, N.J.: Princeton University Press.

Chorley-Smith, Peter. 1989. *White Bears and Other Curiosities: The First*

One Hundred Years of the Royal British Columbia Museum. Victoria: Royal British Columbia Museum.

Christophers, Brett. 1998. *Positioning the Missionary: John Booth Good and the Confluence of Cultures in Nineteenth-Century British Columbia.* Vancouver: University of British Columbia Press.

Clayton, Daniel. 1999. *Islands of Truth: The Imperial Fashioning of British Columbia.* Vancouver: University of British Columbia Press.

Clements, Frederic. 1928. *Plant Succession and Indicators: A Definitive Edition of Plant Succession and Plant Indicators.* New York: H. W. Wilson.

Clifford, James. 1988. *The Predicament of Culture: Twentieth-Century Ethnography, Literature and Art.* Cambridge: Harvard University Press.

———. 1997. *Routes: Travel and Translation in the Late Twentieth Century.* Cambridge: Harvard University Press.

Cole, Douglas. 1985. *Captured Heritage: The Scramble for Northwest Coast Artifacts.* Seattle: University of Washington Press.

———. 2000. The invented Indian/the imagined Emily. *BC Studies* 125–26: 147–62.

Cole, Douglas, and Bradley Lockner (Eds.). 1993. *To the Charlottes: George Dawson's 1878 Survey of the Queen Charlotte Islands.* Vancouver: University of British Columbia Press.

Collis, Septima. 1890. *Woman's Trip to Alaska.* New York: Cassell Publishing Company.

Commoner, Barry. 1971. *The Closing Circle: Nature, Man and Technology.* New York: Knopf.

———. 1990. *Making Peace with the Planet.* New York: Pantheon Books.

Coombe, Rosemary. 1998. *The Cultural Life of Intellectual Properties: Authorship, Appropriation and the Law.* Durham, N.C.: Duke University Press.

Cooper, W. S. 1913. The climax forest of Isle Royale, Lake Superior, and its development, I, II, and III. *Botanical Gazette* 55: 1–44, 115–40, 189–235.

Cosgrove, Denis. 1984. *Social Formation and Symbolic Landscape.* London: Croom Helm.

Cosgrove, Denis, and Stephen Daniels (Eds.). 1988. *The Iconography of Landscape: Essays on the Symbolic Representation, Design and Use of Past Environments.* Cambridge: Cambridge University Press.

Crary, Jonathan. 1988. Modernizing vision. In Hal Foster (Ed.), *Vision and Visuality.* Seattle: Bay Press. 29–50.

———. 1990. *Techniques of the Observer: Vision and Modernity.* Cambridge: MIT Press.

Cronon, William. 1992. A place for stories: Nature, history and narrative. *Journal of American History* 78: 1347–76.

———. 1995. The trouble with wilderness: Or, getting back to the wrong nature. In William Cronon (Ed.), *Uncommon Ground: Toward Reinventing Nature.* New York: Norton. 69–90.

Crosby, Marcia. 1991. Construction of the imaginary Indian. In Stan Douglas (Ed.), *Vancouver Anthology.* Vancouver: Talon Books. 267–91.

Crosby, Thomas. 1914. *Up and Down the North Pacific Coast by Canoe and Mission Ship.* Toronto: Frederick Clarke Stephenson.

Culler, Jonathan. 1988. *Framing the Sign: Criticism and Its Institutions.* Norman: University of Oklahoma Press.

Curtis, Barry, and Claire Pajaczkowska. 1994. "Getting there": Travel, time and narrative. In George Robertson, Melinda Mash, Lisa Tickner, Jon Bird, and Barry Curtis (Eds.), *Traveller's Tales: Narratives of Home and Displacement.* London: Routledge. 199–215.

Curtis, Edward. 1907. *The North American Indian; Being a Series of Volumes Picturing and Describing the Indians of the United States, the Dominion of Canada, and Alaska.* Vol. 1. Seattle: E. S. Curtis.

Darier, Eric. 1999. *Discourses of the Environment.* Oxford: Blackwell.

Davis, Margaret. 1986. Climatic instability, time lags and community disequilibrium. In Jared Diamond and Ted Case (Eds.), *Community Ecology.* New York: Harper and Row.

Dawkins, Heather. 1986. Paul Kane and the eye of power: Racism in Canadian art history. *Vanguard,* September. 24–27.

Dawson, George. 1877. Economic minerals and mines of British Columbia. Report on surveys. *Canadian Pacific Railway Report.* 227–34.

———. 1880a. Note on the Distribution of Some of the More Important Trees of British Columbia. *Canadian Naturalist* 9(6): 1–11.

———. 1880b. *Report on the Queen Charlotte Islands.* Montreal: Geological Survey of Canada.

———. 1896. Canada as a field for mining investment. *National Review.* 242–51.

De Cosmos, Amor. 1863. Diffusion of geological knowledge. *British Colonist,* June 27. 2.

Delgamuukw v. Her Majesty the Queen in Right of the Province of British Columbia and the Attorney General of Canada. 1991. Unreported judgment of Chief Justice Allan McEachern (British Columbia Supreme Court), No. 0843 Smithers.

Deleuze, Gilles. 1983. *Nietzsche and Philosophy.* Trans. Hugh Tomlinson. New York: Columbia University Press.

Deleuze, Gilles, and Claire Parnet. 1987. *Dialogues*. Trans. Hugh Tomlinson and Barbara Habberjam. New York: Columbia University Press.

Deleuze, Gilles, and Félix Guattari. 1977. *Anti-Oedipus: Capitalism and Schizophrenia*. Trans. Robert Hurley, Mark Seem, and Helen R. Lane. New York: Viking Press.

———. 1987. *A Thousand Plateaus: Capitalism and Schizophrenia*. Trans. Brian Massumi. Minneapolis: University of Minnesota Press.

Demeritt, David. 1994. Ecology, objectivity and critique in writings on nature and human societies. *Journal of Historical Geography* 20: 22–37.

Demeritt, David. 1997. Knowledge, nature and representation: Clearings for conservation in the Maine Woods. Ph.D. dissertation, Department of Geography, University of British Columbia, Vancouver.

Derickson, Harold. 1991. *Native Forestry in British Columbia: A New Approach. Final Report*. Victoria: Task Force on Native Forestry.

Derrida, Jacques. 1976. *Of Grammatology*. Trans. Gayatri Chakravorty Spivak. Baltimore: Johns Hopkins University Press.

———. 1982. *Margins of Philosophy*. Trans. Alan Bass. Chicago: University of Chicago Press.

Devall, Bill (Ed.). 1993. *Clearcut: The Tragedy of Industrial Forestry*. San Francisco: Sierra Club Books.

Devall, Bill, and George Sessions. 1985. *Deep Ecology*. Salt Lake City: G. M. Smith.

di Chiro, Giovanna. 1995. Nature as community; the convergence of environment and social justice. In William Cronon (Ed.), *Uncommon Ground: Toward Reinventing Nature*. New York: Norton. 290–320.

———. 2000. Beyond ecoliberal/common futures: Environmental justice, toxic touring, and a transcommunal politics of place. Paper presented at Race, Nature, and the Politics of Difference Workshop, Berkeley, February.

Dilworth, Ira. 1941. Emily Carr—Canadian artist-author. *Saturday Night*, 1 November. 26.

Dirlik, Arlik. 1994. The postcolonial aura: Third world criticism in the age of global capitalism. *Critical Inquiry* 20: 328–56.

Doel, Marcus. 1999. *Poststructuralist Geographies: The Diabolical Art of Spatial Science*. Edinburgh: Edinburgh University Press.

Dorst, Adrian, and Cameron Young. 1990. *Clayoquot: On the Wild Side*. Vancouver: Western Canada Wilderness Committee.

Douglas, Stan (Ed.). 1991. *Vancouver Anthology*. Vancouver: Talon Books.

Dreyfus, Hubert. 1991. *Being-in-the-World: A Commentary on Heidegger's Being and Time, Division I*. Cambridge: MIT Press.

Drinnon, Richard. 1980. *Facing West: The Metaphysics of Indian-Hating and Empire Building.* Minneapolis: University of Minnesota Press.

Drucker, Philip. 1951. *The Northern and Central Nootkan Tribes.* Washington, D.C.: Smithsonian Institution Bureau of American Ethnology, Bulletin 144.

———. 1955. *Indians of the Northwest Coast.* New York: McGraw-Hill.

———. 1965. *Cultures of the North Pacific Coast.* San Francisco: Chandler Publishing Company.

Drushka, Ken, Bob Nixon, and Ray Travers. 1993. *Touch Wood: BC Forests at the Crossroads.* Madeira Park, B.C.: Harbour Publishing.

Eckersley, Robyn. 1992. *Environmentalism and Political Theory: Towards an Ecocentric Approach.* Albany: State University of New York Press.

Ecosummer. 1990. Journeys of Discovery. Summer/fall. (Advertising.)

———. 1994. Journeys of Discovery. Summer/fall. (Advertising.)

Ecotrust Canada. 1997. *Seeing the Ocean through the Trees: A Conservation-Based Development Strategy for Clayoquot Sound.* Vancouver: Ecotrust Canada.

Ehrlich, Paul. 1968. *The Population Bomb.* New York: Ballantine Books.

Elliott, Anthony. 1992. *Social Theory and Psychoanalysis in Transition: Self and Society from Freud to Kristeva.* Oxford: Blackwell.

———. 1996. *Subject to Ourselves: Social Theory, Psychoanalysis and Postmodernism.* Cambridge: Polity Press.

Elton, Charles. 1930. *Animal Ecology and Evolution.* New York: Oxford University Press.

Emberley, Julia. 1996. Simulated politics: Animal bodies, fur-bearing women, indigenous survival. *New Formations* 24: 66–91.

Escobar, Arturo. 1996. Constructing nature: Elements for a post-structural political ecology. In Richard Peet and Michael Watts (Eds.), *Liberation Ecology.* New York: Routledge. 46–68.

Featherstone, Mike. 1991. *Consumer Culture and Postmodernism.* London: Sage.

Featherstone, Mike, and Scott Lash. 1995. Globalization, modernity and the spatialization of social theory: An introduction. In Mike Featherstone, Scott Lash, and Roland Robertson (Eds.), *Global Modernities.* London: Sage. 1–24.

Fisher, Robin. 1977. *Contact and Conflict: Indian-European Relations in British Columbia, 1774–1890.* Vancouver: University of British Columbia Press.

Fitzsimmons, Margaret. 1989. The matter of nature. *Antipode* 21: 106–20.

Fitzsimmons, Margaret, and David Goodman. 1998. Incorporating nature:

Environmental narratives and the reproduction of food. In Bruce Braun and Noel Castree (Eds.), *Remaking Reality: Nature at the Millennium.* London: Routledge. 194–220.

Foehr, Stephen. 1995. True mettle: Gaining confidence and altitude in Colorado's Sangre de Cristo. *Ecotraveller* 1(5): 30–37.

Foreman, David. 1995. Wilderness areas are vital. *Wild Earth* (winter 1994–95): 64–68.

Forest Alliance of British Columbia. 1994. *The Economic Impact of the Forest Industry on British Columbia.* Vancouver: Forest Alliance of British Columbia.

Foucault, Michel. 1970. *The Order of Things: An Archaeology of the Human Sciences.* New York: Vintage Books.

———. 1977. Nietzsche, genealogy, history. Trans. Donald F. Bouchard and Sherry Simon. In Donald F. Bouchard (Ed.), *Language, Counter-Memory, Practice: Selected Essays and Interviews.* Ithaca, N.Y.: Cornell University Press. 139–64.

———. 1979. *Discipline and Punish: The Birth of the Prison.* Trans. Alan Sheridan. New York: Vintage Books.

———. 1980. *The History of Sexuality,* volume 1, *An Introduction.* Trans. Robert Hurley. New York: Vintage Books.

———. 1984. What is enlightenment? In Paul Rabinow (Ed.), *The Foucault Reader.* New York: Pantheon Books. 32–50.

Fraser, Nancy. 1991. Rethinking the public sphere. *Social Text* 8/9: 56–80.

Fulford, Robert. 1993. The trouble with Emily: How Canada's greatest woman painter ended up on the wrong side of the political correctness debate. *Canadian Art* 10: 32–39.

Fulton, Frederick J. 1910. *Final Report of the Royal Commission of Inquiry on Timber and Forestry, 1909–1910.* Victoria: R. Wolfenden.

Fussell, Paul. 1980. *Abroad: British Literary Traveling between the Wars.* New York: Oxford University Press.

Galois, Robert, and Cole Harris. 1994. Recalibrating society: The population geography of British Columbia in 1881. *Canadian Geographer* 38: 37–53.

Gaonkar, Dilip Parameshwar. 1999. On alternative modernities. *Public Culture* 2: 1–18.

Ghandi, Leela. 1998. *Postcolonial Theory: A Critical Introduction.* New York: Columbia University Press.

Gleason, H. A. 1926. The individualistic concept of the plant association. *Bulletin of Torrey Botany Club* 53: 1–20.

Glickman, Joe. 1995. Paddling the unknown. *Ecotraveler* 2: 8–9.

Government of British Columbia. 1993a. *Administrative Fairness of the Process Leading to the Clayoquot Sound Land Use Decision.* Victoria: Office of the Ombudsman.

———. 1993b. *Clayoquot Sound: A Balanced Decision, a Sustainable Future.* Victoria: Queen's Printer.

Grafton, J. J. 1894. *Grafton's Tours to Alaska, 1984, via Colorado, California and British Columbia.* Chicago: Poole Bros.

Green, Nicholas. 1990. *The Spectacle of Nature: Landscape and Bourgeois Culture in Nineteenth Century France.* Manchester: Manchester University Press.

Greenpeace, Canada. N.d. *Cutting Down Canada: The Giveaway of Our National Forests.* Vancouver: Greenpeace, Canada.

Greenpeace, UK. 1994. *Clayoquot Sound: Clearcut Sound.* London: Greenpeace, UK.

Gregory, Derek. 1985. Suspended animation: The stasis of diffusion theory. In Derek Gregory and John Urry (Eds.), *Social Relations and Spatial Structures.* London: Macmillan. 296–336.

———. 1994. *Geographical Imaginations.* Cambridge: Blackwell.

———. 1995. Between the book and the lamp: Imaginative geographies of Egypt, 1849–60. *Transactions of the Institute of British Geographers* 20: 29–57.

Grove, Richard. 1995. *Green Imperialism: Colonial Expansion, Tropical Island Edens and the Origins of Environmentalism.* Cambridge: Cambridge University Press.

Guha, Ramachandra. 1989. Radical American environmentalism and wilderness preservation: A third world critique. *Environmental Ethics* 11: 71–84.

Guntau, M. 1978. The emergence of geology as a scientific discipline. *History of Science* 16: 280–90.

Gupta, Akhil. 1998. *Postcolonial Developments: Agriculture in the Making of Modern India.* Durham, N.C.: Duke University Press.

Gupta, Akhil, and James Ferguson. 1997. Beyond "culture": Space, identity and the politics of difference. In Akhil Gupta and James Ferguson (Eds.), *Culture, Power, Place: Explorations in Critical Anthropology.* Durham, N.C.: Duke University Press, 33–51.

Habermas, Jürgen. 1971. *Toward a Rational Society.* London: Heinemann.

———. 1972. *Knowledge and Human Interests.* London: Heinemann.

———. 1987. *Theory of Communicative Action.* Vol. 2. Boston: Beacon Press.

Hacking, Ian. 1983. *Representing and Intervening: Introductory Topics in the Philosophy of Natural Science.* Cambridge: Cambridge University Press.

Hacking, Norman. 1974. *The Princess Story: A Century and a Half of West Coast Shipping.* Vancouver: Mitchell Press.

Hagen, Joel. 1992. *An Entangled Bank: The Origins of Ecosystem Ecology.* New Brunswick, N.J.: Rutgers University Press.

Hall, Stuart. 1986. Gramsci's relevance for the study of race and ethnicity. *Journal of Communication Inquiry* 10: 5–27.

———. 1996. When was the "post-colonial"? Thinking at the limit. In Iain Chambers and Lidia Curti (Eds.), *The Post-Colonial Question: Common Skin, Divided Horizons.* London: Routledge. 242–60.

Hamilton, Gordon. 1996. Indians tell Greenpeace to get out of Clayoquot. *Vancouver Sun,* 22 June. A1, A24.

Hammond, Herb. 1991. *Seeing the Forest among the Trees: The Case for Wholistic Forest Use.* Vancouver: Polestar Press.

Haraway, Donna. 1989. *Primate Vision: Gender, Race and Nature in the World of Modern Science.* New York: Routledge.

———. 1991. *Simians, Cyborgs and Women: The Reinvention of Nature.* New York: Routledge.

———. 1992. The promises of monsters: A regenerative politics for inappropriate/d others. In Lawrence Grossberg, Cary Nelson, and Paula A. Treichler (Eds.), *Cultural Studies.* New York: Routledge. 295–337.

———. 1994. Game of cat's cradle: Science studies, feminist theory, cultural studies. *Configurations* 1: 59–71.

———. 1997. *Modest Witness@Second Millennium.Female Man© Meets OncoMouseᵀᴹ.* New York: Routledge.

Harley, J. B. 1992. Deconstructing the map. In Trevor Barnes and James Duncan (Eds.), *Writing Worlds: Discourse, Text and Metaphor in Representation of Landscape.* London: Routledge. 231–47.

Harris, Cole. 1997. *The Resettlement of British Columbia: Essays on Colonialism and Geographical Change.* Vancouver: University of British Columbia Press.

———. Forthcoming. *Theory and Practice of Native Land.* Vancouver: University of British Columbia Press.

Harvey, David. 1989. *The Condition of Postmodernity: An Enquiry into the Origins of Cultural Change.* Cambridge: Blackwell.

———. 1996. *Justice, Nature, and the Geography of Difference.* Oxford: Blackwell.

Hatch, Christopher. 1994. The Clayoquot protests: Taking stock one year later. In Tzeporah Berman (Ed.), *Clayoquot and Dissent*. Vancouver: Ronsdale Press, 199–208.

Hayman, John (Ed.). 1989. *Robert Brown and the Vancouver Island Exploring Expedition*. Vancouver: University of British Columbia Press.

Hays, Samuel. 1959. *Conservation and the Gospel of Efficiency*. Cambridge: Harvard University Press.

Hecht, Suzanna, and Alexander Cockburn. 1989. *The Fate of the Forest: Developers, Destroyers and Defenders of the Amazon*. New York: HarperCollins.

Heidegger, Martin. 1962. *Being and Time*. Trans. John Macquarrie and Edward Robinson. New York: Harper and Row.

———. 1977a. Letter on humanism. In D. Krell (Ed.), *Basic Writings*. New York: Harper Collins. 193–242.

———. 1977b. *The Question concerning Technology and Other Essays*. Trans. William Lovitt. New York: Harper and Row.

Hess, David. 1997. *Science Studies: An Advanced Introduction*. New York: New York University Press.

Hinckley, Ted C. 1965. The inside passage: A popular gilded age tour. *Pacific Northwest Quarterly*, April. 67–74.

Holm, Bill. 1965. *Northwest Coast Indian Art: An Analysis of Form*. Seattle: University of Washington Press.

Holm, Bill, and George Irving Quimby. 1980. *Edward S. Curtis in the Land of the War Canoes: A Pioneer Cinematographer in the Pacific Northwest*. Vancouver: Douglas and McIntyre.

Houser, F. B. 1926. *The Canadian Art Movement: The Story of the Group of Seven*. Toronto: Macmillan.

Ingram, Gordon. 1994. The ecology of a conflict. In Tzeporah Berman (Ed.), *Clayoquot and Dissent*. Vancouver: Ronsdale Press. 9–72.

Jackson, Peter. 1992. Constructions of culture, representations of race: Edward Curtis's "way of seeing". In Kay Anderson and Fay Gale (Eds.), *Inventing Places: Studies in Cultural Geography*. London: Longman Cheshire. 89–106.

Jacknis, Ira. 1992. George Hunt, Kwakiutl photographer. In Elizabeth Edwards (Ed.), *Anthropology and Photography: 1860–1920*. New Haven: Yale University Press. 143–51.

Jacobs, Jane. 1996a. *Edge of Empire: Postcolonialism and the City*. London: Routledge.

———. 1996b. Speaking always as geographers. *Environment and Planning D: Society and Space* 14: 379–94.

Jasen, Patricia. 1995. *Wild Things: Nature, Culture and Tourism in Ontario, 1790–1914.* Toronto: University of Toronto Press.

Jay, Martin. 1993. *Downcast Eyes: The Denigration of Vision in Twentieth-Century French Thought.* Berkeley: University of California Press.

Kaplan, Caren. 1996. *Questions of Travel: Postmodern Discourses of Displacement.* Durham, N.C.: Duke University Press.

Katz, Cindi. 1998. Whose nature, whose culture? Private productions of space and the "preservation" of nature. In Bruce Braun and Noel Castree (Eds.), *Remaking Reality: Nature at the Millennium.* London: Routledge. 46–63.

Kennedy, Robert F., Jr. 1993. Logging Clayoquot will strip province of its natural beauty. *Vancouver Sun,* 20 February. B4.

Kimmins, Hamish. 1992. *Balancing Act: Environmental Issues in Forestry.* Vancouver: University of British Columbia Press.

Kirby, Kathleen. 1996. *Indifferent Boundaries: Spatial Concepts of Human Subjectivity.* New York: Guilford Press.

Knapp, Marilyn. 1980. *Carved History: The Totem Poles and House Posts of Sitka National Historical Park.* Sitka: Alaska Natural History Association.

Kollin, Susan. 1997. "The first white women in the last frontier": Writing race, gender and nature in Alaska travel narratives. *Frontiers* 18: 105–24.

Kolodny, Annette. 1975. *The Lay of the Land: Metaphors as Experience and History in American Life and Letters.* Chapel Hill: University of North Carolina Press.

Kuehls, Thom. 1996. *Beyond Sovereign Territory: The Space of Ecopolitics.* Minneapolis: University of Minnesota Press.

Kuletz, Valerie. 1998. *Tainted Desert: Environmental and Social Ruin in the American West.* New York: Routledge.

Laclau, Ernesto, and Chantal Mouffe. 1985. *Hegemony and Socialist Strategy.* London: Verso.

Lamb, J. Mortimer. 1933. A British Columbian Painter. *Saturday Night,* 14 January. 3.

Landau, R. 1987. *From Mineralogy to Geology: The Foundations of a Science, 1650–1830.* Chicago: University of Chicago Press.

Lasch, Christopher. 1979. *The Culture of Narcissism: American Life in an Age of Diminishing Expectations.* New York: Warner Books.

Latour, Bruno. 1986. Visualization and cognition: Thinking with eyes and hands. In Henrika Kulchek and Elizabeth Long (Eds.), *Knowledge and*

Society: Studies in the Sociology of Culture Past and Present. London: JAI Press. 1–40.

———. 1987. *Science in Action: How to Follow Scientists and Engineers through Society*. Cambridge: Harvard University Press.

———. 1988. *The Pasteurization of France*. Trans. Alan Sheridan and J. Law. Cambridge: Harvard University Press.

———. 1993. *We Have Never Been Modern*. Trans. Catherine Porter. Cambridge: Harvard University Press.

———. 1999. *Pandora's Hope: Essays on the Reality of Science Studies*. Cambridge: Harvard University Press.

Lefebvre, Henri. 1991. *The Production of Space*. Trans. Donald Nicholson-Smith. Oxford: Blackwell.

Lenoir, Timothy. 1994. Was the last turn the right turn?: The semiotic turn and A. J. Greimas. *Configurations* 1: 119–36.

Lévi-Strauss, Claude. 1962. *Jean-Jacques Rousseau*. Paris: La Baconnière.

———. 1992. *Triste-Tropiques*. Trans. John Weightman and Doreen Weightman. New York: Penguin.

———. 1966. *The Savage Mind*. Chicago: University of Chicago Press.

Levin, S. A. 1979. Multiple equilibria in ecological models. In *Proceedings of the International Symposium of Mathematical Modelling of Man-Environment Interaction*. 164–239.

Li, Tania. 2000. Articulating indigenous identity in Indonesia: Resource politics and the tribal slot. *Comparative Studies in Society and History* 42: 149–79.

Light, Andrew. 1995. Urban wilderness. In David Rothenberg (Ed.), *Wild Ideas*. Minneapolis: University of Minnesota Press, 195–212.

Limerick, Patricia. 1987. *The Legacy of Conquest: The Unbroken Past of the American West*. New York: Norton.

Linsley, Robert. 1991. Painting and the social history of British Columbia. In Stan Douglas (Ed.), *Vancouver Anthology*. Vancouver: Talon Books.

———. 1995. Yuxweluptun and the west coast landscape. In Scott Watson (Ed.), *Lawrence Paul: Yuxweluptun; Born to Live and Die on Your Colonialist Reservation*. Vancouver: Morris and Helen Belkin Art Gallery. 29–32.

Locke, John. 1967 [1680]. *Two Treaties of Government*. Ed. P. Laslett. Cambridge: Cambridge University Press.

Lutz, Catherine, and Jane Collins. 1993. *Reading National Geographic*. Chicago: University of Chicago Press.

MacCannell, Dean. 1976. *The Tourist: A New Theory of the Leisure Class*. New York: Schocken Books.

———. 1989. Introduction to the 1989 Edition. In Dean MacCannell, *The Tourist*. London: Macmillan. ix–xx.

———. 1992. *Empty Meeting Grounds: The Tourist Papers*. London: Routledge.

MacIsaac, Ron, and Anne Champagne (Eds.). 1994. *Clayoquot Mass Trials: Defending the Rainforest*. Gabriola, B.C.: New Society Publishers.

MacMillan Bloedel. N.d. *Beyond the Cut: MacMillan Bloedel's Forest Management Program*. Vancouver: MacMillan Bloedel.

McClintock, Ann. 1992. The angel of progress: Pitfalls of the term postcolonialism. *Social Text* 31/32: 84–98.

———. 1995. *Imperial Leather: Race, Gender and Sexuality in the Colonial Contest*. New York: Routledge.

McCrory, Colleen (Ed.). 1993. *Brazil of the North: The National and Global Crisis in Canada's Forests*. New Denver: Canada's Future Forest Alliance.

McIntosh, Robert. 1985. *The Background of Ecology*. Cambridge: Cambridge University Press.

———. 1987. Pluralism in ecology. *Annual Review of Ecology and Systematics* 18: 321–41.

McKibben, Bill. 1989. *The End of Nature*. New York: Random House.

McMaster, Gerald, and Lee-Ann Martin (Eds.). 1992. *Indigena: Contemporary Native Perspectives*. Vancouver: Douglas and McIntyre.

McQuire, Scott. 1998. *Visions of Modernity: Representation, Memory, Time and Space in the Age of the Camera*. London: Sage.

Maigan, Loys. 1994. Clayoquot: Recovering from cultural rape. In Tzeporah Berman (Ed.), *Clayoquot and Dissent*. Vancouver: Ronsdale Press. 155–98.

Malkki, Lisa. 1997. National geographic: The rooting of peoples and the territorialization of national identity among scholars and refugees. In Akhil Gupta and James Ferguson (Eds.), *Culture, Power, Place: Explorations in Critical Anthropology*. Durham, N.C.: Duke University Press. 52–74.

Marchak, Patricia. 1983. *Green Gold: The Forest Industry in British Columbia*. Vancouver: University of British Columbia Press.

———. 1995. *Logging the Globe*. Montreal: McGill-Queen's University Press.

———. 1997. A changing global context for British Columbia's forest industry. In Trevor Barner and Roger Hayter (Eds.), *Trouble in the Rainforest: British Columbia's Forest Economy in Transition*. Victoria: Western Geographical Press. 149–66.

Marcuse, Herbert. 1955. *Eros and Civilization: A Philosophical Inquiry into Freud.* Boston: Beacon Press.

———. 1964. *One-Dimensional Man: Studies in the Ideology of Advanced Industrial Society.* Boston: Beacon Press.

Marshall, Yvonne. 1994. A political history of the Nuu-chah-nulth people: A case study of the Mowachaht and Muchalaht tribes. Ph.D. dissertation, Department of Anthropology, Simon Fraser University, Burnaby.

Martin, Emily. 1996. Citadels, rhizomes and string figures. In Stanley Aronowitz, Barbara Martinsons, and Michael Menser (Eds.), *Technoscience and Cyberculture.* New York: Routledge. 97–109.

———. 1998. Fluid bodies, managed nature. In Bruce Braun and Noel Castree (Eds.), *Remaking Reality: Nature at the Millennium.* London: Routledge. 64–83.

Mason, Michael. 1999. *Environmental Democracy.* London: Earthscan.

Massey, Doreen. 1994. *Space, Place, and Gender.* Minneapolis: University of Minnesota Press.

Massey, Vincent. 1948. *On Being Canadian.* Toronto: J. M. Dent.

Matas, Robert. 1993. Clayoquot: the sound and the fury. *Toronto Globe and Mail,* 22 May. A1, A6–A7.

Merchant, Carolyn. 1980. *The Death of Nature: Women, Ecology, and the Scientific Revolution.* San Francisco: Harper and Row.

Mills, Sara. 1991. *Discourses of Difference: An Analysis of Women's Travel Writing and Colonialism.* London: Routledge.

Mitchell, Katharyne. 1997. Different diasporas and the hype of hybridity. *Environment and Planning D: Society and Space* 15: 533–53.

Mitchell, Timothy. 1988. *Colonizing Egypt.* Cambridge: Cambridge University Press.

Monet, Don, and Skan'nu. 1992. *Colonialism on Trial: Indigenous Land Rights and the Gitksan Wet'suwet'en Sovereignty Case.* Philadelphia: New Society Publishers.

Moore, Patrick. 1995. *Pacific Spirit: The Forest Reborn.* West Vancouver, B.C.: Terra Bella.

Moray, Gerta. 1993. Northwest Coast Native culture and the early paintings of Emily Carr. Ph.D. dissertation, Department of Art History, University of Toronto.

Morley, David, and Kevin Robins. 1995. *Spaces of Identity: Global Media, Electronic Landscapes and Cultural Boundaries.* London: Routledge.

Morrison, Ann. 1991. Canadian art and cultural appropriation: Emily Carr and the 1927 *Exhibition of Canadian West Coast Art—Native and Modern.* Unpublished master's thesis.

Mulholland, F. D. 1937. *The Forest Resources of British Columbia.* Victoria: C. F. Banfield.

Munro, John. 1988. Coastal shipping and the development of coastal British Columbia. Paper presented to the Fifth BC Studies Conference, Burnaby.

Munt, Ian. 1994. The "other" postmodern tourism: Culture, travel and the new middle classes. *Theory, Culture and Society* 11: 101–24.

Naess, Arne. 1989. *Ecology, Community and Lifestyle: Outline of an Ecosophy.* Cambridge: Cambridge University Press.

Nash, Roderick. 1967. *Wilderness and the American Mind.* New Haven: Yale University Press.

Nathan, Holly. 1993. Aboriginal forestry: The role of First Nations. In Ken Drushka, Bob Nixon, and Ray Travers (Eds.), *Touch Wood: BC Forests at the Crossroads.* Madeira Park: Harbour Publishing. 137–70.

Nemiroff, Diana. 1992. Modernism, nationalism and beyond: A critical history of exhibitions of First Nations art. In Diana Nemiroff, Robert Houle, and Charlotte Townsend-Gault (Eds.), *Land, Spirit, Power: First Nations at the National Gallery of Canada.* Ottawa: National Gallery of Canada. 15–41.

Neumann, Roderick. 1998. *Imposing Wilderness: Struggles over Livelihood and Nature Preservation in Africa.* Berkeley: University of California Press.

Newcombe, Charles F. 1909. *Guide to Anthropological Collection in the Provincial Museum.* Victoria: British Columbia Legislative Assembly.

Newton, Eric. 1939. Canadian art through English eyes. *Canadian Forum* 28: 344–45.

Nochlin, Linda. 1989. *The Politics of Vision: Essays on Nineteenth-Century Art and Society.* New York: Harper and Row.

Norgaard, Richard. 1994. *Development Betrayed: The End of Progress and a Co-evolutionary Revisioning of the Future.* New York: Routledge.

Noss, Reed. 1995. Wilderness—now more than ever. *Wild Earth* (winter 1994–95): 60–63.

Nuu-chah-nulth Tribal Council. 1990. *The Land Question: Land, Sea and Resources.* Port Alberni: Nuu-chah-nulth Tribal Council.

Oakes, Tim. 1999. Bathing in the far village: Globalization, transnational capital, and the cultural politics of modernity in China. *Positions* 7: 307–42.

O'Connor, James. 1998. *Natural Causes: Essays in Ecological Marxism.* New York: Guilford Press.

Odum, Howard. 1983. *Ecological and General Systems: An Introduction to Systems Ecology.* Niwot: University of Colorado Press.

O'Neill, Wiggs. 1960. *Steamboat Days on the Skeena River.* Kitimat: Northern Sentinal Press.

Pearse, Peter. 1976. *Timber Rights and Forest Policy in British Columbia: Report of the Royal Commission on Forest Resources.* Victoria: Royal Commission on Forest Resources.

Persky, Stan. 1998. *Delgamuukw: The Supreme Court of Canada Decision on Aboriginal Title.* Vancouver: Douglas and McIntyre.

Phillips, Richard. 1997. *Mapping Men and Empire: A Geography of Adventure.* London: Routledge.

Polan, Dana. 1991. Review of Habermas, *The Structural Transformation of the Public Sphere. Social Text* 8/9: 260–87.

Pollock, Griselda. 1988. *Vision and Difference: Femininity, Feminism and Histories of Art.* London: Routledge.

———. 1994. Territories of Desire: Reconsideration of an African Childhood. In George Robertson, Melinda Mash, Lisa Tichner, Jon Bird, and Barry Curtis (Eds.), *Traveller's Tales: Narratives of Home and Displacement.* London: Routledge. 63–92.

Poole, Deborah. 1997. *Vision, Race and Modernity: A Visual Economy of the Andean Image World.* Princeton, N.J.: Princeton University Press.

Porter, Dennis. 1991. *Haunted Journeys: Desire and Transgression in European Travel Writing.* Princeton, N.J.: Princeton University Press.

Porter, Roy. 1977. *The Making of Geology: Earth Science in Britain, 1660–1815.* Cambridge: Cambridge University Press.

Povinelli, Elizabeth. 1999. Settler modernity and the quest for an indigenous tradition. *Public Culture* 11: 19–48.

Prakash, Gyan. 1999. *Another Reason: Science and the Imagination of Modern India.* Princeton, N.J.: Princeton University Press.

Pratt, Mary Louise. 1992. *Imperial Eyes: Travel Writing and Transculturation.* London: Routledge.

Raffles, Hugh. Forthcoming. *On the Nature of the Amazon.* Princeton, N.J.: Princeton University Press.

Rajala, Richard. 1998. *Clearcutting the Pacific Rain Forest: Production, Science and Regulation.* Vancouver: University of British Columbia Press.

Rajchmann, John. 1988. Foucault's art of seeing. *October* 44: 89–117.

Riding, Alan. 1998. Freida Kahlo introduces her husband, Diego Rivera. *New York Times,* July 8. B2.

Robbins, Bruce (Ed.). 1993. *The Phantom Public Sphere.* Minneapolis: University of Minnesota Press.

Rojek, Chris. 1993. *Ways of Escape: Modern Transformations in Leisure and Travel.* London: Macmillan.

Rosaldo, Renato. 1989. *Culture and Truth: The Remaking of Social Analysis.* Boston: Beacon Press.

Rose, Gillian. 1993. *Feminism and Geography: The Limits of Geographical Knowledge.* Minneapolis: University of Minnesota Press.

Rothenberg, David (Ed.). 1995. *Wild Ideas.* Minneapolis: University of Minnesota Press.

Rouse, Joseph. 1987. *Knowledge and Power: Toward a Political Philosophy of Science.* Ithaca, N.Y.: Cornell University Press.

———. 1993. What are the cultural studies of scientific knowledge? *Configurations* 1: 1–22.

Rowell, Andrew. 1996. *Green Backlash: Global Subversion of the Environmental Movement.* London: Routledge.

Rudwick, Martin. 1976. The emergence of a visual language for geological science, 1760–1840. *History of Science* 14: 149–95.

———. 1996. Minerals, strata and fossils. In Nicholas Jardine, Anne Secord and Emma Spary (Eds.), *Cultures of Natural History.* Cambridge: Cambridge University Press. 266–86.

Rushing, W. Jackson. 1995. *Native American Art and the New York Avant-Garde.* Austin: University of Texas Press.

Rushton, Gerald. 1969. Union Steamship Company of British Columbia, a short history. In Ruth Greene (Ed.), *Personality Ships of British Columbia.* West Vancouver: Marine Tapestry.

———. 1974. *Whistle up the Inlet: The Union Steamship Story.* Vancouver: J. J. Douglas.

———. 1980. *Echoes of the Whistle: An Illustrated History of the Union Steamship Company.* Vancouver: Douglas and McIntyre.

Rutherford, Paul. 1999. The entry of life into history. In Eric Darier (Ed.), *Discourses of the Environment.* Oxford: Blackwell. 37–62.

Ryan, Maureen. 1992. Picturing Canada's Native landscape: Colonial expansion, national identity and the image of a dying race. *RACAR* 18: 138–49.

Said, Edward. 1978. *Orientalism.* New York: Vintage Books.

———. 1994. *Culture and Imperialism.* New York: Vintage Books.

Sam, Stanley. 1997. *Ahousaht Wild Side Heritage Trail.* Vancouver: Western Canada Wilderness Committee.

Schivelbusch, Wolfgang. 1986. *The Railway Journey: The Industrialization of Time and Space in the Nineteenth Century.* Berkeley: University of California Press.

Scientific Panel for Sustainable Forest Practices in Clayoquot Sound (B.C.). 1995a. *Report 3: First Nation's Perspectives Relating to Forest Practices in Clayoquot Sound.* Victoria: Clayoquot Sound Scientific Panel.

———. 1995b. *Report 5: Sustainable Ecosystem Management in Clayoquot Sound: Planning and Practices.* Victoria: Clayoquot Sound Scientific Panel.

Schmidt, Alfred. 1971. *The Concept of Nature in Marx.* Trans. Ben Fowkes. London: New Left Books.

Scoones, I. 1999. New ecology and the social sciences: What prospects for a fruitful engagement? *Annual Review of Anthropology* 28: 479–507.

Scott, David. 1995. Habitation sites and culturally modified trees. *Cultural Survival* 18: 19–20.

Scott, James. 1985. *Weapons of the Weak: Everyday Forms of Peasant Resistance.* New Haven: Yale University Press.

Secord, J. 1986. *Controversy in Victorian Geology: The Cambrian-Silurian Dispute.* Princeton, N.J.: Princeton University Press.

Seltzer, Mark. 1992. *Bodies and Machines.* New York: Routledge.

Sessions, George. 1997. Reinventing nature? The end of Wilderness? A response to William Cronon's *Uncommon Ground. Wild Earth* (winter 1996–97): 46–52.

Shadbolt, Doris. 1975. *Emily Carr: A Centennial Exhibition Celebrating the One Hundredth Anniversary of Her Birth.* Vancouver: J. J. Douglas.

———. 1979. *The Art of Emily Carr.* Toronto: Clarke, Irwin.

———. 1990. *Emily Carr.* Vancouver: Douglas and McIntyre.

Shapin, Steven, and Simon Schaffer. 1985. *Leviathan and the Air Pump: Hobbes, Boyle and the Experimental Life.* Princeton, N.J.: Princeton University Press.

Shapiro, Gary. 1993. In the shadows of philosophy: Nietzsche and the question of vision. In David Levin (Ed.), *Modernity and the Hegemony of Vision.* Berkeley: University of California Press. 124–42.

Shohat, Ella. 1992. Notes on the "post-colonial." *Social Text* 31/32: 99–113.

Sierra Club of Western Canada. 1991. *Ancient Rainforests at Risk: An Interim Report by the Vancouver Island Mapping Project.* Victoria: Sierra Club.

Siviramakrishna, K. 1999. *Modern Forests: Statemaking and Environmental Change in Colonial Eastern India.* Stanford, Calif.: Stanford University Press.

Sloan, Gordon. 1945. *Report of the Commissioner Relating to the Forest Resources of British Columbia, 1945.* Victoria: C. F. Banfield.

———. 1957. *Report of the Commissioner Relating to the Forest Resources of British Columbia: 1955–1957.* Victoria: C. F. Banfield.

Slotkin, Richard. 1985. *The Fatal Environment: The Myth of the Frontier in the Age of Industrialization, 1800–1890.* New York: Atheneum.

Smith, Michael Peter. 1994. Transnational migration and the globalization of grassroots politics. *Social Text* 39: 15–33.

Smith, Neil. 1990 [1984]. *Uneven Development: Nature, Capital and the Production of Space.* Oxford: Blackwell.

———. 1996. The production of nature. In George Robertson, Melinda Mash, Lisa Tickner, Jon Bird, Barry Curtis, and Tim Putnam (Eds.), *FutureNatural.* London: Routledge. 35–54.

Solnick, Timothy. 1992. Power, resistance and the law in a British Columbia land title trial. Master's thesis, Department of Geography, University of British Columbia, Vancouver.

Sontag, Susan. 1977. *On Photography.* New York: Farrar, Straus and Giroux.

Sorkin, Michael (Ed.). 1992. *Variations on a Theme Park: The New American City and the End of Public Space.* New York: Hill and Wang.

Sotriffer, Kristian. 1972. *Expressionism and Fauvism.* Trans. Richard Rickett. Vienna: Anton Schroll and Company.

Soule, Michael, and Gary Lease (Eds.). 1995. *Reinventing Nature? Responses to Postmodern Deconstructionism.* Washington, D.C.: Island Press.

Sparke, Matthew. 1995. Between demythologizing and deconstructing the map: Shawnadithit's New-Found-Land and the alienation of Canada. *Cartographica* 32: 1–21.

———. 1998. A map that roared and an original atlas: Canada, cartography and the narration of nation. *Annals of the Association of American Geographers* 88: 463–95.

Spilsbury, Jim. 1990. *Spilsbury's Album: Photographs and Reminiscences of the British Columbia Coast.* Madeira Park: Harbour Publishing.

Spivak, Gayatri Chakravorty. 1988a. Can the subaltern speak? In Carey Nelson and Lawrence Grossberg (Eds.), *Marxism and the Interpretation of Culture.* Urbana: University of Illinois Press. 271–313.

———. 1988b. *In Other Worlds: Essays in Cultural Politics.* New York: Routledge.

———. 1990. *The Post-Colonial Critic: Interviews, Strategies, Dialogues.* Ed. Sarah Harawsym. New York: Routledge.

Sprugal, Douglas. 1991. Disturbance, equilibrium, and environmental variability: What is "natural" vegetation in a changing environment? *Biological Conservation* 58: 1–18.

Stafford, Robert A. 1990. Annexing the landscapes of the past: British imperial geology in the nineteenth century. In J. MacKenzie (Ed.), *Imperialism and the Natural World.* New York: St. Martin's Press. 67–89.

Sterritt, Neil. 1999. The Nisga'a treaty: Competing claims ignored. *BC Studies* 120: 73–98.

Stewart, Kathleen. 1996. *A Space on the Side of the Road: Cultural Politics in the "Other" America.* Princeton, N.J.: Princeton University Press.

Stewart, Susan. 1984. *On Longing: Narratives of the Miniature, the Gigantic, the Souvenir, the Collection.* Baltimore: Johns Hopkins University Press.

Stoddart, David. 1995. Darwin and the seeing eye: Iconography and meaning in the *Beagle* years. *Earth Sciences History* 14: 3–22.

Sturgeon, Noel. 1997. *Ecofeminist Natures: Race, Gender, Feminist Theory and Political Action.* New York: Routledge.

Tagg, John. 1988. *The Burden of Representation: Essays on Photographies and Histories.* Minneapolis: University of Minnesota Press.

Tennant, Paul. 1990. *Aboriginal Peoples and Politics: The Indian Land Question in British Columbia, 1849–1989.* Vancouver: University of British Columbia Press.

Thom, Ian M. 1991. *Emily Carr in France: Vancouver Art Gallery, June 22 to September 22, 1991.* Vancouver: Vancouver Art Gallery.

Thomas, Nicholas. 1994. *Colonialism's Culture: Anthropology, Travel and Government.* Princeton, N.J.: Princeton University Press.

Thompson, John. 1995. *Media and Modernity: A Social Theory of the Media.* Cambridge: Polity Press.

Tippett, Maria. 1979. *Emily Carr: A Biography.* Toronto: Oxford University Press.

Todd, Loretta. 1990. Notes on appropriation. *Parallelogramme* 16: 24–33.

Townsend-Gault, Charlotte. 1995. The salvation art of Yuxweluptun. In Scott Watson (Ed.), *Lawrence Paul: Yuxweluptun; Born to Live and Die on Your Colonialist Reservations.* Vancouver: Morris and Helen Belkin Art Gallery. 7–28.

———. 1999. Hot dogs, a ball gown, adobe and words: The modes and materials of identity. In W. Jackson Rushing III (Ed.), *Native American Art in the Twentieth Century.* New York: Routledge. 113–33.

Turnbull, David. 1997. Reframing science and other local knowledge traditions. *Futures* 29: 551–62.

Turner, Robert. 1977. *The Pacific Princesses: An Illustrated History of the Canadian Pacific Railway's Princess Fleet on the North Coast.* Victoria: Sono Nis Press.

Urry, John. 1990. *The Tourist Gaze.* London: Sage.

Valie, Smadar, and Ted Swedenburg. 1996. Introduction: Displacement, diaspora and geographies of identity. In Smadar Valie and Ted

Swedenburg (Eds.), *Displacement, Diaspora and Geographies of Identity.* Durham, N.C.: Duke University Press. 1–25.

Vancouver, George. 1798. *A Voyage of Discovery to the North Pacific Ocean, and Round the World.* 3 vols. London: G. G. and J. Robinson.

Walker, Stephanie. 1996. *This Woman in Particular: Contexts for the Biographical Image of Emily Carr.* Waterloo: Wilfred Laurier University Press.

Watson, Scott. 1995. The modernist past of Lawrence Paul Yuxweluptun's language allegories. In Scott Watson (Ed.), *Lawrence Paul: Yuxweluptun; Born to Live and Die on Your Colonialist Reservations.* Vancouver: Morris and Helen Belkin Art Gallery. 62–74.

Watt, A. S. 1947. Pattern and process in the plant community. *Journal of Ecology* 35: 1–22.

Western Canada Wilderness Committee. 1994. Clayoquot: A heritage worth protecting. In *Western Canada Wilderness Committee Educational Report* 13: 1.

Whatmore, Sarah. 1997. Dissecting the autonomous self: Hybrid cartographies for a relational ethics. *Environment and Planning D: Society and Space* 15: 37–53.

Whitlock, Cathy. 1992. Vegetation and climatic history of the Pacific Northwest during the last twenty thousand years: Implications for understanding present-day biodiversity. *Northwest Environmental Journal* 8: 5–28.

Willers, Bill. 1997. The trouble with Cronon. *Wild Earth* (winter 1996–97): 59–61.

Williams, Raymond. 1973. *The Country and the City.* New York: Oxford University Press.

Wilson, Alexander. 1991. *The Culture of Nature: North American Landscape from Disney to the Exxon Valdez.* Toronto: Between the Lines Press.

Wolff, Janet. 1985. The impossible flaneuse: Women and the literature of modernity. *Theory, Culture and Society* 2: 37–46.

Worster, Donald. 1988. *Nature's Economy: A History of Ecological Ideas.* Cambridge: University of Cambridge Press.

———. 1990. The ecology of order and chaos. *Environmental History Review* 14: 1–18.

Wu, Jianguo, and Orie Loucks. 1995. From balance of nature to hierarchical patch dynamics: A paradigm shift in ecology. *Quarterly Review of Biology* 70: 439–66.

Young, Robert. 1990. *White Mythologies: Writing History and the West.* London: Routledge.

Yuxweluptun (Lawrence Paul). 1995. Artist's statement. In Scott Watson (Ed.), *Lawrence Paul: Yuxweluptun; Born to Live and Die on Your Colonialist Reservations.* Vancouver: Morris and Helen Belkin Art Gallery. 1–2.

Zaslov, Morris. 1975. *Reading the Rocks: The Story of the Geological Survey of Canada.* Ottawa: Information Canada.

Zeller, Suzanne. 1987. *Inventing Canada: Early Victorian Science and the Idea of a Transcontinental Nation.* Toronto: University of Toronto Press.

Žižek, Slavoj. 1989. *The Sublime Object of Ideology.* London: Verso.

Zukin, Sharon. 1991. *Landscapes of Power: From Detroit to Disney World.* Berkeley: University of California Press.

Index

Adorno, Theodor, 138

Adventure: and desire, 126–27, 134; and metropolitian culture, 133–34; reasons for, 126

Adventure travel: and the Ahousaht, 121–23, 147–53; and anachronistic space, 120–21, 134; and authenticity, 131, 134; and civilization, 115, 129; in Clayoquot Sound, 121–23, 141; and Clayoquot Sound Land Use Plan, 143–44; and colonialism, 118–20, 145–46; as colonizing, 143, 154; commodification of, 113–14, 117; economic aspects of, 121, 288 n.14; and environmental politics, 145, 146; and forestry, 144; and identity, 126–27; impacts of, 142–46; as masculinist, 133; modernity and, 111–12, 123, 134–39, 137, 290 n.27, 291 n.35; modern subject and, 112, 133, 142; and Native Canadians, 117–20, 121–23, 145–46; and nostalgia, 131, 136, 291 n.33; politics of, 121, 124, 151–54; primitivism and, 116–17, 131, 292 n.38; and production of nature, 124, 139, 143; and territorialization, 138, 141–42, 154–55; visual logic of, 117, 143, 151. *See*

also Ahousaht Wild Side Heritage Trail; Ecosummer Expeditions

Ahousaht: and adventure travel, 121–22, 147–53; and Ahousaht Wild Side Heritage Trail, 147–53; economic conditions, 152–53; Ma-Mook Development Corporation, 268; Marktosis, 120–23, 147, 151; Meares Island study, 97–99, 101; modernity of, 151; and tourism, 150–51; women, 152–53

Ahousaht Wild Side Heritage Trail: and Ahousaht culture, 150; and forest politics, 153; gender and, 152; history, 147–49; and land claims, 149–50; and local politics, 151–53; and tourist gaze, 151; and Western Canada Wilderness Committee, 149

Anachronistic space: and adventure travel, 120–21, 134; British Columbia coast as, 120–21; Clayoquot Sound as, 86; Native Canadians placed in, 61; in work of George Dawson, 54–57, 60. *See also* Primitivism

Anthropocentrism, 257, 273 n.13, 273 n.14

Anthropology, 103–4, 189–91. *See also* Imperialist nostalgia

Anticolonialism: and colonialism,

333

Bruce Braun is assistant professor of geography at the University of Minnesota. He has written extensively on the cultural politics of nature, science and government, colonialism and environmentalism, and social theory and nature. He is coeditor, with Noel Castree, of *Remaking Reality: Nature at the Millennium* and *Social Nature: Theory, Practice, and Politics.*